UNIVERSITY OF STRATHCLYDE
30125 00490256 4

KU-214-042

ANDERSONIAN LIBRARY
WITHDRAWN FROM LIBRARY STOCK
UNIVERSITY OF STRATHCLYDE

Books are to be returned on or before
the last date below.

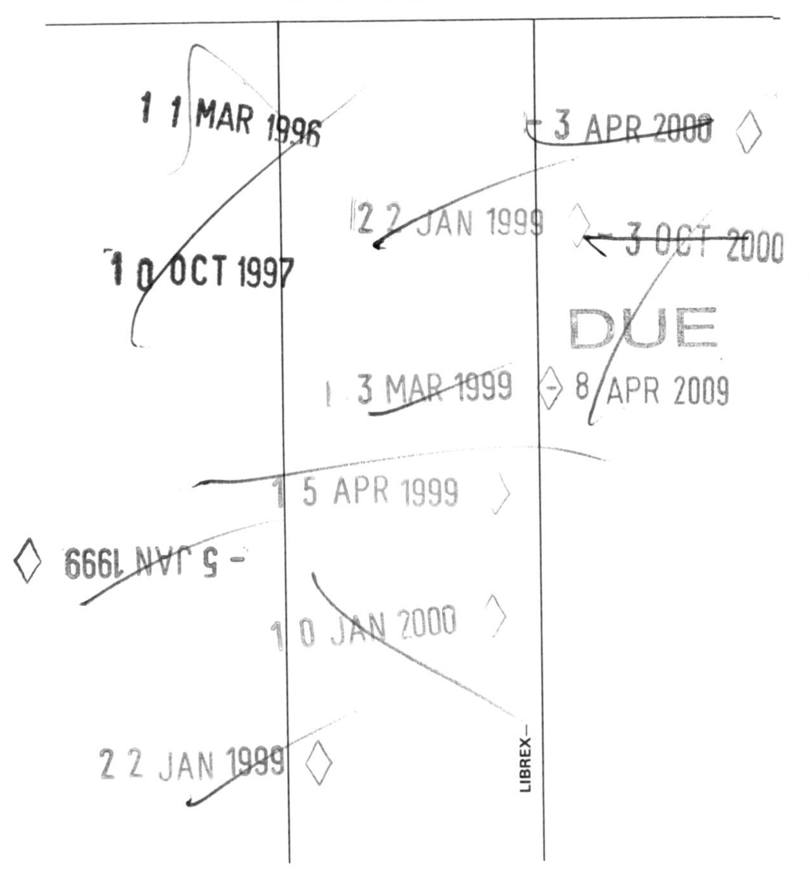

MULTI PHASE PRODUCTION

This volume consists of papers presented at the 6th International Conference on Multi Phase Production, 16–18 June 1993, held in Cannes, France.

ACKNOWLEDGEMENTS

The valuable assistance of the Technical Advisory Committee and panel of referees is gratefully acknowledged.

TECHNICAL ADVISORY COMMITTEE

Prof A Wilson OBE (Chairman)	Total Oil Marine plc
Dr A P Burns	Elf UK
Dr J Finnegan	Conoco (UK) Ltd
M. J-F Giannesini	Institut Français du Pétrole, France
Prof G Hewitt	Imperial College
Mr S Sarchar	BHR Group Limited
Mr J Tiratsoo	Pipes and Pipelines International
Dr A Wilcockson	AEA Petroleum Services

CORRESPONDING MEMBERS

Prof P Andreussi	Universitá degli Studi di Pisa, Italy
Mr I Brandt	SINTEF, Norway
Dr J P Brill	The University of Tulsa and TUFFP, USA
Dr E F Caetano	Petrobrás R & D Centre, Brazil
Prof J Fabre	Institut National Polytechnique de Toulouse, France
Dr P Fuchs	Statoil, Norway
Dr E Guevara	INTEVEP, Venezuela
Dt Ing S Brendsund Hansen	Norsk Hydro A/S, Norway
Dr R V-A Oliemans	Shell Research BV, The Netherlands

Organized and sponsored by BHR Group Limited
Co-sponsored by the Institut Français du Pétrole (IFP)
and the Tulsa University Fluid Flow Projects (TUFFP)

8633941

6TH INTERNATIONAL CONFERENCE ON

Multi Phase Production

Edited by Professor A Wilson

BHR Group Conference Series
Publication No. 4

Papers presented at the *6th International Conference on Multi Phase Production*, organized and sponsored by BHR Group Limited, and held in Cannes, France, on 16–18 June 1993

Mechanical Engineering Publications Limited
LONDON

© BHR Group Limited 1993

This publication is copyright under the Berne Convention and the International Copyright Convention. All rights reserved. Apart from any fair dealing for the purpose of private study, research, criticism, or review, as permitted under the Copyright, Designs and Patents Act, 1988, no part of this publication may be reproduced, stored in a retrieval system, or transmitted in any form or by any means without the prior permission of the copyright owners.

ISBN 0 85298 872 9

A CIP catalogue record for this book is available from the British Library.

Other titles in the BHR Group Conference Series

1 *Offshore Loss Prevention – a Systematic Approach*
Edited by C P A Thompson

2 *Fluid Power – The Future for Hydraulics*
Edited by N Way

3 *Effective Membrane Processes – New Perspectives*
Edited by R Paterson

D
622.338
INT

Produced by Technical Communications (Publishing) Ltd.

Printed by Information Press Ltd, Oxford, England.

CONTENTS

6th International Conference on
MULTI PHASE PRODUCTION
Cannes, France: 16–18 June 1993

ANALYSIS OF SLUG FLOW

EXPERIMENTAL INVESTIGATIONS I – Horizontal Flow Loops

EXPERIMENTAL INVESTIGATIONS II – Vertical Flow Loops

MULTIPHASE METERING

MULTIPHASE BOOSTING

FOREWORD

Although the term Multiphase Production has figured in discussions concerning hydrocarbon production for many years, much still remains to be understood. Today it is less of a black art than it was twenty years ago, but it should be remembered that one aspect, hydrate formation, was defined at the turn of the century. Today, ninety years on, major projects to better understand this one phenomenon within multiphase production are being carried out with participation by almost all major oil companies as well as many universities and research associations.

The state of the art of multiphase production changes continually, and in Europe and North America more than one hundred related projects are underway. The availability of powerful computer models has helped considerably to better understand the results obtained from research work but the total cost of investment for these tools is very high. Funding is an ongoing problem.

Governments world-wide are taking a greater interest in this form of production. For them, it offers, in the future, lower investment costs and operating costs, whilst simplifying the very costly conventional installations that are needed to produce hydrocarbons, particularly from deep water offshore reservoirs. Oil companies also see these techniques as a considerable help in boosting production world-wide, whilst avoiding high costs in a market that has reduced earnings, in constant dollar terms, against an increasingly difficult exploration background. Thus, Government and industry funding is increasingly important in progressing this area of long term development.

The Community of European Countries has also appreciated the importance of understanding correctly, and in depth, the various phenomena associated with multiphase production. Hence, through their various commercial and technical aid schemes they have invested heavily in a wide variety of relevant projects to everyone's benefit.

This, the 6th International Conference organised by the British Hydromechanics Research Group Limited, with sponsorship from the Institut Français du Pétrole, the University of Tulsa and the Tulsa University Fluid Flow Projects, provides a forum to examine recent progress in this most important subject. Activities such as slug flow analysis, flow modelling, multiphase metering and boosting, are all examined in depth and will lead to a further improvement in the overall level of knowledge. The conference papers and discussion will reveal the many areas still in need of detailed evaluation and understanding.

May I, on behalf of the Technical Advisory Committee, wish you all welcome to Cannes and hope that all of you will benefit from the wide range of excellent papers that are being presented.

Professor Alan Wilson
Chairman
Technical Advisory Committee

6th International Conference on
MULTI PHASE PRODUCTION
Cannes, France : 16-18 June 1993

OPENING ADDRESS

Mr Chairman, Ladies and Gentlemen,

It is a great pleasure to open this 6th Conference on Multiphase Production. It is also difficult to introduce such a large subject!

So let me start with words you have probably already heard a hundred times, but which will nevertheless be useful to keep in mind during these few days to stir up our discussions.

Oil industry is facing new challenges. New exploration areas are limited, production from major fields is on the decrease and companies are forced to extensively explore their "own garden" around their existing facilities. Consequently, the size of discoveries is generally smaller. At the same time, new technical challenges have come up : high pressure, high temperature, reservoir fluids in critical state, deep water...

Taking all these elements into consideration and in order to cash in on these discoveries or existing portfolios under the current rather depressed economic conditions, governments and oil companies have to stimulate innovation and promote cheaper production concepts.

Naturally Petroleum Engineering should be on the first line. Petroleum Engineering aims to maximize recovery from the reservoir and at the same time to optimize the use of its proper downhole energy. This requires to look at it from a global perspective so as to overstep the present-day limits of flow conditions in, let's say, the reservoir, the well or the pipeline system.

From the reservoir to export facilities, we handle a multiphase product which behaves acording to thermodynamic laws. Multiphase production is therefore the essence of Petroleum Engineering. More specifically, it applies to the simultaneous handling of significant proportions of gas, oil, water and sometimes solids. And this is a challenge in itself...

First it is a challenging domain for **modelling**. We must remember that Bernouilli introduced hydrodynamics more than two centuries ago. Reynolds wrote down his unforgettable number in the middle of the 19th century and we have all played with it...

We have started from an empirical approach. During the fifties, data became available from test facilities and from field production. The idea was to extrapolate from these laws that describe one-phase fluid. Multiphase fluids were considered as mixtures of homogeneous components. Attention was focused on slippage effects and the frist flow pattern maps were constructed.

The first mechanistic models for slug flow and flow pattern prediction appeared in the '70 s.

In the early '80 s the personal computer phenomenon spread extensively and with it, the use of software to evaluate pressure drops. But all this was based on beloved old-time correlations and extensive use of existing models and correlations that were far from being perfect.

To improve prediction accuracy it was necessary to have a better knowledge of basic physical mechanisms... New sensors and instrumentation asociated with computer hardware and software helped obtain a better data acquisition and thereby a better understanding of these phenomena.

Nodal analysis was also introduced and engineers were able to optimize, for instance, the operation of gas-lift supply networks.

New equations and experimental data are continuously introduced into flow models. Steady state two-phase flow models were first developed, then commonly used.

Recently, transient configurations have been evaluated. Operational events, as shutting down or opening a well, changing the outlet pressure, are now simulated with a fair degree of reliability, which is to be further improved.

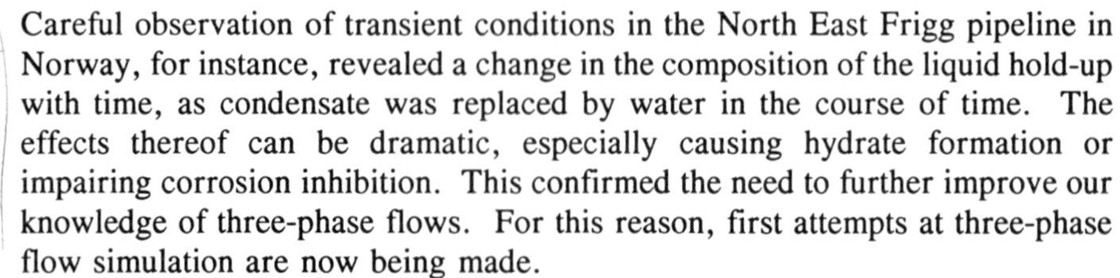

Careful observation of transient conditions in the North East Frigg pipeline in Norway, for instance, revealed a change in the composition of the liquid hold-up with time, as condensate was replaced by water in the course of time. The effects thereof can be dramatic, especially causing hydrate formation or impairing corrosion inhibition. This confirmed the need to further improve our knowledge of three-phase flows. For this reason, first attempts at three-phase flow simulation are now being made.

As far as multiphase technology is concerned considerable progress has been made too.

To minimize back pressure effects on the formation, engineers imagined to implement subsea separators. Test programs have been recently performed in Europe and US. It can be concluded that conventional subsea separation is technically feasible but not yet sufficiently reliable.

On the other hand, engineers worked on downhole pumping.

A lot of efforts were made to increase the gas void fraction acceptance at the downhole pump suction. The hydraulic cells of pumps were modified so as to handle increasing quantities of gas. Such pumps could even be installed downstream the well choke... The Poséidon concept was born. It had moved from the bottom to the surface.

Different pumping systems were studied. Screw pumps proved to be feasible while promising results were obtained for reciprocating pumps.

On the metering side, the 88 IMPEL study has led to an impressive list of theoretical concepts which could be used to build multiphase meters. Some of them have been selected and some others are now being tested in laboratories or through field test loops.

Let me say a few words about **physico-chemistry** in multiphase transportation. There are two main areas of concern today. Different from a thermodynamic point of view, they are nevertheless similar in terms of operational consequences. I am talking of hydrates and high pour point or waxy crudes.

For a while, it has been thought that hydrate plugging could be prevented by using conventional methods such as alcohol inhibition or thermal isolation to keep the flow system away from the thermodynamic hydrate area. But new production conditions constrain such methods in certain economic and environmental limits. New methods for hydrate inhibition are being developed in laboratories. Though their field of application may not be universal, they can however be complementary.

In the field of high pour point or waxy crudes, much progress has been made, too. Crudes, regarded as non producible 10 years ago, are now flowing through pipelines at temperature below their pour point.

This rapid overview of the last decades shows how knowledge and technology have progressed in the multiphase domain.

If we look back at the list of contents of the first multiphase production conference which took place in Coventry in April '83 under the official mane of "The International Conference on the Physical Modelling of Multiphase Flow" and if we compare it with the present-day Conference in Cannes, ten years later, we can measure the whole progress that has been made since then. The '83 Conference was mainly dedicated to multiphase flow modelling and the '91 Conference included reports on multiphase technology and on operational experiences.

Even if we do not see yet many pages about the physico-chemistry phenomena, we must keep in mind that more challenging developments will require a more global approach from operators which will have to face them.

Mr Chairman, Ladies and Gentlemen, I am sure this 6th Multiphase Production Conference will contribute to the promotion of new ideas and new applications, and I thank you for your kind attention.

M. Jacques Duquesne
Directeur Délégué aux Techniques Industrielles
Elf Aquitaine, Pau, France

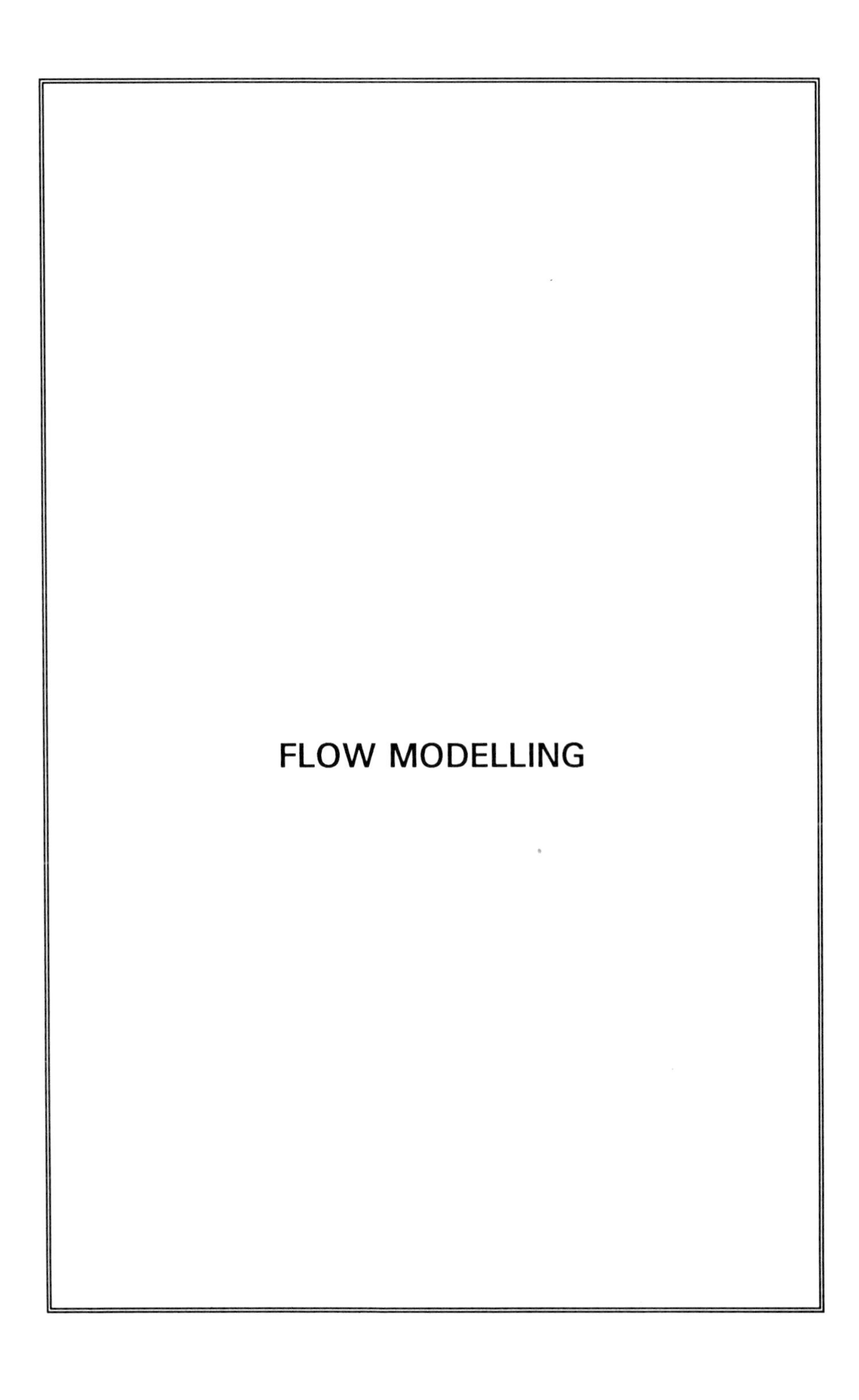

FLOW MODELLING

A MECHANISTIC MODEL FOR THE FLOW PHENOMENA IN JET-LIFTING AND DISPLACEMENT OF WELL FLUIDS WITH COILED TUBING AND NITROGEN

L.C.B. Bianco, PETROBRÁS S.A., Brazil
F.A.A. Barbuto, PETROBRÁS R&D Centre, Brazil
E.F. Caetano, PETROBRÁS R&D Centre, Brazil
A.C. Bannwart, UNICAMP, Brazil

ABSTRACT

Surgence induction and displacement of well fluids with coiled tubing and nitrogen is a very common technique in the oil industry. Although effective, it is also a quite expensive operation. Bianco (1990) developed a full mechanistic and comprehensive model to predict fluid rates and recovery in order to minimize costs related to this operation. Laboratory experiments were carried out to supply reliable data regarding the validation of these model. A reasonable agreement between the lab data and the numerical scheme was observed. It was also verified that the frictional pressure drops plays a very important role in both predicted fluid recovery and total pressure gradients. Moreover, the authors verified that the gradients due to accelerational effects must not be neglected.

NOMENCLATURE

A_{an}	Cross sectional area of the coil tubing-production tubing annulus
A_c	Cross sectional area of the production tubing, internal
A_g	Area occupied by gas in the annulus cross section
A_i	Area of the orifice in the ending nozzle of the coiled tubing
A_L	Area occupied by liquid in the annulus cross section
A_T	Cross sectional area of the coiled tubing
d	Distance between the centres of the coiled tubing and production tbg.
D_C	Internal diameter of the production tubing
$D_{e,fl}$	External diameter of the coiled tubing (O.D.)
$D_{i,fl}$	Internal diameter of the coiled tubing (I.D.)
D_{eq}	Equivalent diameter of the annular section
f_{an}	Fanning friction factor for annular geometries
f_T	Fanning friction factor for tubular geometries
g	Acceleration of gravity
L	Length of the annular section
L_e	Length of production tubing below the ending nozzle of the coil tubing.
L_S	Length of prod. tbg. below the static level of fluid
$\dot{M}_L$	Mass flow rate of the liquid phase
$\dot{M}_G$	Mass flow rate of the gas phase
P_{atm}	Atmospheric pressure
P_i	Pressure at the coiled tubing outlet

P_1	Pressure at the bottom of the production tubing
P_2	Pressure at the top of the prod. tubing (back pressure)
SUB	Submergence factor or fraction of the production tbg. filled with liquid
U_{SL}	Superficial velocity of the liquid phase
ε	Eccentricity, $2d/(D_C-D_{i,fl})$
ρ_L	Specific weight of the liquid phase
$\bar{\rho}_G$	Specific weight of the gas phase, local averaged value
ξ_L	Coefficient of irreversibilities at the nozzle's region
$\left(\frac{dP}{dz}\right)_{f,L}$	Frictional gradient, local, one phase, liquid
$\left(\frac{dP}{dz}\right)_{f,G,L}$	Frictional gradient, local, two-phase (L+G)

1. INTRODUCTION

The loss of surgence in a well requires specific operations to improve the reservoir's capability to keep the well flowing longer. Among these acidizing and fracturing may be cited. Both acidizing and fracturing require the previous dampening of the well with completion fluid. Sometimes, the well "kills" itself due to bad production practices, and the remaining dead oil avoids surgence. In each of these cases the well fluids must be displaced to re-start production.

Clean-up, surgence induction and fluid displacement operations in oil wells as they are currently done were implemented since they were successfully employed in 1963 at the Barataria field, Louisiana, USA. A coil tubing with 0.0344 m O.D. (1.315") and minimum wall thickness of 0.003175 m (0.125") formed by 300 welded pieces of 15.24 m (50 ft) each was then used. Nitrogen was employed as displacing medium, with flow rates ranging from 5.66 m^3 Std./min (200 scf/min) to 25.4 m^3 Std./min (900 scf/min). Nitrogen pumping pressures were set at the level of 62 MPa (9000 psi).

The surgence induction operation was ever conducted in a rather subjective basis. Injection rates and flow times were controlled by the operator's crew; no criteria was set to minimize costs and equipment downtime. Moreover, no account was taken on the characteristics of the complex flows of well fluids and nitrogen in annular geometry. So, the predicted and observed operational results were hardly ever close. Most service companies commonly employ empirical or semi-empirical-based guiding tables to execute the job. These tables try to adjust the theory for two-phase flow in circular ducts to the actual flow in an annulus, and coarse errors in the prediction of pressure gradients have been produced so, contributing to the standing of empiricism in the execution of this operation.

This work is bounded to study the mechanisms of fluid motion in jet-lifting operations, based on a set of lab experiments carried out to collect data in upward gas-liquid two-phase flows in annular geometries similar to that found in the field. A simplified numerical scheme to simulate jet-lifting was developed to optimize fluid recovery by means of a better prediction of both total injected volume and flow rate of nitrogen.

2. PREVIOUS WORK

In the technical literature there exists a good amount of published works on this subject. The authors selected the most contributing literature to this present work.

Slator & Hanson(1965) presented in one of the earliest papers on this subject the advantages of jet-lifting in completion and workover operations.

In a similar way, Weeks(1970) pointed out the advantages and low costs of jet-lifting for some clean-up, stimulation, recompletion and kick-off operations. This author has also presented a diagram for frictional pressure losses due to the flow of xylene and nitrogen as a function of the coil length only. He identified as common flow rates figures in the order of 5.66 scm/min (200 scf/min) for nitrogen, and surface pressures such as 27.6 MPa (4000 psi).

Cashion(1973) associated a correlation for friction factors in a theoretical development, generating diagrams to predict frictional pressure drops for several flow rates and fluid types. He pointed out that flow rates of 25.4 scm/min (900 scf/min) at 34.5 MPa (5000 psi) were very common in well treating, and that in the case of jet-lifting the rate of 14.1 scm/min (500 scf/min) was rarely exceeded, and with returning flow of recovered liquid less than 0.00044 m^3/s (10 bbl/h).

Many researchers had developed empirical works aiming to predict pressure behavior and pressure gradients in two-phase flows in circular pipes. Poettman & Carpenter(1952), Duns & Ros(1963) and Hagedorn & Brown(1965) - among many others - must be cited by their outstanding contributions to the study of these flows.

For non-circular geometries, the hydraulic diameter concept is largely employed, specially to determine friction factors, although this concept should rather be applied for highly turbulent flows. Bird *et al*(1976) developed an analytical solution for velocity profiles and friction factors to laminar single-phase flows in concentric annular geometries.

A very few works for upward two-phase flows in annular geometries have been published so far. Among these remarks shall be given to Winkler(1968), Sadatomi *et al*(1982), Kelessidis(1986) and Kelessidis & Dukler(1989).

Finally, Caetano(1986) gathered the works of Tosun(1984) and Snyder & Goldstein(1965) in order to develop an analytical solution for the velocity profiles for two-phase flows through concentric and non-concentric annular flows. Caetano also presented a complex modeling to predict pressure gradients in these flows.

3. EXPERIMENTAL FACILITY

The experimental lab facility was conceived and built at PETROBRÁS R&D Centre, Rio de Janeiro, Brazil. The experimental conditions were set to simulate the typical conditions of surgence induction operations: for a given gas flow rate established at the surface (entering the coiled tubing) a specific volume of liquid shall be recovered as a function of the feeding system's productivity index and losses for each specific experimental run.

The lab facility had three basic systems: fluids feeding, return for fluids and test column. An air-water mixture was employed throughout the experiments.

Figure 1 shows an scheme of the lab facility. The equipment associated to each fluid stream is described below:

Liquid stream: storage tanks, pumping, flow rate metering station, flow rate control valve and pipes.

Gaseous stream: air compressors, pressure control valve, flow rate metering skid and pipes.

The system for fluids return is also shown in Fig. 1: return lines and a gas/liquid separator.

The test string and related instruments is shown in Figure 2. It was composed of a N-80 tubing column with 0.073 m O.D. with two transparent plexiglas sections with a coiled tubing inside it. The coil had 0.0254 m O.D. with a standard ending nozzle. The coiled tubing was fixed only at the top of the tubing string. This last one was conceived to allow an inclination between -15 to 90 degrees with the horizontal.

The column was operated by filling it up with liquid and injecting gas at its base through the described coil. The two-phase mixture so formed raises the column up to the outlet at the top and then is sent to the separation station through the returning system. The liquid feeding is kept constant by means of the flow control valve.

The P1 station measures pressure to the liquid rate control valve. P2 and P4 stations were used to take friction factors for one-phase liquid flows as well as two-phase flow pressure gradients. The P5 station was used to control the pressure levels at the outlet of the string, avoiding siphoning. Figure 3 shows the scheme used to measure single-phase friction factors for air inside the coiled tubing. T1 and T2 stations were assigned to take temperatures. Orifice plates were used to measure gas flow rates; liquid rates were taken through a turbine-type meter.

4. EXPERIMENTAL PROCEDURES

The following steps were adopted as experimental basis: calibration of all instruments; setting of a fixed gas rate; stabilizing liquid and gas rates; flow pattern characterization; reading of temperature, differential and gauge pressures and recovered liquid rates.

Two-phase flow patterns were detected by visual inspection through a transparent section located at the top of the column. The following criteria was adopted to identify the flow pattern transitions:

Bubble to Slug Flow: the onset of slug flow occurs whenever the small bubbles coalesce giving rise to a greater (Taylor) bubble which fills up most of the view-through section of the string; this bubble has a bullet-shaped format and a velocity greater than that experienced by the remaining smaller bubbles.

Slug to Churn Flow: Chaos takes place in the flow and distorted wrecks of Taylor bubbles are observed. A liquid film of constant profile and thickness around the just broken Taylor bubbles are no longer seen.

Churn to Annular Flow: the flow assumes a "foggy" pattern and both liquid and gas streams cannot be identified anymore.

Properties of both liquid and gas streams were determined by the correlation due to Miller(1983). A typical theoretical flow pattern map is presented in figure 4.

5. EXPERIMENTAL RESULTS

Table 1 shows experimental results for two-phase vertical flows whilst Tables 2A, 2B and 2C show the results found for two-phase inclined flows.

Figure 5 shows the experimental graph for Fanning friction factors for one-phase liquid or gas flows in annular geometry. It is also shown a curve set for theoretical friction factors for the lab procedures carried out by Bianco (1990) based on Caetano's (1986) work for several eccentricities (ε) and considering a relative roughness of 0.0045 and a diameter ratio of 0.409. It can be verified that the averaged eccentricity for the experimental system lies around 0.4 for a Reynolds number up to 50,000. From this Reynolds number on the eccentricity deviates from this average value due to the turbulence and vibrations in the coiled tubing inside the tubing. These variations were not taken into account in the theoretical modeling of jet-lifting flow.

Figure 6 shows measured Fanning friction factors for air inside the coiled tubing associated with the theoretical figures given by the Moody's diagram for a relative roughness of 0.003. The good agreement between the measured and predicted data is remarkable.

Figure 7 shows a flow pattern map built according Caetano(1986) and Kelessidis(1986) for experimentally observed superficial velocities in upward flow.

Figure 8 shows the relationship between the imposed gas flow rate and the flow rate of recovered liquid. Figure 9 shows the pressure at the column base (station P1) against gas flow rate. With the aid given by figure 8 one can verify in figure 9 that to a maximum liquid rate corresponds a minimum pressure value, for a same flow rate. In terms of mass the gas flow rate that maximizes the liquid recovery ranges around 0.024 kg/s which corresponds to a gas-liquid ratio of 0.020 ($\simeq$ 17 Nm^3/m^3 or $\simeq$ 95 scf/bbl).

There is a relationship between the existence of a region of maximum liquid production and two-phase flow pressure gradients. A reasonable explanation to this fact may be given as being this region where the pressure losses in two-phase flow are minimum. For gas rates close to the maximizing rates the liquid production is maximum coupled with optimized gas consumption. As long as the gas flow rates are increased beyond the maximizing rate the energy losses in the system also raise up without any corresponding increase in recovered liquid. Figure 10 shows the relationship between the recovered liquid rates and pressure at the column base, and an operational curve for the system is established. This curve shall be employed later to develop a liquid recovery model for jet-lifting.

In figure 11 it is presented the relationship between the total experimental pressure gradient and the gas rate imposed to the system. It can be seen three reasonably well defined regions where the behavior in the alignment of points can be explained from the analysis on different influences under which each region is submitted. The first one corresponds to slug flow near the bubble flow region; here the proeminent factor is the effect of decreasing specific mass of the mixture due to the incorporating gas mass to the liquid medium. In the second region the flow patterns varies from slug to chaotic (churn) and a balance between gravitational, frictional and accelerational forces is observed; the total gradient remains stable with a low decrease where the liquid phase gives its place to the gas phase which has a displacement with increasing velocities. The third region corresponds to annular flow where the gas phase takes a continuous phase format; with the increase in gas velocities smaller volumes of liquid are incorporated to the two-phase flow, adhering to the tube walls as a thin film with a quite smaller velocity when compared with the gaseous core velocity. Being so both gravitational and frictional forces decrease, and rising accelerational forces are observed.

Finally, figure 12 shows the effect on the string inclination in the liquid recovery by jet-lifting. It can be verified that the increase in inclination implies in a fall in the system performance, since a greater ability for gas segregation is allowed, lowering the drag of the liquid phase.

6. NUMERICAL MODELING

The experimental facility is pictorially represented in figure 2. Taking a control volume delimited by the tube walls and cross sections relative to the top and the bottom of the string as also as to the nozzle's region, the balance for momentum quantity in the vertical direction yields:

$$P_1 A_C + \frac{\dot{M}_L^2}{(\rho_L A_C)} = P_i A_T + \rho_L A_C g L_e + \frac{\dot{M}_G^2}{(\rho_{G,i} A_i)} + P_2 A_{an} + \int_1^2 (A_G \rho_G + A_L \rho_L) g \, dz + $$

$$+ \frac{\dot{M}_G^2}{(\rho_{G,2} A_{G,2})} + \frac{\dot{M}_L^2}{(\rho_L A_{L,2})} - \int_1^i \left(\frac{dP}{dz}\right)_{f,L} A_C \, dz - \int_i^2 \left(\frac{dP}{dz}\right)_{f,G,L} A_{an} \, dz \tag{1}$$

For the control volume limited by sections 1 and *i* yields:

$$P_1 A_C + \frac{\dot{M}_L^2}{(\rho_L A_C)} = P_i A_C + \rho_L A_C g L_e + \frac{\dot{M}_G^2}{(\rho_{G,i} A_i)} + \frac{\dot{M}_G^2}{(\rho_{G,i} A_{G,i})} + \frac{\dot{M}_L^2}{(\rho_L A_{L,i})} - \int_1^i \left(\frac{dP}{dz}\right)_{F,L} A_C dz \tag{2}$$

In order to simulate the behavior of the liquid feeding system we shall take into account that all irreversibilities upstream the section 1 are represented by the coefficient ξ_L. So it can be written:

$$P_1 = P_{atm} + \rho_L g\left(L_e + L_s\right) - \frac{1}{2}\xi_L \frac{\dot{M}_L^2}{\left(\rho_L A_C^2\right)} \tag{3}$$

Substituting eq. (3) into eq. (2) and grouping terms:

$$\rho_L g L_S - \int_i^2 \left(\alpha_G \rho_G + \alpha_L \rho_L\right) g\, dz = \left(P_2 - P_{atm}\right) - \int_1^i \left(\frac{dP}{dz}\right)_{f,L} dz - \int_i^2 \left(\frac{dP}{dz}\right)_{f,L} dz +$$

$$+\frac{\dot{M}_L^2}{\rho_L A_{an}^2}\left[\frac{1}{\alpha_{L,2}} - \frac{A_T}{\alpha_{L,i} A_C} + \frac{A_{an}^2}{A_C^2}\left(0.5\xi_L - 1\right)\right] + \frac{\dot{M}_G^2}{\rho_{G,2} A_{an}^2}\left[\frac{1}{\alpha_{G,2}} - \frac{\rho_{G,2} A_T}{\rho_{G,i}\alpha_{G,i} A_C} + \frac{\rho_{G,2} A_{an}^2}{\rho_{G,i} A_i A_C}\right] \tag{4}$$

Equation (4) represents the existing relationship between mass flow rates of liquid and gas respectively in the employed apparatus, since correlations for void fractions and frictional pressure losses are available. Todoroki *et al* (1973) had presented similar results for apparatus with gas injection at one side.

For small segments of pipe one could assume the following simplifications:

$$\rho_{G,2} \cong \rho_{G,i} \cong \bar{\rho}_G$$

$$\alpha_{G,2} = \rho_{G,i} = \bar{\alpha}_G = 1\text{-}H_L$$

$$\alpha_{L,2} = \rho_{L,i} = \bar{\alpha}_L = H_L$$

With the aid given by these three assumptions one could rewrite eq. (4) in the following manner:

$$\rho_L g L \left[\left(1 - H_L\right)\left(1 - \frac{\bar{\rho}_G}{\rho_L}\right) - \left(1 - \frac{L_S}{L}\right)\right] = \frac{\dot{M}_L^2}{\rho_L A_{an}^2}\left[\frac{A_{an}}{H_L A_C} + \frac{A_{an}^2}{A_C^2}\left(\frac{\xi_L}{2} - 1\right)\right] +$$

$$+ \frac{\dot{M}_G^2}{\bar{\rho}_G A_{an}^2}\left[\frac{A_{an}}{(1\text{-}H_L)A_C} + \frac{A_{an}^2}{A_i A_C}\right] + \left(P_2 - P_{atm}\right) - \overline{\left(\frac{dP}{dz}\right)}_{f,L} L_e - \overline{\left(\frac{dP}{dz}\right)}_{f,G,L} L \tag{5}$$

where "L" is the length of the annular section and L_S/L is the submergence of the coiled tubing inside the column.

Separating the liquid mass flow rate term from eq. (5) one can obtain the relationship between the volumetric flow rates of liquid and gas:

$$\bar{Q}_L = \left[\frac{C_1(H_L) - C_2(Q_G, H_L) + C_3(Q_L) + C_4(Q_L, Q_G, H_L) - C_5}{C_6(H_L)}\right]^{1/2} \tag{6}$$

where

$$C_1(H_L) = \rho_L g L \left[\left(1 - H_L\right)\left(1 - \frac{\bar{\rho}_G}{\rho_L}\right) - \left(1 - \frac{L_S}{L}\right)\right] \tag{7}$$

$$C_2(Q_G,H_L) = \frac{\rho_{Gs}\bar{Q}_G^2}{\bar{\rho}_G A_C A_{an}}\left[\frac{1}{(1-H_L)} - \frac{A_{an}}{A_i}\right] \tag{8}$$

$$C_3(Q_L) = \overline{\left(\frac{dP}{dz}\right)}_{f,L} L_e = -2f_T\,\rho_L \frac{\bar{Q}_L^2}{D_C\,A_T^2} L_e \tag{9}$$

$$C_4(Q_L,Q_G,H_L) = \overline{\left(\frac{dP}{dz}\right)}_{f,G,L} L \tag{10}$$

$$C_5 = \left(P_2 - P_{atm}\right) \tag{11}$$

$$C_6(H_L) = \frac{\rho_L}{A_{an}A_C}\left[\frac{1}{H_L} + \frac{A_{an}}{A_C}\left(\frac{\xi_L}{2} - 1\right)\right] \tag{12}$$

$$H_L = H_L(Q_L,Q_G) \tag{13}$$

Equations (10) and (13) are defined from a two-phase flow model.

Todoroki *et al* (1973) have also presented a relationship between the so-called "system's dimensionless kinetic energy" and the volumetric fraction of gas, which takes the following form in the present case:

$$\frac{\bar{Q}_L^2}{A_{an}^2 gL} = \frac{C_1(H_L)-C_2(Q_G,H_L)+C_3(Q_L)+C_4(Q_L,Q_G,H_L)-C_5}{2A_{an}^2 gLC_6(H_L)} \tag{14}$$

The above equation suggests the existence of an optimum value for the void fraction which maximizes the kinetic energy of the system and is also a function of the coiled tubing submergence.

7. APPLICATION OF A TWO-PHASE FLOW MODEL

Aiming to verify the behavior of the proposed model the correlation of Zuber & Findlay(1965) to predict the average volumetric fraction of liquid in two-phase flow will be adopted:

$$H_L = 1 - \frac{\dot{M}_G}{C_{G,L}\left(\dot{M}_G+\dot{M}_L\frac{\bar{\rho}_G}{\rho_L}\right) + \bar{\rho}_G A_{an} U_S} \quad (15)$$

where $C_{G,L}$ is refined experimentally and

$$U_S = 0.345\sqrt{g(D_{i,T}+D_{e,fl})} \quad (16)$$

is the slip velocity between phases for the slug flow pattern in annular geometry according to Sadatomi *et al* (1982).

It will be also assumed that the two-phase frictional pressure gradient can be represented by the expression due to Akagawa (1957):

$$\overline{\left(\frac{dP}{dz}\right)}_{f,G,L} = H_L^{-k}\overline{\left(\frac{dP}{dz}\right)}_{f,L} = -2f_{an}\frac{H_L^{-k}\left(\dot{M}_L+\dot{M}_G\right)^2}{D_{eq}\rho_L A_{an}^2} \quad (17)$$

where "k" is a factor fitted by experiments.

Substituting eqs. (15) and (17) in equations (7) to (13) one can now have all elements to predict the recovered liquid flow rate, given: gas flow rate, back pressure in the returning line, physical properties of both liquid and gas streams, dimensions of both column and coiled tubing and depths.

Table 3 presents the comparative results between lab data for liquid flow rates and the values generated by the numerical simulator. The experimental fitted constants were: $C_{G,L}$ = 1.20 and k = 2.10. The average Fanning friction factor for the test column is equal to 0.0075. This factor was employed throughout the numerical simulation. The coefficient for irreversible losses was adjusted through figure 10 and set as being ξ_L=457.0.

Figure 13 presents a graph on the existing relationship between the imposed gas flow rates and liquid rates as calculated by the numerical simulator. Figure 14 presents a diagram with dimensionless kinetic energy as a function of gas/liquid ratio. In these two graphs it can be clearly verified the existence of points of maximum values assuring the optimization of the system.

8. KEYS FOR OPERATIONAL PROCEDURES

Todoroki *et al* (1973) suggested that the system's maximum efficiency during air lifting is reached under the following conditions:

$0.4 < U_{SL} < 0.6$

where U_{SL} is the superficial velocity of the liquid phase in an annulus as given by the expression below:

$$U_{SL} = \frac{Q_L}{A_{an}} \tag{18}$$

where

$$A_{an} = \frac{\pi}{4}\left(D^2_{int,T} - D^2_{ext,ft} \right) \tag{19}$$

$$SUB = \frac{L_s}{L} \quad , \ 0.6 < SUB < 0.8 \tag{20}$$

Being so it is a must to know and control the following parameters for a better planning of jet-lifting operations:

* The static level of fluid inside the well;
* Maximum depth (length) of the coiled tubing;
* In the case of composed columns, the depth in which the diameter changes;
* Inside diameter of the column and external coiled tubing diameter.

With all these information the depth to position the end of the coiled tubing can be defined (eq. (20)) before starting the displacement of nitrogen. It can also be known which liquid flow rate yields (eq. (18)) for an optimum gas consumption. The adjustment shall be done by varying the nitrogen flow rate up to the point in which the specified liquid velocity is reached.

It shall be kept in mind that *any* changes in the flow parameters demand a lag time to stabilize the flow. This is due to a new equilibrium level onset between phase equilibrium and flow rates. Sharp and sudden variations in nitrogen pumping rates must be avoided so.

9. CONCLUSIONS

An experimental facility was designed, built and employed to generate and collect data in fluid displacement with air and water. This facility showed a good performance in the job to which it had been bounded for: the evaluation of the main parameters of jet-lifting flows with coiled tubing and the existing relationships between imposed gas flow rates and recovered liquid flow rates.

One-phase friction factors for liquid (water) and gas (air) in annular geometries similar to those found in the field were experimentally determined and checked. The one-phase friction factor for air inside a coiled tubing with 0.0254 m O.D. were also collected.

An optimum gas flow rate for maximum liquid recovery was experimentally found. It was verified that this optimum rate is quite below the values usually found in field. For example, for the vertical column with 11 m high/0.073 m O.D. a gas mass flow rate of 0.024 kg/s was found as the maximizing rate. It was also observed that the inclination of the system causes a decrease in the system's performance, since the gas stream flows preferably through the mixture at the inner top of the annular space between the column and the coiled tubing, decreasing the drag on the liquid phase. With a 45 degree inclination the performance dropped approximately 30%.

A numerical model for upward flows to simulate jet-lifting operations was developed. Comparisons of this tool against models such as Zuber & Findlay and Akagawa (1957) were performed. The results showed a good agreement with lab data and with the results published by Todoroki *et al* (1973).

Specifically in relation with the experiments, the model showed an error of 0.2% and a standard deviation of ± 5.7 for the measured recovered liquid flow rate. This fact demonstrates the ability of the numerical model in the simulations of jet-lifting situations for the employed experimental facility.

Output printings during the numerical simulation showed that accelerational pressure drops are very high for some studied cases, and accelerations gradients greater than 5% of the total pressure drop were observed in some cases.

ACKNOWLEDGEMENTS

The authors wish to express their full gratitude for all the support given by PETROBRÁS S.A. and by PETROBRÁS R&D Centre in all the steps of this work. Thanks shall also be given to all the people who has made this job possible.

REFERENCES

AKAGAWA, T. - *Trans. Japan Soc. Mech. Eng.*,24, 128, p. 292, (1957)

BIANCO, L.C.B. - "Estudo de Otimização das Operações com Tubo Flexível e Nitrogênio", M.Sc. Thesis, Universidade Estadual de Campinas, Brazil (1990).

BIRD, R., STEWART, W. & LIGHTFOOT, E. - "Transport Phenomena", John Wiley & Sons, (1976).

CAETANO, E.F. - "Upward Vertical Two-phase Flow through an Annulus", Ph.D. Thesis, The University of Tulsa (1985).

CASHION, J.L. - "Use of Nitrogen and Coiled Tubing in Deep Wells", *Petr. Eng.*, 45, 3, (1973).

DUNS, Jr, H. & ROS, N.C.J. - "Vertical Flow of Gas and Liquid Mixture in Wells", *Proceedings of the 6th World Pet. Congress*, 451, (1963).

HAGEDORN, A.R. & BROWN, K.E. - "Experimental Study of Pressure Gradients Occuring During Continuous Two-Phase Flow in Small-Diameter Vertical Conduicts", *J. Pet. Tech.*, pp: 475-484, (1965).

KELESSIDIS, V.C. - "Vertical Upward Gas-Liquid Flow in Concentric and Eccentric Annuli", Ph.D. Thesis, University of Houston, Texas, (1986).

KELESSIDIS, V.C. & DUKLER, A.E. - "Modeling Flow Pattern Transitions for Upward Gas-Liquid Flow in Concentric and Eccentric Annuli", *Int. J. of Multiphase Flow*, 15, 2, pp: 173-191, (1989).

MILLER, R.W. - "Flow Measurement Engineering Handbook", McGraw-Hill Books, (1983).

POETTMAN, F.H. & CARPENTER, P.G. - "The Multiphase Flow of Gas, Oil and Water Through Vertical Flow Strings with Application to the Design of Gas-lift Installations", *Drill and Prod. Prac.*, API, (1952).

SADATOMI, M., SATO, Y. & SARUWATARI, S. - "Two Phase Flow in Vertical Noncircular Channels", *Int. J. Mult. Flow*, 8, 6, pp: 641-655, (1982).

SLATOR, D.T. & HANSON Jr., W.E. - "Continuous String Light Workover Unit", *J. Pet. Tech.*, 17, pp:39-44 (1965).

SNYDER, W.T. & GOLDSTEIN, G.A. - "An analysis of Fully Developed Laminar Flow in an Eccentric Annulus", *AIChE J.*, 11, pp:462-467, (1965).

TODOROKI, I., SATO, T., & HONDA, T. - "Performance of Air-lift Pump", *Bulletin JSME*, 16:94, pp:733-741 (1973).

TOSUN, I. - "Axial Laminar Flow in a Eccentric Annulus: an Approximate Solution", *AIChE J.*, 30, PP:877-878, (1984).

WEEKS, S.G. - "Coil Tubing, Nitrogen Cut Workover Costs", *World Oil*, 70, 2, pp:29-32, (1970)

ZUBER, N. & FINDLAY, J. - "Average Volumetric Concentration in Two-phase Systems", *ASME J. Heat Transfer*, 87, pp:453-468, (1965).

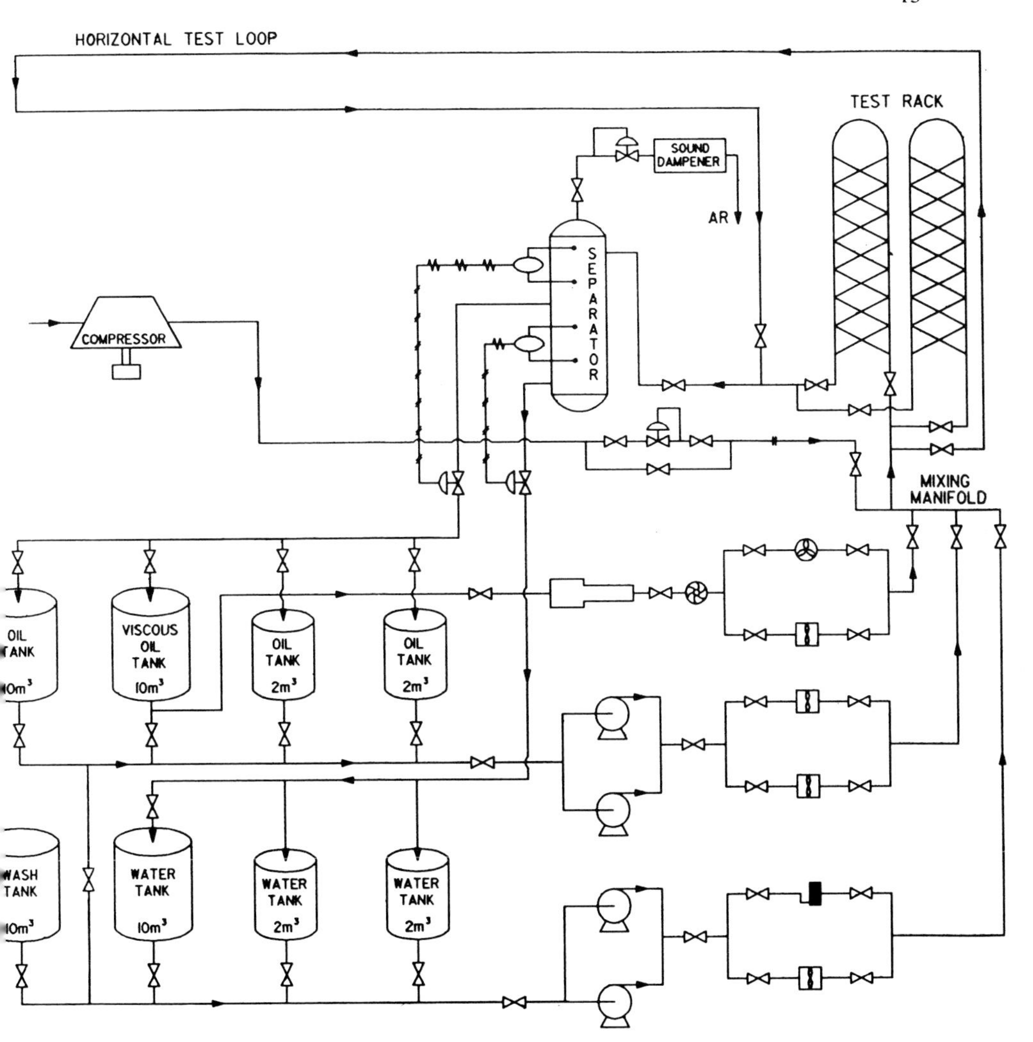
HORIZONTAL TEST LOOP
TEST RACK
SOUND DAMPENER
AR
SEPARATOR
COMPRESSOR
MIXING MANIFOLD
OIL TANK
VISCOUS OIL TANK 10m³
OIL TANK 2m³
OIL TANK 2m³
WASH TANK
WATER TANK 10m³
WATER TANK 2m³
WATER TANK 2m³

Figure 1

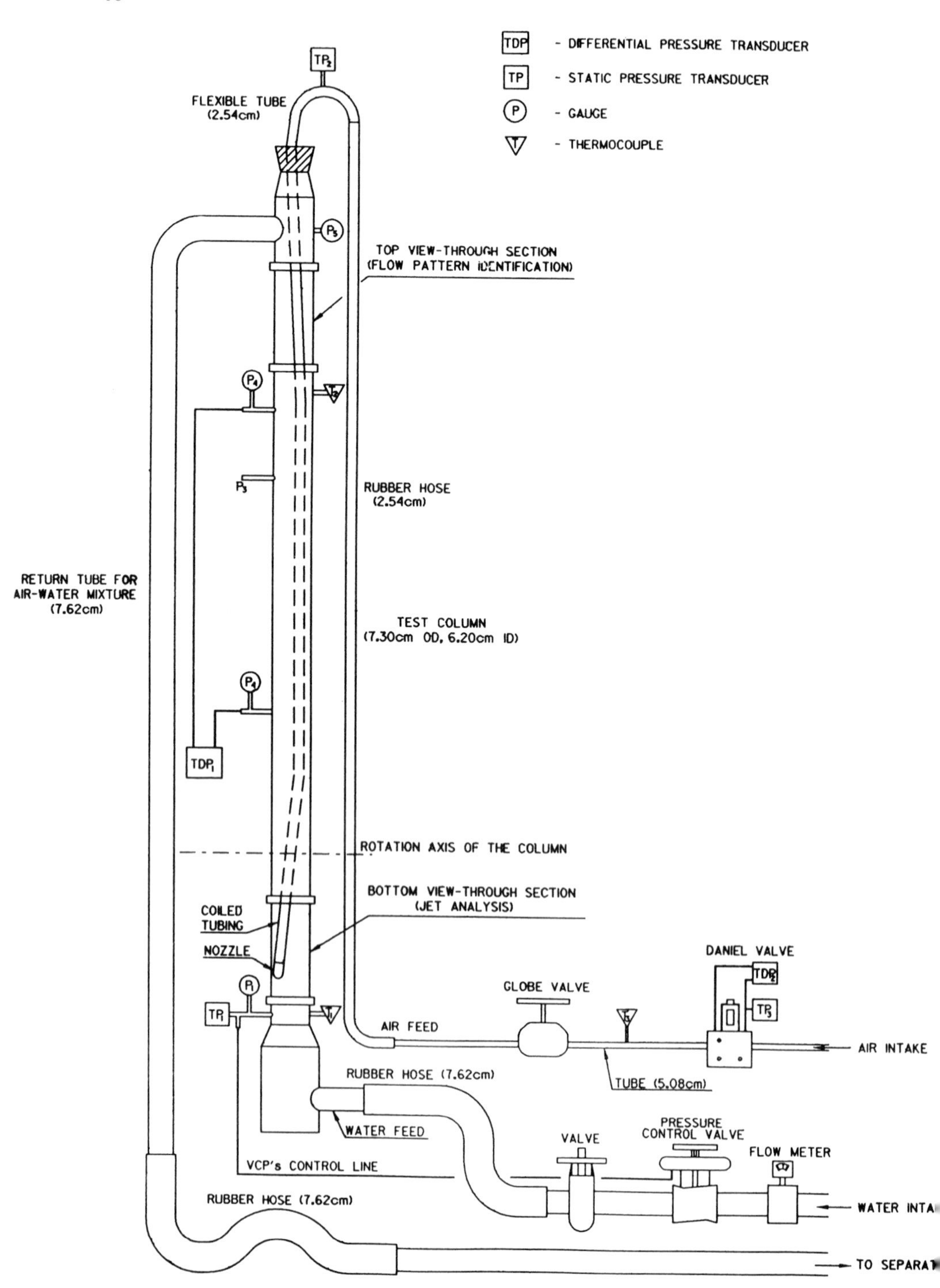
TDP - DIFFERENTIAL PRESSURE TRANSDUCER
TP - STATIC PRESSURE TRANSDUCER
P - GAUGE
T - THERMOCOUPLE
FLEXIBLE TUBE (2.54cm)
TOP VIEW-THROUGH SECTION (FLOW PATTERN IDENTIFICATION)
RUBBER HOSE (2.54cm)
RETURN TUBE FOR AIR-WATER MIXTURE (7.62cm)
TEST COLUMN (7.30cm OD, 6.20cm ID)
ROTATION AXIS OF THE COLUMN
BOTTOM VIEW-THROUGH SECTION (JET ANALYSIS)
COILED TUBING
NOZZLE
DANIEL VALVE
GLOBE VALVE
AIR FEED
AIR INTAKE
TUBE (5.08cm)
RUBBER HOSE (7.62cm)
WATER FEED
VALVE
PRESSURE CONTROL VALVE
FLOW METER
VCP's CONTROL LINE
RUBBER HOSE (7.62cm)
WATER INTA
TO SEPARAT

Figure 2

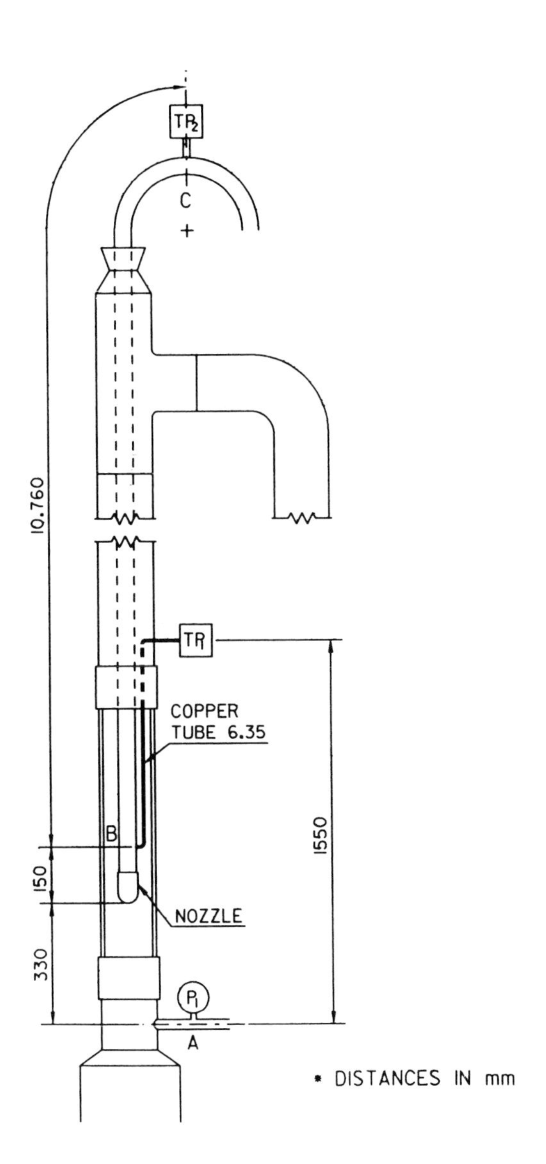
TP2
C
10.760
TR
COPPER
TUBE 6.35
B
150
NOZZLE
330
1550
P1
A
* DISTANCES IN mm

Figure 3

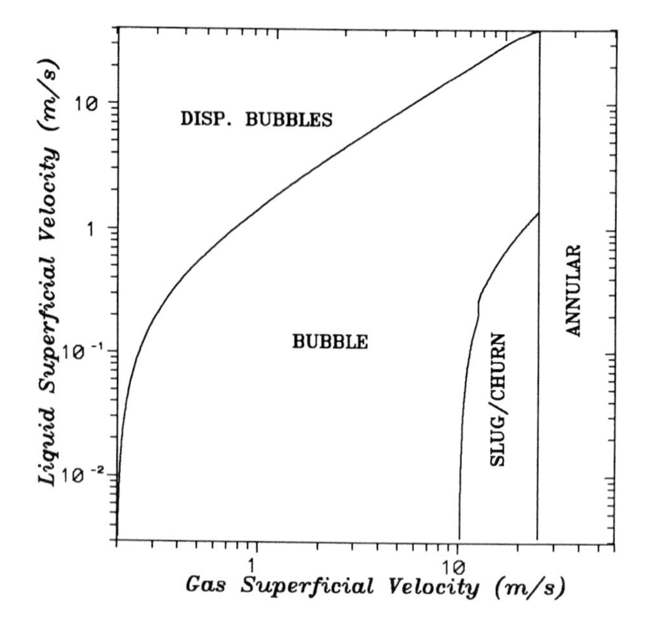

Figure 4

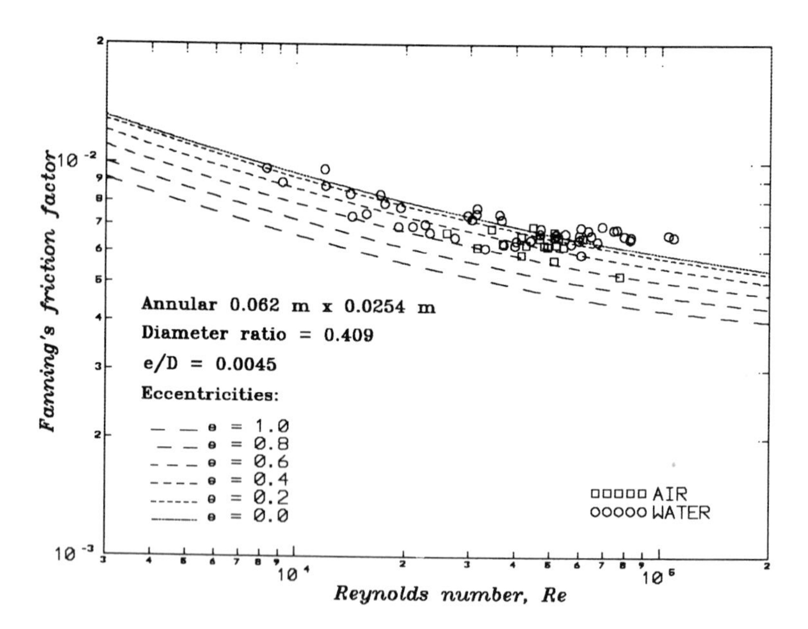

Figure 5

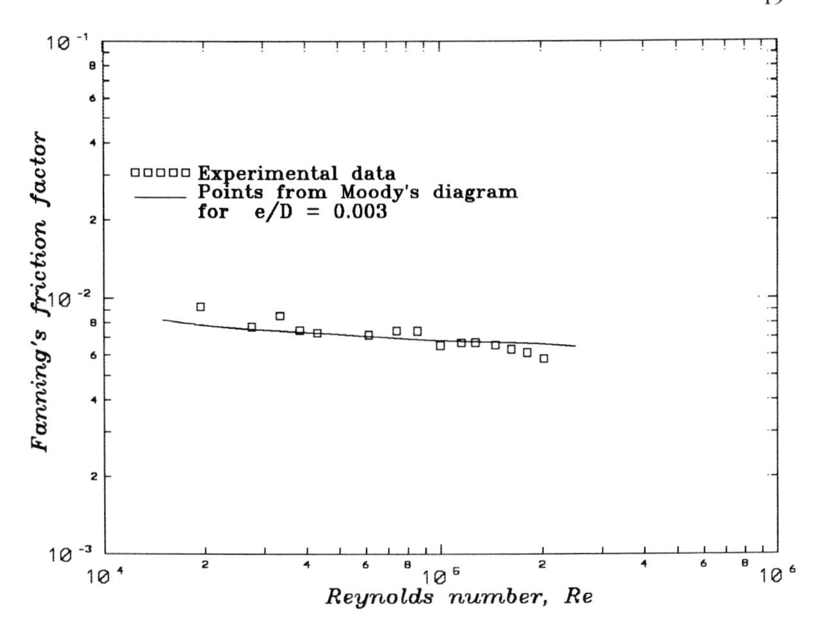

Figure 6

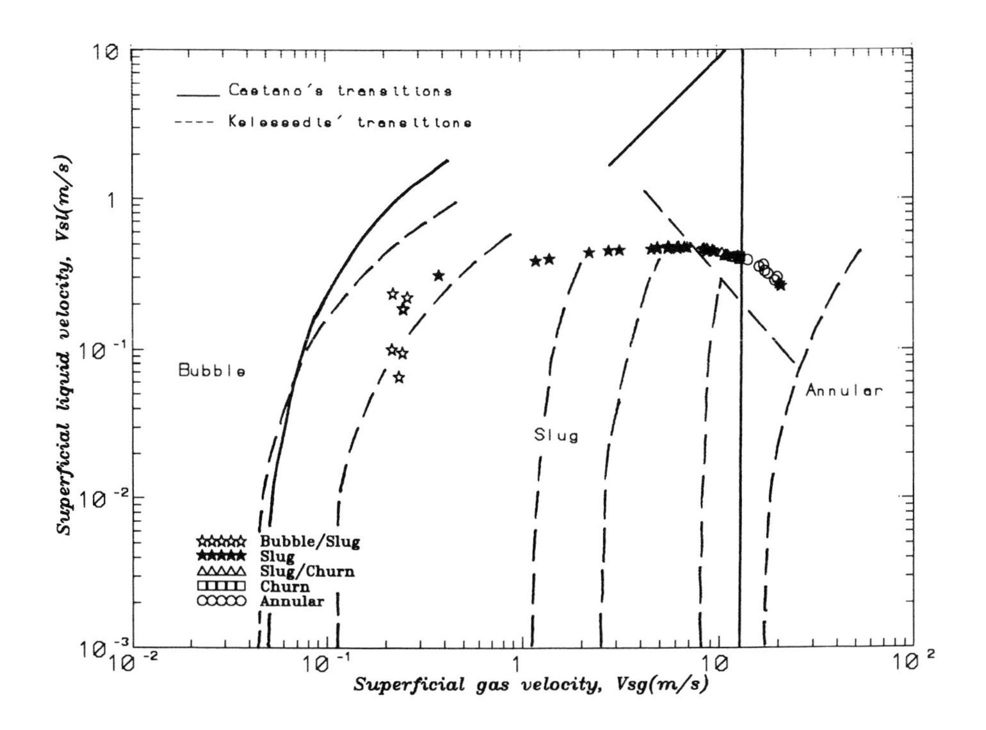

Figure 7

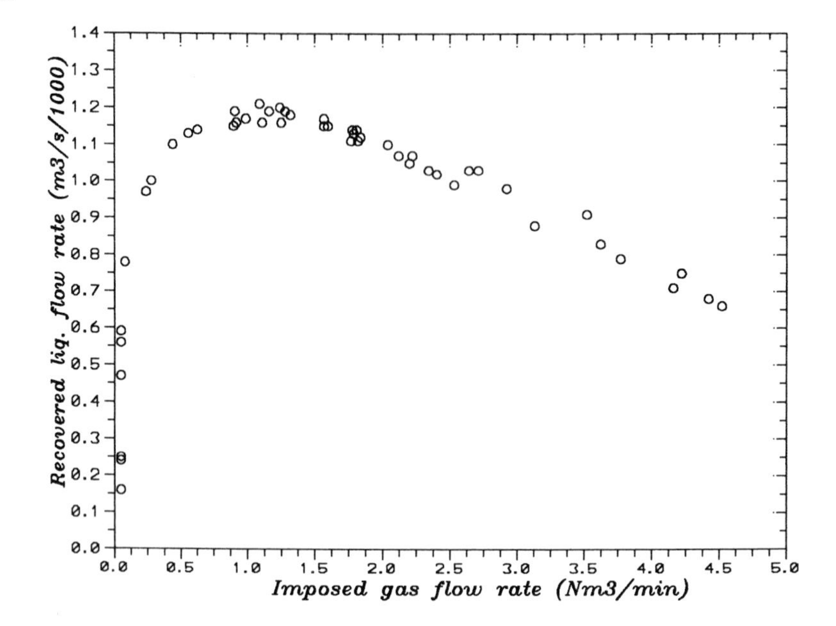

Figure 8

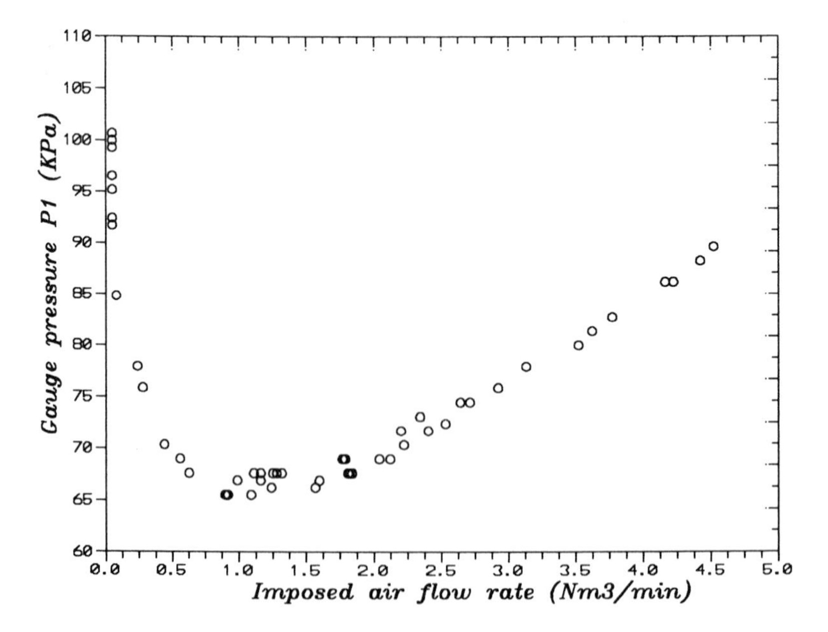

Figure 9

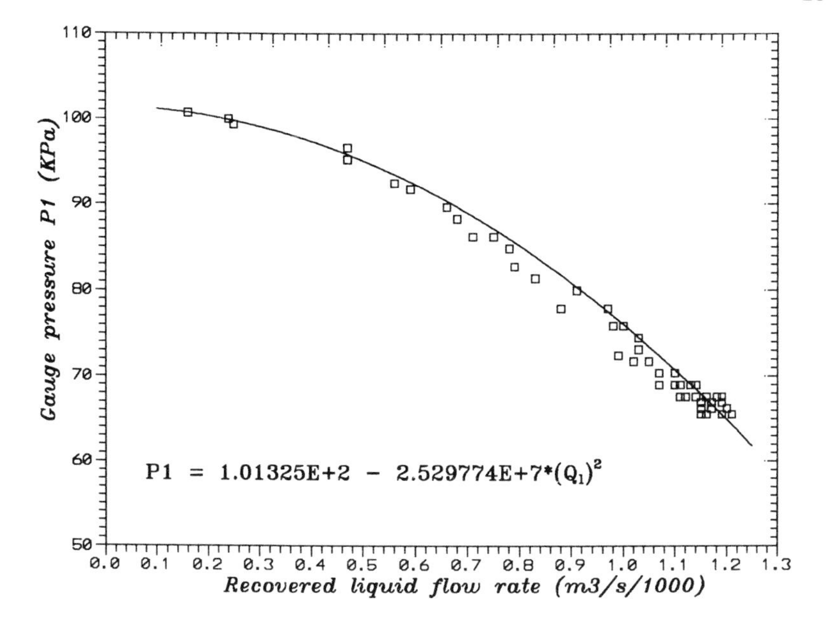

Figure 10

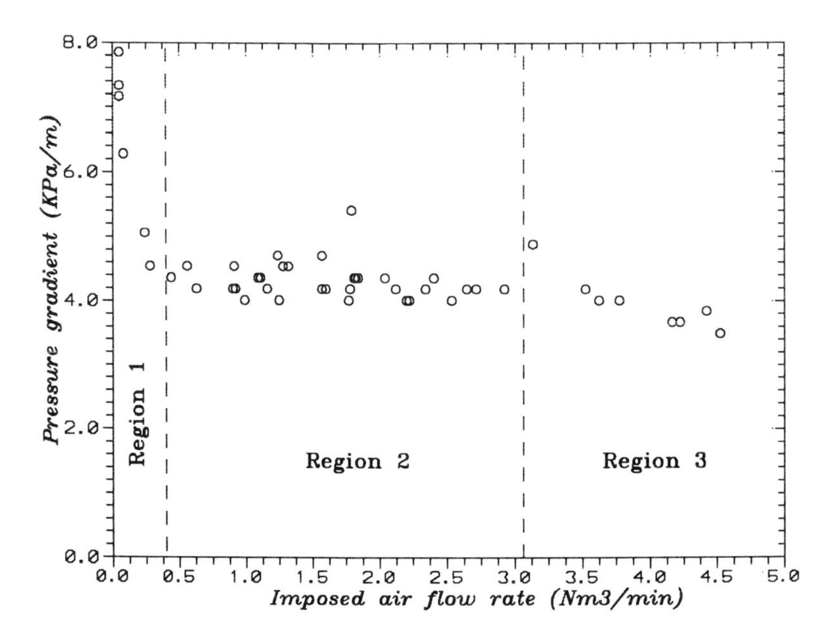

Figure 11

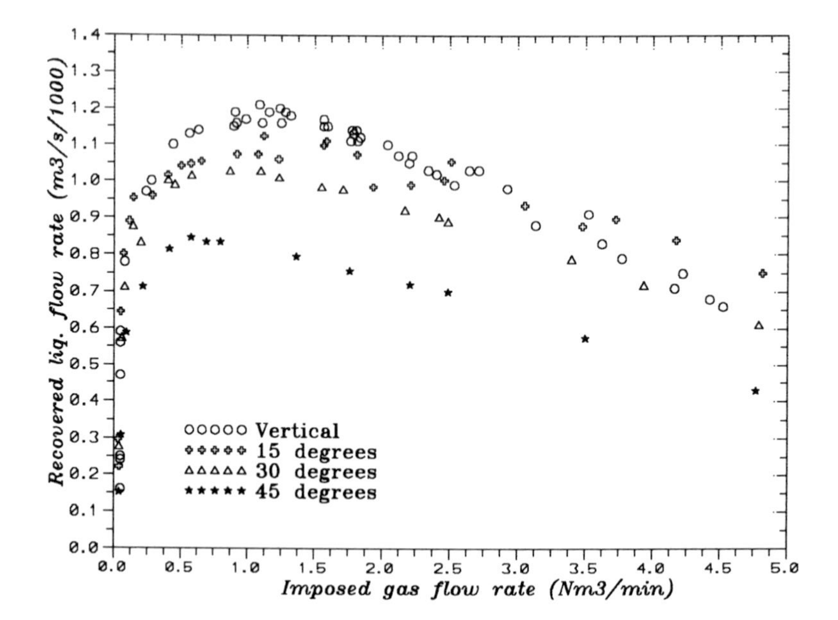

Figure 12

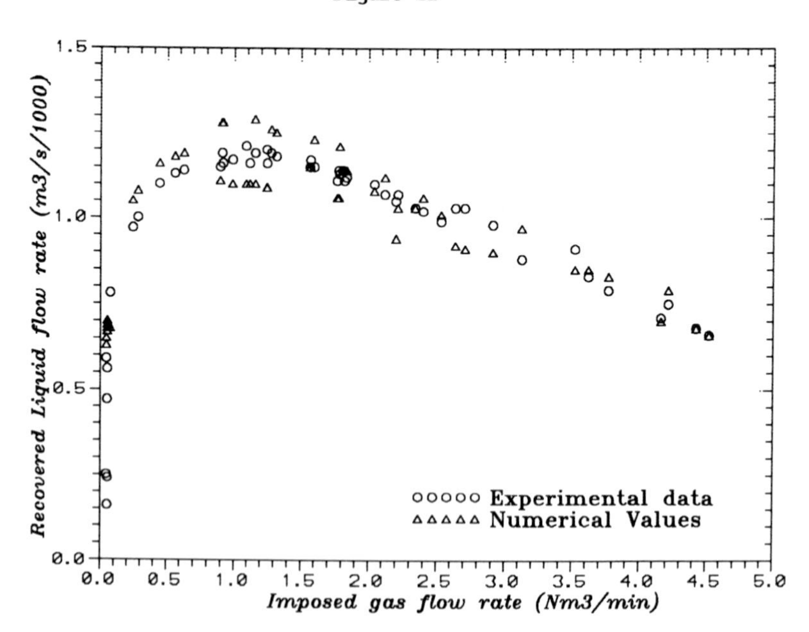

Figure 13

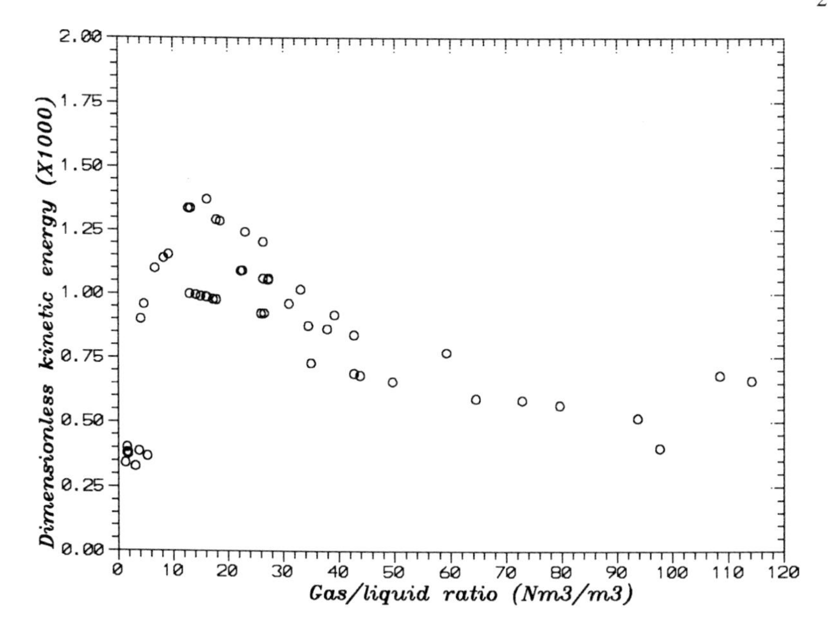
Dimensionless kinetic energy (X1000)
2.00
1.75
1.50
1.25
1.00
0.75
0.50
0.25
0.00
0
10
20
30
40
50
60
70
80
90
100
110
120
Gas/liquid ratio (Nm3/m3)

Figure 14

TABLE 1

VERTICAL STRING

OBS	Qgas scf/min	Qliq gpm	TP1 psi	P1 psi	P2 psi	P4 psi	P5 psi	TP2 psi	TDP psi	Tcol dg C	FLOW PATT.	JET cm
1	45.1	18.8	10.1	9.8	6.1	3.5	1.7	15.2	2.3	28.0	SLUG	15.0
2	32.2	18.8	9.8	9.5	5.6	3.0	1.7	12.8	2.1	28.0	SLUG	13.0
3	55.5	18.5	9.8	9.6	5.6	3.2	2.8	17.8	1.8	29.0	SLUG	17.0
4	63.9	18.0	10.1	9.8	6.4	3.9	2.8	20.0	1.8	29.0	SLUG	18.0
5	72.2	17.5	10.3	10.0	6.6	4.1	3.3	22.0	1.8	28.0	SLUG	19.0
6	78.2	17.0	10.5	10.2	6.8	4.5	3.8	24.2	1.8	28.0	SLUG	20.0
7	82.8	16.3	10.9	10.6	6.8	4.4	3.8	26.8	2.1	25.0	SL/CH	20.0
8	89.5	15.7	10.8	10.5	6.6	4.3	3.8	28.8	1.8	26.0	SL/CH	20.5
9	128.0	13.1	12.1	11.8	7.7	5.4	4.9	44.0	1.6	26.0	CHURN	23.0
10	64.4	17.6	10.1	9.8	6.4	3.9	2.8	20.0	1.8	26.0	SLUG	18.5
11	32.4	18.4	9.8	9.5	5.6	3.2	1.7	12.6	1.9	26.0	SLUG	13.0
12	110.5	14.0	11.6	11.3	6.4	3.6	3.8	36.0	2.1	28.0	CHURN	22.0
13	55.4	18.3	9.9	9.6	5.0	2.3	2.8	17.0	2.3	28.0	SLUG	17.0
14	147.0	11.3	12.8	12.5	8.2	6.1	6.0	53.0	1.4	29.0	CH/AN	24.0
15	133.2	12.6	12.3	12.0	8.0	5.7	4.9	47.0	1.5	29.0	CHURN	23.5
16	64.9	17.7	10.1	9.8	6.4	3.9	2.8	19.8	2.0	30.0	SLUG	18.5
17	84.8	16.2	10.7	10.4	6.8	4.3	3.3	26.0	1.8	29.0	SL/CH	19.5
18	148.9	11.9	12.8	12.5	8.0	5.9	4.9	50.5	1.5	30.0	CH/AN	24.0
19	159.7	10.4	13.3	13.0	8.6	6.6	6.0	58.5	1.3	30.0	ANNUL	25.0
20	46.5	18.7	10.1	9.8	6.1	3.5	1.7	15.0	1.8	30.0	SLUG	16.0
21	156.3	10.8	13.1	12.8	8.6	6.4	6.0	50.9	1.4	29.0	ANNUL	25.0
22	75.0	17.0	10.3	10.0	6.6	4.2	2.8	23.0	1.8	30.0	SLUG	19.0
23	56.5	18.2	9.9	9.7	6.3	3.9	1.7	17.2	1.8	30.0	SLUG	17.0
24	40.8	18.8	10.1	9.8	5.6	3.2	1.3	13.5	1.8	29.0	SLUG	15.0
25	31.8	18.3	9.8	9.5	6.5	4.1	3.8	11.2	1.9	27.0	SLUG	13.0
26	38.4	19.2	9.9	9.5	6.6	4.1	3.8	12.7	1.9	29.0	SLUG	14.0
27	44.0	19.0	10.0	9.6	6.9	4.2	3.8	13.8	1.9	29.0	SLUG	15.0
28	2.7	12.4	12.6	12.3	7.7	4.1	2.8	11.1	3.1	30.0	SLUG	0.5
29	1.8	7.4	14.0	13.8	8.8	4.7	1.8	12.7	3.8	31.0	BU/SL	0.5
30	1.6	9.4	13.6	13.3	9.3	4.8	1.8	13.1	4.0	30.0	BU/SL	0.5
31	1.9	8.9	13.6	13.4	8.7	4.6	1.8	12.4	3.7	30.0	BU/SL	0.5
32	1.8	7.5	13.7	14.0	8.8	4.7	1.8	12.9	3.8	31.0	BU/SL	0.5
33	1.6	4.0	14.7	14.4	9.3	4.8	1.8	13.3	4.1	30.0	BU/SL	0.5
34	1.9	3.8	14.8	14.5	9.8	5.7	1.8	14.0	3.6	28.0	BU/SL	0.5
35	1.8	2.6	14.9	14.6	9.7	5.5	1.8	13.8	3.5	28.0	BU/SL	0.5
36	8.5	15.4	11.6	11.3	7.7	4.8	2.8	11.0	2.4	30.0	SLUG	3.5
37	19.8	17.9	10.4	10.0	7.4	4.8	2.8	11.1	2.0	30.0	SLUG	9.0
38	15.6	17.4	10.6	10.2	7.3	4.8	2.8	10.9	2.1	31.0	SLUG	7.0
39	9.9	15.8	11.3	11.0	7.4	4.8	2.8	10.9	2.3	31.0	SLUG	4.0
40	44.1	18.4	10.1	9.8	7.3	5.0	3.8	14.4	1.9	31.0	SLUG	15.5
41	34.9	18.6	10.0	9.7	7.2	4.9	3.8	12.9	1.9	31.0	SLUG	13.5
42	22.2	18.1	10.1	9.8	7.2	4.8	2.8	11.1	2.0	32.0	SLUG	9.5
43	77.8	16.7	10.8	10.4	7.5	5.2	4.9	23.7	1.8	31.0	SLUG	19.5
44	62.5	17.6	10.4	10.0	7.4	5.1	3.8	18.0	1.8	31.0	SLUG	18.0
45	39.4	18.4	10.2	9.8	7.3	4.8	3.8	12.8	1.9	31.0	SLUG	14.0
46	124.5	14.4	12.1	11.6	7.8	5.4	4.9	42.0	1.8	31.0	CHURN	22.0
47	103.0	15.5	11.4	11.0	7.7	5.3	4.9	30.7	1.8	30.0	SL/CH	20.5
48	62.8	18.1	10.4	10.0	7.2	4.8	3.8	17.6	1.8	30.0	SLUG	17.5
49	93.2	16.4	11.1	10.8	7.9	5.5	4.9	25.9	1.8	27.0	SL/CH	20.5
50	95.7	16.3	11.1	10.8	7.8	5.4	4.9	27.0	1.8	27.0	SL/CH	20.5
51	63.1	17.9	10.4	10.0	7.6	4.5	1.8	18.0	1.8	27.0	SLUG	18.5
52	40.9	18.8	10.0	9.7	7.4	5.0	3.8	13.0	1.9	26.0	SLUG	14.5

TABLE 2A

EXPERIMENTAL DATA FOR AIR-WATER TWO-PHSE FLOWS IN ANNULAR GEOMETRY.
UPWARD INCLINATION WITH VERTICAL: 15 DEGREES.

OBS	Qgas scf/min	Qliq gpm	TP1 psi	P1 psi	P2 psi	P4 psi	P5 psi	TP2 psi	TDP psi	T dgC	FLOW PATT.	JET cm
1	78.1	15.7	10.9	10.4	7.4	4.5	4.9	22.0	0.0	24	CHURN	19.5
2	68.3	15.6	10.7	10.1	7.3	4.3	6.0	20.2	0.0	24	CHURN	19.0
3	56.1	17.6	10.7	10.0	7.4	4.3	4.9	16.1	0.0	24	SL/CH	17.5
4	39.7	17.8	10.5	10.0	7.3	4.3	4.9	12.8	0.0	25	SLUG	15.0
5	55.4	17.4	10.8	10.0	7.3	4.2	4.9	16.2	0.0	25	SL/CH	17.0
6	23.1	16.7	9.1	10.0	7.4	4.3	4.9	10.8	0.0	26	SLUG	10.0
7	32.7	17.0	9.5	10.0	6.9	4.5	4.9	12.0	0.0	27	SLUG	14.0
8	38.1	17.0	9.5	10.0	7.2	4.2	4.9	13.0	0.0	26	SLUG	15.5
9	43.5	16.8	9.5	10.1	7.4	4.2	6.0	14.1	0.0	26	SLUG	16.0
10	20.2	16.6	10.8	10.0	6.9	4.2	4.9	11.3	2.5	26	SLUG	9.5
11	17.6	16.5	10.9	10.0	6.8	4.2	4.9	11.2	2.4	27	SLUG	8.0
12	14.2	16.1	11.1	10.1	6.9	4.1	4.9	11.1	2.5	27	SLUG	7.0
13	10.2	15.2	11.4	10.2	7.2	4.1	4.9	11.0	2.7	27	SLUG	4.5
14	1.4	3.5	14.9	14.1	9.4	4.6	4.9	13.7	4.3	31	SLUG	0.5
15	1.1	4.7	15.1	14.1	9.4	4.5	4.9	14.0	4.3	31	SLUG	0.5
16	1.8	10.2	13.7	13.0	8.6	4.5	4.9	12.8	3.7	31	SLUG	0.5
17	2.6	12.7	11.4	11.5	7.5	4.3	4.9	11.0	3.3	30	SLUG	2.0
18	4.1	14.1	10.9	11.0	7.4	4.1	4.9	10.6	2.9	29	SLUG	2.5
19	5.1	15.1	10.7	10.8	7.3	4.1	4.9	10.8	2.8	29	SLUG	3.0
20	64.1	17.0	10.3	10.1	7.2	4.2	6.0	18.9	2.4	29	SL/CH	17.5
21	86.8	15.9	10.6	10.3	7.3	4.5	6.0	26.5	2.4	30	CHURN	20.5
22	107.7	14.8	10.9	11.1	7.4	4.3	6.0	33.8	2.4	29	CHURN	21.5
23	122.7	13.9	11.4	11.3	7.4	4.7	6.0	40.4	2.3	27	CH/ANN	23.0
24	170.0	11.9	10.2	12.2	7.9	5.1	6.0	.0	2.3	30	ANN	25.0
25	147.3	13.3	12.6	12.9	7.6	4.9	6.0	60.0	2.3	29	ANN	23.0
26	131.5	14.2	12.0	11.8	7.5	4.8	6.0	50.0	2.3	30	CH/ANN	22.0
27	88.5	16.7	11.6	10.8	7.5	4.5	6.0	43.0	2.4	30	CHURN	21.0

TABLE 2B

EXPERIMENTAL DATA FOR AIR-WATER TWO-PHSE FLOWS IN ANNULAR GEOMETRY.
UPWARD INCLINATION WITH VERTICAL: 30 DEGREES.

OBS	Qgas scf/min	Qliq gpm	TP1 psi	P1 psi	P2 psi	P4 psi	P5 psi	TP2 psi	TDP psi	T dgC	FLOW PATT.	JET cm
1	169.1	9.7	11.6	11.4	7.1	4.8	6.0	59.3	1.9	31	ANNUL	26.0
2	138.9	11.4	11.0	11.1	6.9	4.6	6.0	46.2	2.1	31	CHURN	23.0
3	87.7	14.1	10.0	10.2	6.8	4.2	6.0	24.5	2.3	31	SL/CH	25.0
4	119.9	12.5	10.7	10.5	6.9	4.5	6.0	39.4	2.1	32	CHURN	24.0
5	85.4	14.3	10.1	10.1	6.8	4.2	6.0	25.2	2.2	32	SL/CH	21.0
6	60.5	15.5	9.5	9.8	6.8	4.0	6.0	17.2	2.2	32	SLUG	19.0
7	76.6	14.6	9.9	10.0	6.8	4.1	6.0	22.4	2.3	31	SL/CH	20.0
8	54.9	15.6	9.5	9.6	6.6	4.0	5.5	16.0	2.3	31	SLUG	18.5
9	38.7	16.3	9.2	9.2	6.3	3.8	5.5	12.2	2.2	31	SLUG	14.0
10	43.6	16.0	9.3	9.2	6.3	3.8	6.0	13.8	2.3	31	SLUG	17.0
11	30.8	16.3	9.2	9.1	6.3	3.9	5.5	11.4	2.2	31	SLUG	12.5
12	16.0	15.7	9.4	9.4	6.6	3.9	5.5	9.5	2.3	31	SLUG	6.0
13	20.4	16.1	9.3	9.1	6.4	3.9	5.5	10.5	2.2	30	SLUG	9.0
14	14.3	15.9	9.3	9.2	6.5	3.9	5.5	9.7	2.2	30	SLUG	6.5
15	7.1	13.2	9.8	9.8	6.9	4.0	5.5	10.0	2.6	30	SLUG	2.5
16	5.1	13.9	9.7	9.8	6.8	3.9	5.5	9.8	2.6	30	SLUG	3.0
17	2.9	11.3	11.0	10.9	7.5	4.1	5.5	10.5	3.1	31	SLUG	1.0
18	2.2	9.1	11.7	11.5	8.0	4.1	5.5	11.1	3.5	31	SLUG	0.5
19	1.4	4.4	12.4	12.5	8.4	4.2	4.9	11.7	3.8	32	SLUG	0.5

TABLE 2C

EXPERIMENTAL DATA FOR AIR-WATER TWO-PHSE FLOWS IN ANNULAR GEOMETRY.
UPWARD INCLINATION WITH VERTICAL: 45 DEGREES.

OBS	Qgas scf/min	Qliq gpm	TP1 psi	P1 psi	P2 psi	P4 psi	P5 psi	TP2 psi	TDP psi	T dgC	FLOW PATT.	JET cm
1	3.3	9.3	9.0	9.0	6.4	3.7	6.0	8.8	2.4	33	SLUG	2.0
2	1.9	4.9	9.9	9.9	6.9	3.8	6.0	9.5	2.7	32	SLUG	1.0
3	1.4	2.4	10.1	10.0	6.7	3.9	6.0	9.6	2.8	32	SLUG	0.5
4	7.7	11.3	8.3	8.3	6.0	3.5	6.0	8.2	2.1	35	SLUG	2.5
5	14.6	12.9	7.8	7.9	5.6	3.4	6.0	8.4	1.7	33	SLUG	8.0
6	20.3	13.4	7.9	7.8	5.6	3.5	6.0	9.0	1.7	33	SLUG	10.0
7	28.3	13.2	7.9	7.9	5.7	3.6	6.0	9.0	1.7	33	SLUG	9.0
8	48.1	12.6	8.0	8.0	6.0	3.5	6.0	13.5	1.8	33	SLUG	19.0
9	62.3	12.0	8.2	8.1	5.8	3.6	6.0	17.3	1.8	32	SLUG	21.0
10	24.4	13.2	7.8	7.8	5.6	3.6	6.0	8.7	1.7	33	SLUG	10.0
11	78.0	11.4	8.6	8.8	6.0	3.8	6.0	22.2	1.9	33	SLUG	25.0
12	87.8	11.1	8.9	9.0	6.1	3.7	6.0	24.5	1.9	33	SLUG	26.0
13	123.5	9.1	9.2	9.2	6.0	4.0	6.0	39.0	1.6	31	CHURN	27.5
14	168.3	6.8	9.8	9.8	5.9	4.1	6.5	62.0	1.4	31	CH/AN	29.0

TABLE 3

COMPARISON BETWEEN EXPERIMENTAL DATA AND NUMERICAL MODEL
CK = 2.10 CGL = 1.20 f = 0.0075
STRING INCLINATION = 0.0 DEGREES ECCENTRICITY = 0.4

OBS	HLC	QG Nm3/min	QL l/s	QLC l/s	ERROR(%)	DIM. KIN. ENERGY(10**3)
1	.256	1.278	1.19	1.26	-6.0	1.292
2	.287	.912	1.19	1.28	-7.8	1.337
3	.238	1.572	1.17	1.15	1.1	1.089
4	.229	1.809	1.14	1.14	-.2	1.059
5	.222	2.045	1.10	1.08	1.8	.962
6	.218	2.218	1.07	1.03	3.6	.875
7	.216	2.344	1.03	1.03	.1	.863
8	.212	2.534	.99	1.01	-2.4	.840
9	.197	3.624	.83	.85	-2.4	.585
10	.230	1.822	1.11	1.14	-2.5	1.059
11	.287	.917	1.16	1.28	-10.2	1.337
12	.203	3.128	.88	.97	-9.9	.770
13	.239	1.569	1.15	1.15	.0	1.090
14	.191	4.161	.71	.70	1.9	.400
15	.195	3.772	.79	.83	-4.7	.566
16	.228	1.837	1.12	1.14	-1.7	1.054
17	.214	2.401	1.02	1.06	-3.7	.918
18	.192	4.217	.75	.79	-5.7	.515
19	.195	4.522	.66	.00	100.0	.000
20	.253	1.318	1.18	1.25	-6.3	1.286
21	.195	4.425	.68	.00	100.0	.000
22	.220	2.122	1.07	1.12	-4.0	1.017
23	.238	1.600	1.15	1.23	-7.4	1.242
24	.265	1.155	1.19	1.29	-9.2	1.371
25	.285	.899	1.15	1.11	4.3	.999
26	.266	1.087	1.21	1.10	9.1	.990
27	.254	1.245	1.20	1.09	8.8	.978
28	.673	.075	.78	.68	12.8	.381
29	.746	.052	.47	.68	-46.1	.380
30	.759	.046	.59	.65	-9.0	.342
31	.739	.054	.56	.70	-24.8	.401
32	.747	.051	.47	.68	-43.4	.376
33	.763	.046	.25	.63	-151.1	.328
34	.743	.054	.24	.69	-187.0	.387
35	.750	.051	.16	.67	-309.0	.368
36	.479	.241	.97	1.05	-7.9	.899
37	.344	.560	1.13	1.18	-4.5	1.139
38	.377	.442	1.10	1.16	-5.6	1.098
39	.453	.279	1.00	1.08	-8.6	.958
40	.254	1.247	1.16	1.09	5.8	.977
41	.274	.989	1.17	1.10	6.0	.996
42	.327	.629	1.14	1.19	-4.0	1.153
43	.216	2.202	1.05	.94	10.4	.729
44	.230	1.769	1.11	1.06	4.3	.924
45	.264	1.115	1.16	1.10	5.3	.988
46	.197	3.524	.91	.85	6.4	.591
47	.204	2.916	.98	.90	8.3	.657
48	.229	1.778	1.14	1.06	7.0	.923
49	.209	2.639	1.03	.92	11.3	.688
50	.207	2.709	1.03	.91	11.3	.681
51	.232	1.787	1.13	1.21	-7.4	1.203
52	.262	1.158	1.19	1.10	7.4	.986

TACITE:

A COMPREHENSIVE MECHANISTIC MODEL FOR TWO-PHASE FLOW

Christian Pauchon (IFP)

Hasmukh Dhulesia (TOTAL)

Denis Lopez (EAP)

Jean Fabre (IMFT)

Abstract

The TACITE code is a steady state as well as transient multiphase flow simulation tool, for the design and control of oil and gas production networks. A unified hydraulic model has been developed which is applicable to any slope and diameter of pipeline. It also handles most of the flow regimes encountered in practice. The new modelling concept differentiates two types of flow patterns: the separated flow pattern (stratified or annular) and the dispersed flow pattern. The intermittent flow pattern (slug, churn flow) is a combination of these two patterns. The same concept has been successfully applied for the transition criteria, insuring continuity of the model variables across the transitions. This requirement is very important to simulate properly transient phenomena. The transient resolution is achieved by an explicit method. The advantages of this method are; its ability to follow wave front propagation, an easy implementation for the resolution of complex pipeline networks and an easy maintenance of the code in view of the evolution of the physical model and numerical scheme. The performance of the unified hydraulic model is demonstrated using a large number of test loop data, and in particular the industrial test loop of Boussens. The first tests performed on field data give encouraging results, especially for gas-condensate pipelines operating at low flowrates.

1 INTRODUCTION

The TACITE code is intended to simulate steady state as well as transient multiphase-phase flow of hydrocarbon mixtures in pipeline networks. Its development results from the collaboration between IFP, TOTAL and ELF in the framework of the EvE project. It is developed with the assistance of the "Institut de Mécanique des Fluides de Toulouse" (IMFT) for the physical modelling, the "Ecole Normale Supérieure de Lyon" (ENSL) for the numerical scheme, and the "Université de Marseille" for the thermodynamic package. IFP, with the help of ELF and TOTAL, is in charge of the integration of these aspects into an industrial code.

The TACITE code is based on the numerical resolution of a drift-flux type model applying to any situation encountered in multiphase production with respect to slope, fluid properties and flow pattern. The model solves a set of four conservation equations, one for the mass of each phase, one for the mixture momentum and one for the mixture energy. The missing information about the slip between phases is restored by a steady state closure relationship depending on the flow regime. In order to identify the regimes the assumption was made that each of them is a space/time combination of two basic patterns: separated flow (stratified and annular) and dispersed flow (Fabre et al. 1989). Then intermittent flow is considered as a combination of these two basic patterns, characterised by the fraction of separated flow β which insures the continuity of the closure laws.

One of the main objectives of TACITE is to predict accurately the propagation of large liquid slugs. Thus, the need for an accurate numerical scheme with good shock preserving capabilities was proposed (Caussade et al. 1989) and improved (Bouvier et al.1991). The time advancing scheme is explicit with second order accuracy in time and space to limit numerical diffusion. The choice of an explicit scheme presents several advantages which are identified in this paper.

The determination of physical properties has also been given special attention. A thermodynamic package is being developed, which takes into account the presence of water, glycol, methanol and dissolved salts.

In the present paper, we give an overview on the hydrodynamic model in TACITE including the transport equations, closure laws and solution algorithm. The time advancing scheme is briefly described. The numerical scheme for regular points and boundary points is described. The transient response of the code is demonstrated on the basis of a test case which shows the different wave systems propagating along the line.

The steady state response of the hydrodynamic model is compared with experimental data and data obtained on fields operated by ELF Aquitaine and TOTAL. Comparisons with the Boussens data bank are detailed for different slopes ranging from -3° to 90° and different flow regimes. The model is also tested against stratified flow data obtained in the Tulsa University Fluid Flow Projects and the data obtained in the MPE project of BHRG.

2 HYDRODYNAMIC MODEL

In this section, we present the set of transport and closure equations and we discuss the concept of flow regime transition which is used. The originality of the whole model lies on:

- the set of transport equations which insures the continuity of the model across flow regime transitions,
- the choice of a limited set of closure laws which have been qualified against experimental results and which has been made continuous with respect to slope and fluid properties,
- the concept of flow pattern transitions based on the continuity of the calculated variables.

2.1 Transport equations

The TACITE model is a drift flux type model with one mass conservation equation for each phases (Eq. 1&2), one mixture momentum equation (Eq. 3) and one mixture energy equation (Eq.4).

$$\frac{\partial}{\partial t}[\rho_G R_G]+\frac{\partial}{\partial x}[\rho_G R_G U_G]=m \tag{1}$$

$$\frac{\partial}{\partial t}[\rho_L R_L]+\frac{\partial}{\partial x}[\rho_L R_L U_L]=-m \tag{2}$$

$$\frac{\partial}{\partial t}[\rho_L R_L U_L+\rho_G R_G U_G]+\frac{\partial}{\partial x}[\rho_L R_L U_L^2+\rho_G R_G U_G^2+P+M_c]= T^w-(\rho_G R_G+\rho_L R_L)g.\sin\theta \tag{3}$$

$$\frac{\partial}{\partial t}\left[\rho_L R_L\left(H_L+\frac{U_L^2}{2}\right)+\rho_G R_G\left(H_G+\frac{U_G^2}{2}\right)-P+\frac{M_c}{2}\right]+ \frac{\partial}{\partial x}\left[\rho_L R_L U_L\left(H_L+\frac{U_L^2}{2}\right)+\rho_G R_G U_G\left(H_G+\frac{U_G^2}{2}\right)+E_c\right]= -Q^w-(\rho_G R_G U_G+\rho_L R_L U_L)g.\sin\theta \tag{4}$$

with $M_c=\beta(1-\beta)\sum_{k=L,G}\left(\frac{\rho_k^S R_k^S \rho_k^D R_k^D}{\rho_k R_k}\left(U_k^S-U_k^D\right)^2\right)$

and $E_c = \beta(1-\beta)\sum_{k=L,G}\left(\frac{\rho_k^S R_k^S \rho_k^D R_k^D}{\rho_k R_k}\left(U_k^S - U_k^D\right)\left(H_k^S - H_k^D + \frac{\left(U_k^S - U_k^D\right)\left(U_k + U_k^S + U_k^D\right)}{2}\right)\right)$

The subscript k stands for the phase (G for the gas phase and L for the liquid phase), the superscripts S, D for the two basic regimes (Separated flow and Dispersed flow). ρ, R, U and H are the density, the volume fraction, the velocity and the enthalpy respectively. θ is the inclination angle with respect to horizontal. T^W and Q^W are the contributions of the wall friction and heat transfer across the wall.

This problem involves five main unknown variables:

- R_G the gas volume fraction,
- P the pressure,
- T the temperature,
- U_G the gas velocity,
- U_L the liquid velocity.

Thus, an additional equation is required to close this system of 4 partial differential equations:

$$Hydro\left[U_G, U_L, R_G, P, T\right] = 0 \tag{5}$$

This equation reduces to a slip equation in the case of dispersed flow. It is a macroscopic momentum balance in stratified flow and it is a set of algebraic equations in the case of intermittent flow. Thus, it is flow regime dependent, and we will see how it can be treated to avoid any discontinuities when flow regime transitions occur.

In addition, two complementary terms M_C and E_C appear in the momentum and energy equations. These terms vanish for dispersed flow (ß=0), and for separated flow (ß=1); in intermittent flow, they account for the non homogeneous distribution of void and velocities in the separated and dispersed parts. They are expressed versus the secondary variables U_k^S, U_k^D, R_k^D and β. Moreover, the wall friction and heat fluxes may be written:

$$T^w = \beta T^{wS} + (1-\beta)T^{wD} \tag{6}$$

$$Q^w = \beta Q^{wS} + (1-\beta)Q^{wD} \tag{7}$$

in which the contributions of separated and dispersed parts appear. A unified formulation is used for the gas and liquid wall friction and the interfacial friction. That is, the form of these closure laws is independent of the flow regime and has the same structure as a single phase flow relationship. Furthermore, we will show that it is possible to obtain continuous solutions of the hydrodynamic model across flow regime transitions, even though the analytical form of the hydrodynamic model for the two basic regimes is very different.

2.2 Closure laws

In this section we will not detail the closure laws which are used in the model. We will rather insist on the solution algorithm used for the general case (0<ß<1) corresponding to intermittency, and for the two degenerated cases ß=0 (dispersed flow) and ß=1 (separated flow).

2.2.1 Dispersed flow (ß=0)

The closure equation (5) reduces to a drift flux relation expressing the gas velocity versus the mixture velocity U^D and the bubble drift velocity V_B.

$$U_G^D = C_0 U^D + V_B\left(d_B, R_G^D, \theta\right) \tag{8}$$

The problem is solved implicitly provided the bubble diameter is known. Hinze (1955) has shown that the bubble diameter results from a balance between the work of surface tension and the turbulent dissipation. Using the turbulent dissipation rate for two-phase flow of Colin (1990), yields:

$$d_B = 1,15\left(R_L^{D2}\frac{\sigma^3}{\rho_L}\right)^{\frac{1}{5}}\left[\left|\frac{4}{D}T^{wD}U_L^D\right| + \left|(\rho_G - \rho_L)g\sin\theta R_G^D\left(U_G^D - U_L^D\right)\right|\right]^{-\frac{2}{5}} \tag{9}$$

The bubble velocity is a continuous function of the bubble diameter. In the viscous stokes regime, the bubbles are considered spherical and the bubble velocity results from a balance between buoyancy and viscosity. In the distorted bubble regime, the bubble diameter is larger, and surface tension effects become important. Of these two regimes, the one corresponding to the minimum bubble velocity is considered to prevail.

The resolution algorithm for the dispersed flow regime is given in Figure 1.

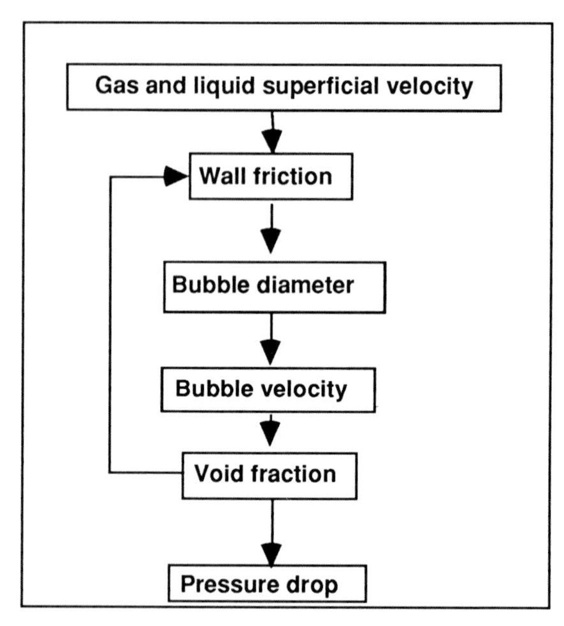

Fig. 1: Resolution algorithm for dispersed flow

2.2.2 Separated flow (ß=1)

For separated flow, Eq. (5) is a combination of the gas and liquid momentum equations from which the pressure gradient has been eliminated:

$$\frac{T_L^{wS}}{R_L^S} - \left(\frac{1}{R_L^S} + \frac{1}{R_G^S}\right)T^i - \frac{T_G^{wS}}{R_G^S} - (\rho_L - \rho_G)g\sin\theta = 0 \tag{10}$$

The iterative solution of this equation as described in Figure 2, is obtained from the expressions of the wall and interfacial friction, T_L^{wS}, T_G^{wS}, T^i. The key point in the solution of Eq. (10) lies in the expression of the interfacial friction. It was found after comparisons with existing data that the best choice is the law proposed by Andritsos and Hanratty (1987). This relation has been modified to take into account the wave velocity with respect to the liquid C_S:

$$T^i = -\frac{1}{2} f^{iS} \rho_G \left(U_G^S - U_L^S - C_S\right)\left|U_G^S - U_L^S - C_S\right| \frac{P^i}{S} \tag{11}$$

where P and S are the interfacial perimeter and the cross section area respectively.

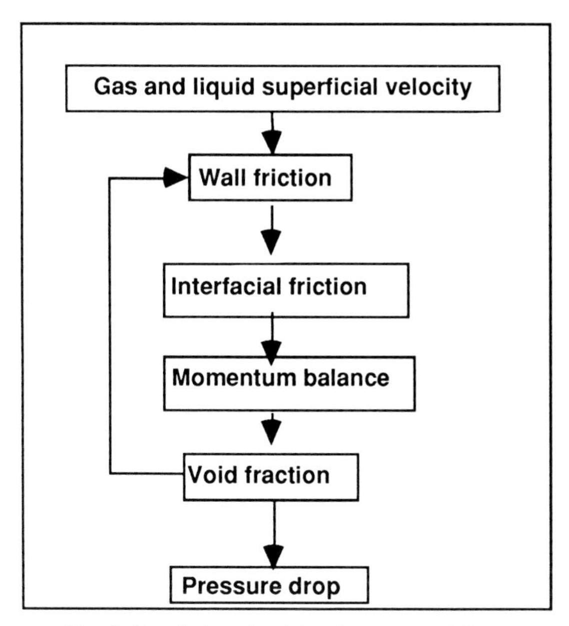

Fig. 2: Resolution algorithm for separated flow

2.2.3 Intermittent flow ($0<\beta<1$)

In intermittent flow, we have a periodic structure of dispersed and separated flows.
The solution of the model depends on an accurate prediction of the slug velocity. An extended expression of the law derived by Nicklin et al.(1962) is used:

$$V = C_0(Re,Bo,Fr,\theta)U^D + C_\infty(Re,Bo,Fr,\theta)\sqrt{\frac{\Delta\rho}{\rho_L}gD} \tag{11}$$

where the coefficients C_0 and C_∞ are function of the inclination angle, the Bond number, Reynolds number and Froude number defined as:

$$Bo = \frac{\Delta\rho g D^2}{\sigma}, \quad Re = \frac{|U^D|D}{\nu_L}, \quad Fr = \frac{|U^D|}{\sqrt{(\Delta\rho/\rho_L)gD}}$$

In order to solve the model, the knowledge of the distribution of gas between the long bubbles and the liquid slugs is needed. Usually, this problem is overcome by using a closure law for the void fraction in the liquid slugs. The accuracy of the model appears to be very sensitive to it. As none of the existing laws gives satisfactory results from a slope ranging between 0 to 90°, the correlation proposed by Bendiksen and Andreussi (1989) has been used. Their proposed correlation can be put in the following dimensionless form

$$R_G^D = \frac{Fr - Nf}{Fr + N_0} \tag{12}$$

with $Nf = 2{,}6\left\{1 - 2\left(\frac{d_o}{D}\right)^2\right\}$ and $N_o = 2400\left\{1 - \frac{\sin\theta}{3}\right\}Bo^{-3/4}$

provided that the critical bubble diameter d_0 is given as:

$$d_o = 12\sqrt{\frac{\sigma}{\Delta\rho g}} \tag{13}$$

Eqs. (12) and (13) underestimate the void fraction in the liquid slugs for vertical upward flow. In order to improve the model, the development of a new correlation is actually on the way.

The solution algorithm for intermittent flow is given in Figure 3.

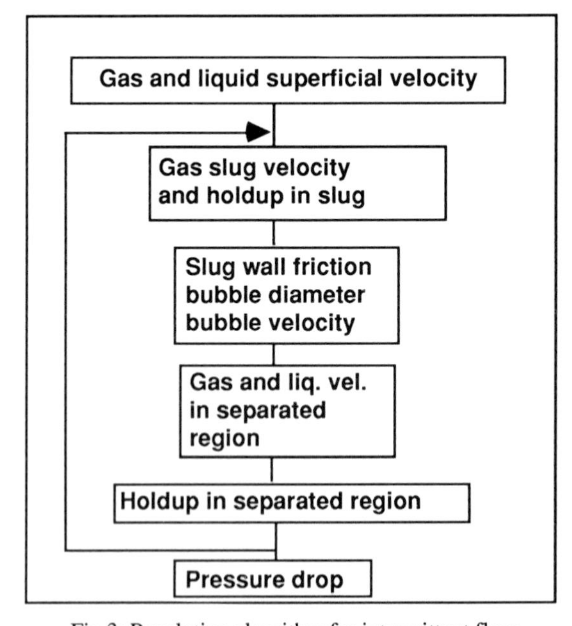

Fig.3: Resolution algorithm for intermittent flow

2.3 Flow regime transitions

The algorithms presented in the previous sections correspond to a calculation with imposed flow regime. In this section, we show how transition criteria lead to discontinuities in the calculated variables, why this is incompatible with the averaged nature of the equations of motion, and how we proceed to avoid these discontinuities.

2.3.1 Why transition criteria lead to discontinuities

A standard approach to determine flow regimes is to apply a set of transition criteria. However, if these criteria is not closely coupled with the models for the individual flow regimes, they lead to discontinuities in the calculated variables. This is best illustrated in Figure 4 where we have represented the hydrodynamic models for flow regimes 1 and 2. This graph represents schematically the dependence of one of the variables (e.g. the gas volume fraction R_G) as a function of the other dependent variables (e.g. the superficial liquid velocity U_{LS}). Figure 4 shows that unless the transition criterion crosses the intersection between the two hydrodynamic models, a discontinuity in one of the dependent variables will occur at the transition.

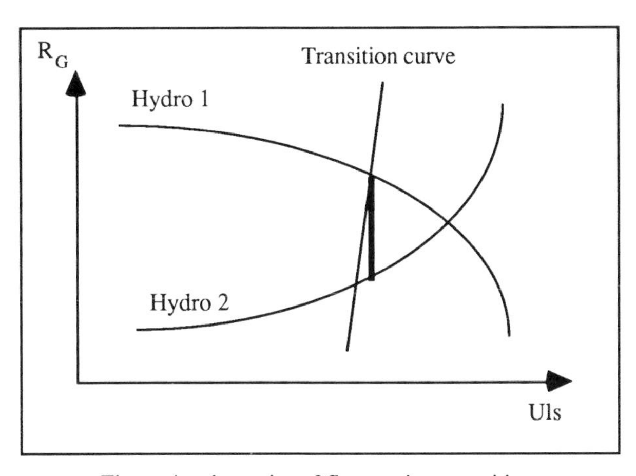

Figure 4: schematics of flow regime transition

2.3.2 Why transitions should be continuous

The averaged equations of motion are based on the assumption that the length scale of the flow is small with respect to the length of the control volume over which the equations are averaged. Similarly, the flow regime must be understood as an averaged property of the flow structure in the control volume. Thus, even if locally, the void fraction and slip velocity may present a discontinuous behaviour, these variables cannot be discontinuous in an averaged sense.

2.3.3 The treatment of flow regime transitions in TACITE

The constraint that the calculated variables be continuous across the transitions, implies that the hydrodynamic models representing the different flow regimes, though very different in form, should lead to continuous solutions across transition boundaries. To obtain this, the following rules are implemented in the physical model:

- The basic flow regime is the intermittent flow regime. The parameter β defined previously ranges from 0. to 1, and is a result of the calculation.

- When $\beta = 0.$, the hydrodynamic model reduces to the dispersed flow model, that is a slip equation in the liquid slug. The transition between dispersed and intermittent flow occurs when the void fraction in the liquid slug is equal to the void fraction of pure dispersed flow.

- When $\beta = 1.$ the hydrodynamic model reduces to the stratified flow model, that is , a momentum balance in the separated region. To force the continuity of the hydrodynamic model between separated and intermittent flows, we assume that the prevailing flow regime is the one with the highest R_G. Figure 4 illustrates that such condition insures the continuity between the two flow regimes. In practice, this implies that there exists a maximum long bubble velocity above which the flow regime must be separated. This velocity is given as:

$$V_{max} = \frac{R_G U_G - R_G^D U_G^D}{R_G^{Sep} - R_G^D} \tag{14}$$

where R_G^{Sep} is the gas fraction in the separated regime.

Figure 5 shows a fairly good prediction of the configuration density β on data from the Boussens loop. Comparisons of this transition algorithm with experimental data along with other transition methods available in the literature (mainly based on the classical Kelvin Helmholtz instability criteria) is under way and will be the subject of a separate detailed publication.

3. NUMERICAL SCHEME

The numerical scheme used in TACITE was proposed originally by Lerat (1981), it was first implemented in TACITE by Benzoni-Gavage (1991). It is a three point predictor corrector scheme. The time advancing scheme in TACITE is explicit, with a second order accuracy. This choice is dictated by the following arguments:

- A good accuracy for the description of transients is needed in order to predict precisely the volume of liquid slugs travelling in the pipeline system and reaching the pipeline outlet.
- Explicit schemes are easy to implement.

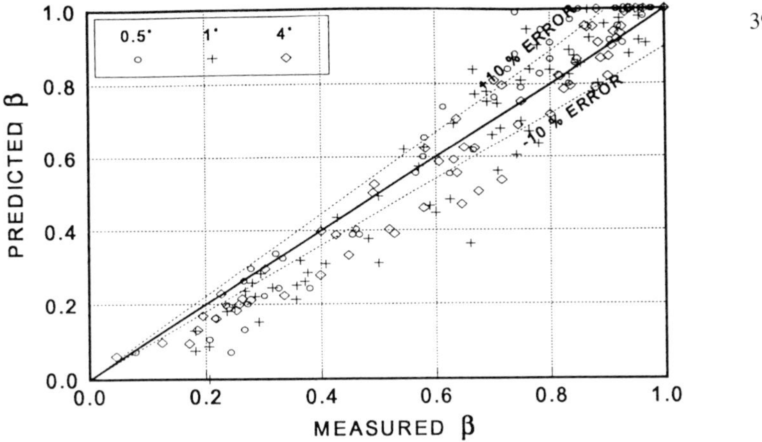

FIG. 5 **COMPARISON BETWEEN MEASURED AND CALCULATED FRACTION OF SEPARATED FLOW FOR DIFFERENT ANGLES OF INCLINATIONS**
(β=1 for separated pattern and β=0 for dispersed pattern)

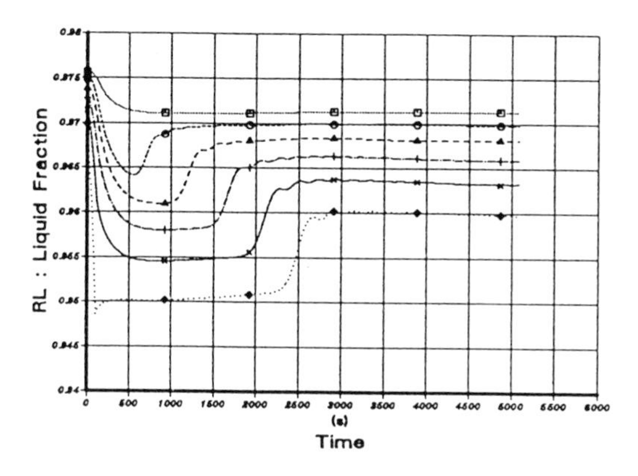

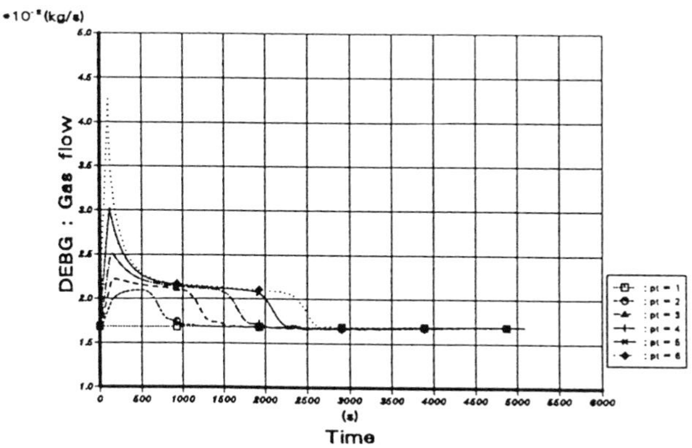

FIG. 6 **EXAMPLE OF PIPELINE DEPRESSURISATION**
(The pipeline outlet pressure is reduced from 10 to 5 bars. Point 1 corresponds to the inlet while Point 2 corresponds to the outlet. Intermediate points are located every 2 km)

- Explicit schemes give more flexibility for the treatment of complex transient networks.

3.1 Time advancing scheme, spatial discretization

We seek the solution of a hyperbolic set of partial differential equations of the form given in (1):

$$\frac{\partial}{\partial t}W(U)+\frac{\partial}{\partial x}F(U)=Q(U) \tag{15}$$

together with the algebraic constraint:

$$Hydro(U)=0 \tag{16}$$

$U = {}^{T}(R_G,U_G,U_L,P,T)$ is the set of dependent variables. At each time step, W is computed by the time advancing scheme, and the dependent variables are calculated with the help of the hydrodynamic model (Eq. 16). The time advancing scheme is explicit, so that the state of the system at time $n+1$ is calculated on the basis of the flux term F and source terms Q evaluated at time n..

Explicit schemes are stable provided that the Courant Friedriech Levy criterion is satisfied. In practice this condition imposes a limitation on the time step Δt depending on the speed of the fastest wave propagating in the system c_{max}:

$$\Delta t \le \frac{\Delta x}{c_{max}} \tag{17}$$

With the evolution towards ever faster computer processors, this limitation is bound to become less of a constraint.

On the other hand, the choice of an explicit time advancing scheme presents substantial advantages compared to implicit or semi-implicit methods:

- For transient simulation of complex networks it allows for the solution of individual network components sequentially instead of requiring their simultaneous solution.

- It has also the advantage of a clear separation between the hydrodynamic calculation module, namely the function *Hydro*, and the treatment of the numerical scheme. This allows an independent evolution of the physical model and the numerical scheme.

The numerical scheme gives second order accuracy in time and space. Thus, it allows good front tracking capabilities. This will be very important when estimating the size of large liquid slugs reaching the pipeline outlet.

Figure 6 shows a typical test case of horizontal pipe 10 km long. The pressure at the outlet is decreased in 100 seconds from 10 bars to 5 bars. Figure 6a shows that the pressure wave propagates very fast towards the inlet to generate a void fraction front which propagates to the outlet in 2500 seconds. Figure 6 illustrates the importance of having a numerical scheme able to track accurately wave fronts propagating in the system.

3.2 Treatment of boundary conditions

The hyperbolic nature of the problem makes it very important to treat precisely the transport of information through the pipe, and in particular the information going in and out of the system. Four types of waves are identified in the analytical model:

- Void fraction waves travelling downstream.
- Pressure waves travelling downstream.
- Pressure waves travelling upstream.
- Enthalpy waves travelling with the mean flow.

The treatment of boundary conditions takes into account the information going out of the system through a set of compatibility relations, which must be solved together with the imposed boundary conditions. Thus, it depends on the direction of propagation of the different wave systems. In normal subsonic cocurrent flowing conditions, the information going out of the system is as shown in Figure 7. In these conditions, the solution of the following set of equations gives a rigorous treatment of the boundary conditions:

At the inlet :

- Mass flowrate of gas
- Mass flowrate of liquid
- Temperature
- Compatibility condition for pressure waves travelling upstream
- Hydrodynamic function

At the outlet :

- Pressure
- Compatibility condition for enthalpy propagation
- Compatibility condition for pressure waves travelling downstream
- Compatibility condition for void fraction waves
- Hydrodynamic function

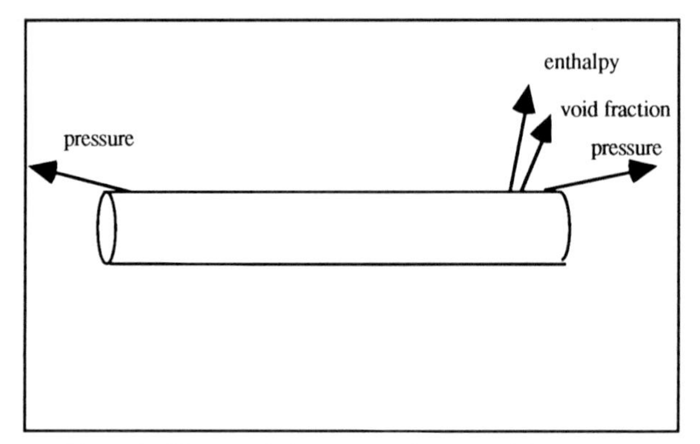

Figure 7: Characteristic waves going out of the system

4. STEADY STATE VALIDATION

4.1 Test loop data

(a) Boussens loop

An early collaboration between IFP, TOTAL and ELF in the field of two-phase flow started in 1974 with the construction of the semi-industrial Boussens test loop. Data from this loop were used for the development of the steady state simulation models PEPITE (for horizontal and slightly inclined pipes) and WELLSITE (for vertical wells).

The hydrodynamic module of TACITE has been tested extensively on the Boussens data. These data cover a wide range of inclinations (-3°, -.5°, 0°, 1°, 4°, 15°, 45°, 75° and 90°) with 3 and 6" piping. Diesel oil and condensate have been used alternatively together with natural gas as working fluids. The fluid properties were measured in the laboratory. Pressures up to 50 bars were investigated over a wide range of flowrates. The horizontal and slightly inclined data bank contains about 1750 points of which 900 are in stratified flow, 600 in slug flow, and 250 in dispersed flow. The data bank on vertical and highly inclined flow consists of about 700 measurement points (260 for bubbly flow, 400 for slug flow and 40 for annular flow). More details on this loop are given by Corteville et al.(1983). Figure 8 shows pressure drop and liquid holdup predictions of TACITE against experimental data for slopes of -3°, -.5°, 0°, 1° and 4°. Figure 9 shows comparisons of the model predictions for inclinations of 15°, 45°, 75° and 90°.

(b) TUFFP stratified flow data

Data from the Tulsa University Fluid Flow Projects (TUFFP) were used to test the TACITE hydrodynamic model in stratified flow conditions. These data cover slopes between 1° and -8°. Diameters of 1", 1.5", 2", 2.5", 3", 4" and 6" were used, with pressures ranging from 1 to 10 bars. The fluids used were air/water and air/kerosene.

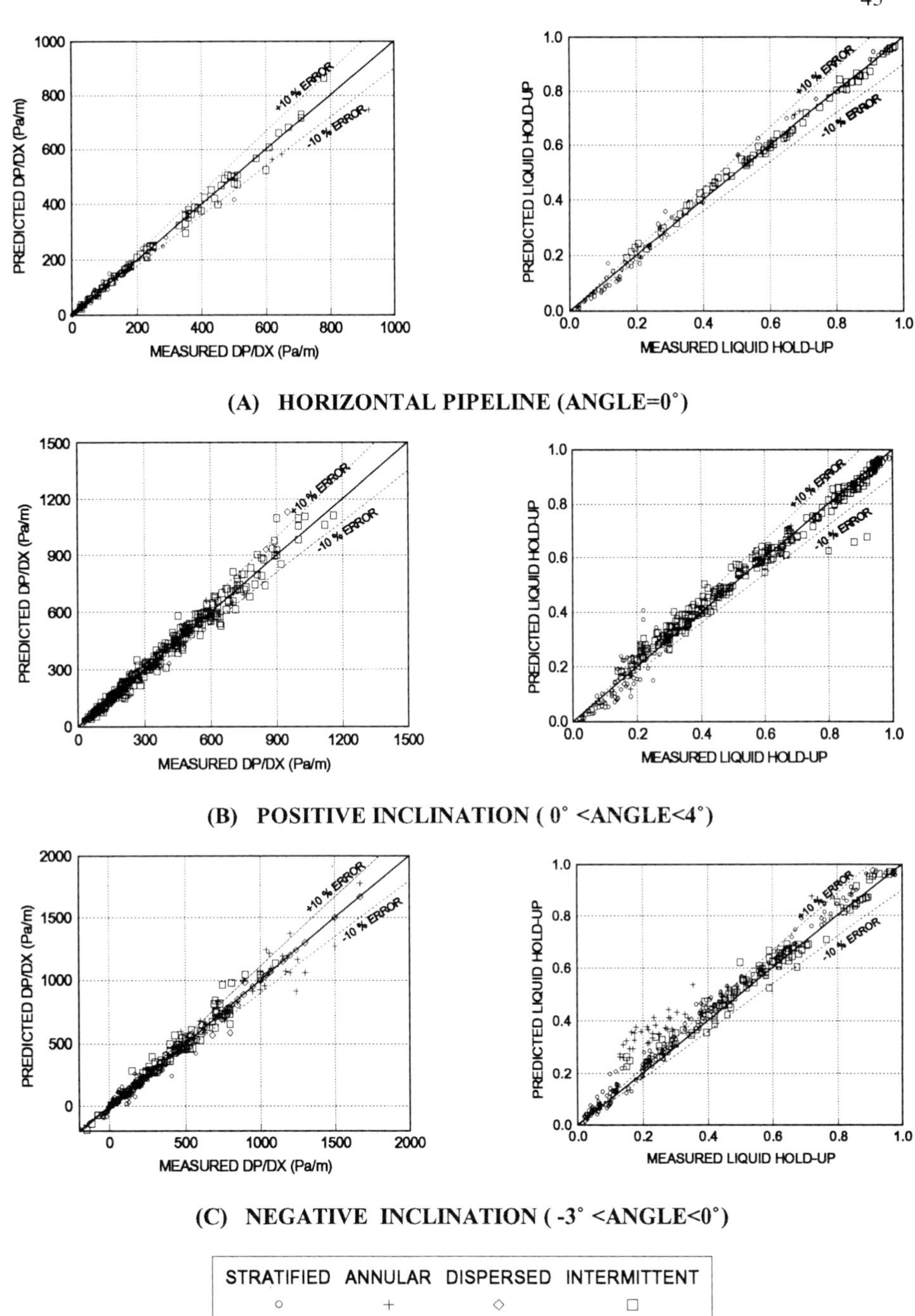

FIG. 8 PREDICTION BY TACITE FOR BOUSSENS DATA (HORIZONTAL AND SLIGHLY INCLINED PIPELINE)

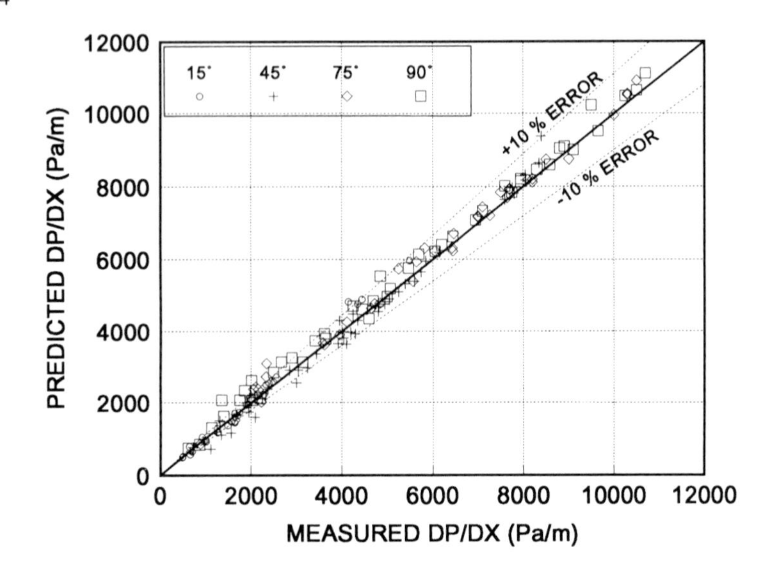

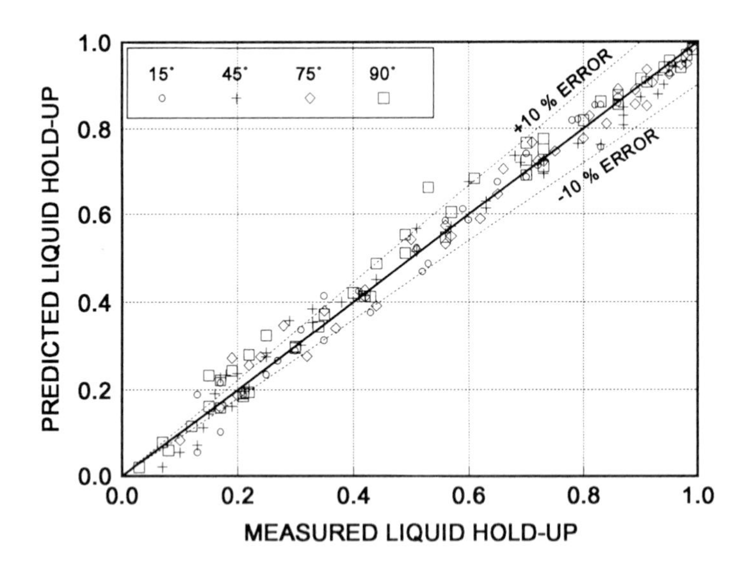

FIG. 9 PREDICTION BY TACITE FOR BOUSSENS DATA (VERTICAL AND HIGHLY INCLINED PIPELINE)

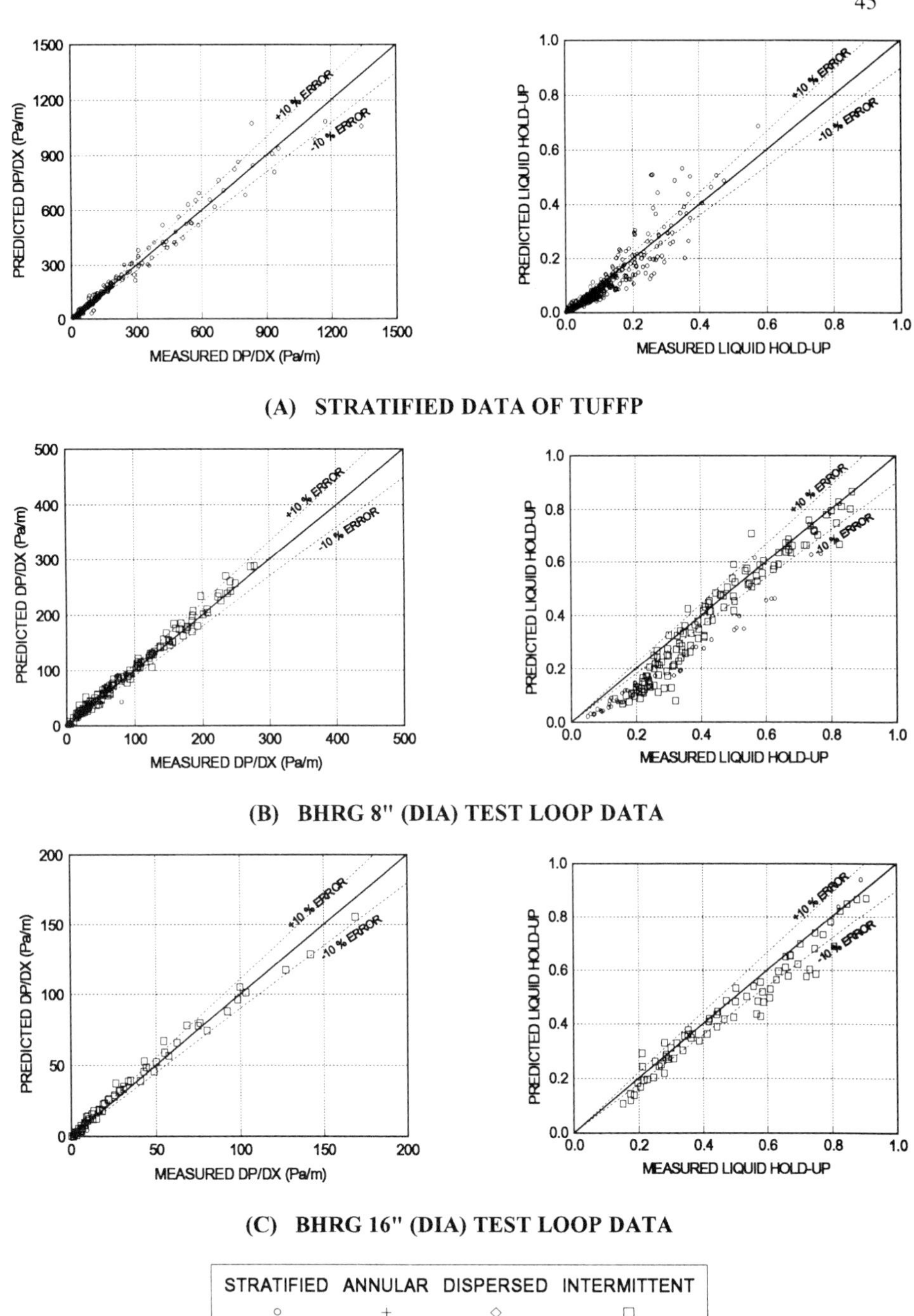

(A) STRATIFIED DATA OF TUFFP

(B) BHRG 8" (DIA) TEST LOOP DATA

(C) BHRG 16" (DIA) TEST LOOP DATA

FIG. 10 PREDICTION BY TACITE FOR TUFFP AND BHRG LOOP DATA

Figure 10-a shows the comparison between measured and predicted pressure drop and liquid hold-up.

(c) BHRG data

Data from the MPE project of BHRG concern air/water experiments in near horizontal flow at atmospheric pressure. 8" and 16" diameters were used for these experiments. Figure 10-b illustrates the comparison between measured and predicted Pressure drop and liquid hold-up. A similar comparison for 16" diameter is shown in Figure 10-c.

4.2 Field data

(a) North Alwyn flowline

TACITE was tested in steady state conditions against field data acquired on the North Alwyn flowline operated by TOTAL. This line located in the North sea is 5 km long, approximately horizontal, followed by a riser 160 meter high leading to the process platform. The pipe diameter is 6", and the flow line is buried at a depth of 1.5 meter. The departure pressure is about 180 bar. The fluid is oil with associated gas, thus the important prameter to predict is the pressure drop. Figure 11 shows the comparison between predicted and calculated pressure drop.

To estimate the fluid composition, liquid and gas samples were taken from the test separator (27 barg, 52°C) and recombined at the measured GOR to reconstitute the wellhead effluent. This effluent was flashed to atmospheric conditions (1 barg, 15°C) to measure the composition.

(b) North East Frigg pipeline

These data were recorded under the supervision of Elf Aquitaine Norge. The NEF-TCP2 line is a 16" pipe, 18 km long. Gas and associated condensates are flowing through the line at a nominal flowrate which is about 5 10^6 Sm^3/d. When operated around this stabilised gas flowrate, the line cannot generate significant transients. But during 1989, the line had to be operated far below its nominal flowrate (around 1.5 10^6 Sm^3/d).The change from this low flowrate to the nominal flowrate gave significant transient phenomena and a series of steady state conditions.

The nominal flowrate was reached through six steps: each step consisting of an increase in gas flowrate. Between each step, a sufficient period of time was allowed to insure that a new steady state equilibrium was reached.

Several parameters were registered:

- upstream pressure,
- upstream temperature
- downstream temperature (just before and after the outlet choke),
- outlet gas flowrate,
- the level of liquid in the outlet separator (to calculate the outlet liquid flowrate),

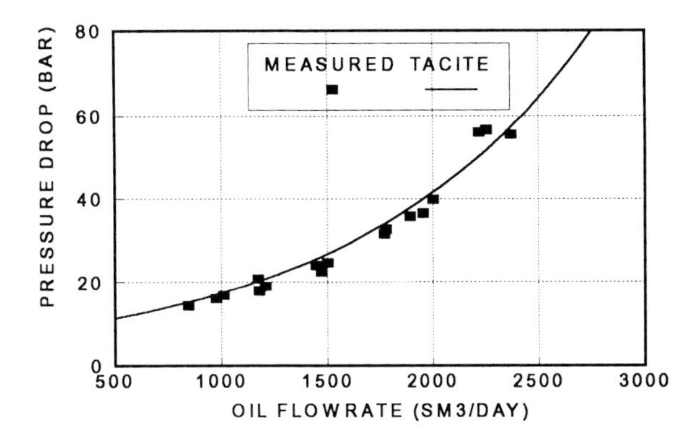

FIG. 11 PREDICTION BY TACITE FOR ALWYN PIPELINE

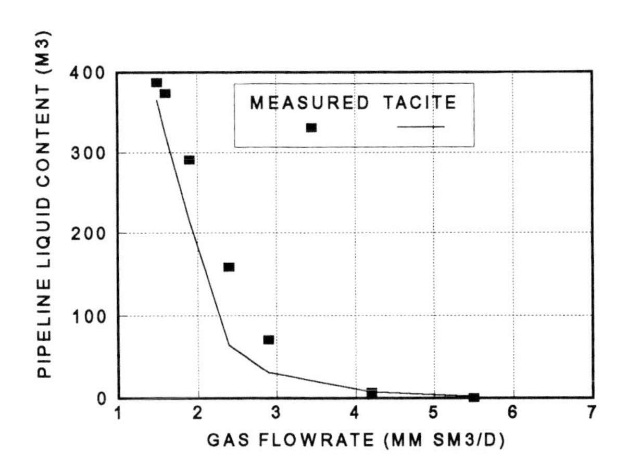

FIG. 12 PREDICTION BY TACITE FOR NORTH EAST FRIGG PIPELINE

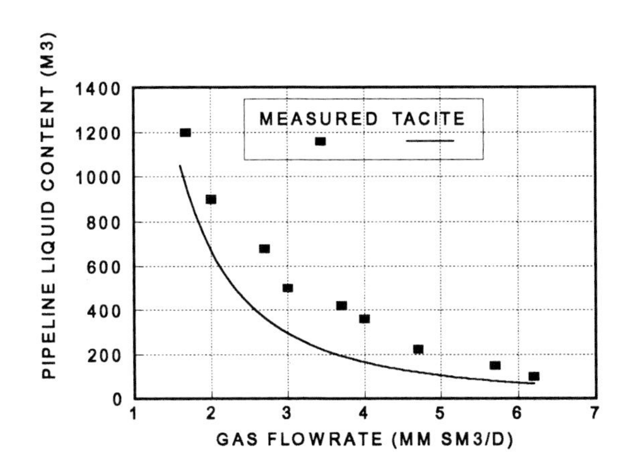

FIG. 13 PREDICTION BY TACITE FOR ZUIDWAAL HARLINGEN PIPELINE

- the mean value of the specific gravity of the outgoing fluid (by using a gamma densitometer which was fixed onto the line just upstream the separator to detect liquid slugs).

The total liquid content of the line at the end of each step after reaching steady state conditions (i.e. when the outlet liquid flowrate is constant and equal to the inlet flowrate) was estimated as follows:

- calculating the cumulated liquid flowrate based on the measured gas flowrate at the outlet during the transient , and taking into account the pressure and temperature at the separator and the molar composition of the fluid,
- substracting from the above quantity the measured outlet liquid recovery during the step.

Figure 12 shows the comparison between the measured and predicted total liquid content using TACITE.

c) Zuidwaal Harlingen pipeline

The Zuidwaal-Harlingen pipeline is 20 km long with a 20" diameter. The upstream gas flowrate is measured with an orifice flowmeter (between 1.7 and 6.5 10^6 Sm^3/d), and the upstream liquid flowrate is equal to the injection flowrate of DEG (between 5.5 and 11.9 M^3/hr). The downstream gas flowrate is measured by the commercial metering, corrected by simulating the process within the gas treatment line. The downstream liquid flowrate is measured by integrating the level variations in the slug catcher (up to 150 m^3/h). The pressure is measured close to the inlet and the outlet of the pipe, free of any valve or choke. The downstream pressure ranges from 90 to 101 bara and the upstream pressure ranges from 95 to 104 bara, depending on the operating conditions.

The total liquid holdup in the line is calculated by substracting liquid recoveries between low and high gas flowrates (when the gas is devoid of liquid). The maximum liquid holdup reached during these experiments was on the order of 25% of the total sea line volume. Figure 13 shows comparisons of the measured versus predicted liquid content for different gas flowrates.

Both Figure 12 and 13 indicate that TACITE slightly underestimates the liquid holdup at low flowrate for gas condensate pipeline. It is believed that the interfacial friction plays an important part in these conditions. Consequently a research project is under way to obtain better correlations for the interfacial friction using a multidimensional approach.

5. CONCLUSION

The objective of this paper is to give an overview of the TACITE code stressing its originalities:

- The physical model integrates a number of existing closure laws into a unified model working for all slopes and flow regimes. Some of these closure laws have been improved and evaluated using a large number of experimental data.

- A new approach to the flow regime transitions based on a single configuration parameter β is presented, which leads to a continuity in the calculated variables across the transitions.

- The numerical scheme, explicit in time, gives a second order accuracy. Its ability to follow front propagations without numerical diffusion is demonstrated on a test case

- It is worth stressing the advantages of an explicit scheme:
  - for the numerical treatment of complex transient networks,
  - for the ease of evolution of the physical model, and the numerical scheme each contained in a single module.

- The steady state validation of the model shows that it gives a good prediction of experimental data taken over a wide range of slopes, fluids and operating conditions.

- Comparisons with field data on gas-condensate lines (North East Frigg, Zuidvaal) show encouraging predictions of the total liquid content in the line. The prediction of pressure drop in the case of the Alwyn oil field is also good.

6. ACKNOWLEDGEMENT

The authors are grateful to the Tulsa University Fluid Flow Projects (TUFFP) and to the MPE Project of BHRG for allowing publication of comparison of TACITE with their experimental data. The contribution of ELF and TOTAL are also gratefully acknowledged for authorising field data acquired on their sites to be used in this work.

7. REFERENCES

Andreussi, P., Bendiksen, K., 1989, An Investigation of void fraction in liquid slugs for horizontal and inclined gas-liquid flow, *Int. J. Multiphase Flow*, Vol. **15**, 2 pp 937-946.

Andristos, N., Hanratty, T.J., 1987, Influence of interfacial waves in stratified gas-liquid flows, *AIChE J.*, Vol. **33**, pp 444-454

Barnea, D., Shoham, O. & Taitel, Y., 1985, Gas-liquid flows in inclined tubes: flow pattern transition for upward flows., *Chem. Eng Sci.*, Vol. **40**, pp 131-136.

Bouvier,V., Ferschneider, G, Fabre, J., Gilquin, H, 1992, A practical hyperbolic problem: the unsteady gas-liquid flow in pipe, *4th Int. Conf. on Hyperbolic Problems*, Taormina , Italy, April 3-8.

Bendiksen, K., Espedal, M., 1992, Onset of slugging in horizontal gas-liquid pipe flow, *Int. J. Multiphase Flow*, Vol. **18**, 2 pp 237-248.

Benzoni-Gavage, S., 1991, *Analyse des modèles hydrodynamiques d'écoulements diphasiques instationnaires dans les réseaux de production pétrolière*, Thèse. Lyon 1, France.

Caussade, B., Fabre, J., Jean, C., Ozon, P., Théron, B., 1989, Unsteady phenomena in horizontal gas-liquid slug flow, *Int. Conference on Multi-Phase Flow*, Nice, France, Cranfield, BHRA.

Colin, C., 1990, *Ecoulements diphasiques à bulles et à poches en micropesanteur*, Institut National Polytechnique, Toulouse.

Corteville, J., Grouvel, J.M., Roux, A., & Lagière, M., 1983, Expérimentation des écoulements diphasiques en conduites pétrolières: boucle d'essais de Boussens, *Rev. Inst. Français du Pétrole*, Vol **38**, pp 143,151.

Fabre, J., Ferschneider, G., Masbernat, L., 1983, Intermittent gas-liquid flow in horizontal or slightly inclined pipes, *Int. Conference on the Physical Modelling of Multi-Phase Flow*, Coventry, England, pp 233, 254

Fabre, J., Liné, A., Péresson L., 1989, Two fluid/two flow pattern model for transient gas liquid flow in pipes, *Int. Conference on Multi-Phase Flow*, Nice, France, pp 269, 284, Cranfield, BHRA.

Hinze, J.O., 1955, Fundamentals of the hydrodynamic mechanism of splitting in dispersion processes. *AIChE J.*, Vol **1**, pp 289-295.

Lerat, A., 1981, *Sur le calcul des solutions faibles des systèmes hyperboliques de lois de conservation à l'aide de schémas aux différences* , Thèse. ONERA, France.

Niklin, D.J., Wilkes, J.O., Davidson, J.F., 1962, Two phase flow in vertical tubes, *Trans. Inst. Chem. Engrs.*, Vol **40**, pp 497-514.

CAPABILITY OF THE OLGA COMPUTER CODE TO SIMULATE MEASURED DATA FROM AGIP OIL FIELD

A. MAZZONI, M. VILLA (Agip) - M. BONUCCELLI (C.P.R-TEA.)
G. DE TOMA, D. MAZZEI, T. CRESCENZI (Enea),

ABSTRACT

This paper presents and analyses the comparison between the OLGA code simulation and the measured data collected in two Agip oil fields.

The first set of data is related to the Trecate-Villafortuna on-shore oil field, which has the main characteristic of being a very extended network with 17 branches connecting 15 wells with a full development of the transportation lines of about 20 Km. The fluid is a light oil (700 - 750 kg/m3 ; 1 - 3 cp.) with a very low water cut (< 2%). Pressure and temperature are measured at the inlet and at the outlet of each well flow line as well as at the slug catcher located at the Oil Centre, where the volumetric flow of each phase and the liquid level are also monitored. All the measured parameters at the Oil Centre and most of the field variables are recorded in order to know their trends. The field data were collected and simulated before, during and after an operational transient of shut-down and start-up.

The second set of data is related to the Prezioso-Perla off-shore field, which is developed with two platforms (Prezioso and Perla) and is characterised by a transportation line (29 km long - 12" diameter) with three risers (two downward and one upward) about 80 meters high and has a heavy oil (900 - 950 kg/m3) with a low water cut (< 3%) and with a viscosity strongly dependent on the pressure and temperature conditions ranging between 50 and 1500 cp. .Pressure and temperature were measured at the transportation line inlet and outlet as well as in the intermediate platform of Perla. The single phase flow rates were measured at the separator in the Gela Oil Centre and verified with the well testing data. Since all the measured parameters have not been recorded on a computer up to now the simulation was performed in steady state conditions.

1. INTRODUCTION

The analysis described in this paper has been performed by AGIP S.p.A. in association with ENEA (Ente per le Nuove teconologie, l'Energia e l'Ambiente) and C.P.R. (Consorzio Pisa Ricerche) and it is aimed at field data collection under controlled conditions and at the OLGA code qualification.

The OLGA code/3/, /4/ is a fluid dynamic, one-dimensional transient system code developed by IFE/SINTEF in Norway.

Among the main objectives of these activities there are the following:

1. Field data collection in controlled conditions.
2. Evaluation of the code capabilities and limits in reproducing typical steady state and transient conditions in hydrocarbon transportation lines.
3. Definition of the basic criteria for a correct code utilisation in the design and analysis of a pipeline transportation system in order to help the user in the code application.

For this scope two AGIP's fields were selected for the field data collection and for the code qualification:

- TRECATE-VILLAFORTUNA on-shore network;
- PREZIOSO-PERLA-GELA off-shore sealine.

The activities described in this paper are related to:

- the simulation of an operational transient in the TRECATE-VILLAFORTUNA field;
- the simulation of the steady state conditions with severe slugging phenomena in the PREZIOSO field.

Part 1 - OPERATIONAL TRANSIENT ANALYSIS IN THE TRECATE-VILLAFORTUNA FIELD

2. DESCRIPTION OF THE FIELD

The field has been developed with three satellites areas (see Figs. 1 and 2) where the flows, coming from several wells, are joined, cooled and delivered to the main flowlines. In each satellite area there are two manifolds, one three phase test separator and a given number of air cooler banks depending on the total flow rate to be cooled. One manifold is used as the test manifold and it is connected to the test separator which allows the measurement of single well flow rates (oil, water, gas). The other is the production manifold where the wells' flows are joined and delivered to the main flow lines through commingling flow lines after having been cooled by the air coolers. Three main flow lines

(10",12",12") deliver the production to the Oil Centre for the oil and gas treatment. At the moment fifteen wells are producing by draining two reservoirs with slightly different fluid compositions. For the simulation an average composition has been used.

The pipelines profile of the whole network is mainly horizontal, except in the areas where streets and channels are crossed. The total flow rates of oil, gas and water as well as the pressure, the temperature and the liquid level are continuously monitored at the slug catcher at the Oil Centre. The single wells' flow rates are measured at the field test separators. Pressures and temperatures are also measured at the inlet and outlet of each well flow line, respectively after the choke valve and before the manifolds and at the inlet and outlet of the air coolers in the satellite areas. At present, the temperatures after the choke valves, the pressure at the well flowlines outlet (manifold in the satellites areas), the temperatures at the air coolers outlet and all the above mentioned parameters measured at the Oil Centre are transmitted and recorded on a PC located in the control room of the Oil Centre while all the remaining parameters are measured locally and are recorded by hand.

The following tables summarise the main characteristics of the field (Tab. 1), the fluids average composition (Tab. 2) and the geometrical characteristic of the flowlines (Tab.3).

3. OPERATIONAL TRANSIENT ANALYSIS.

The operational transient analysed in this paper is related to the shut-down and start-up of the Trecate-Villafortuna production plant.

It started at 00.00 of 29th of October, 1992 with the closure of the of the first well and ended at 11.00 of 30th of October after the reopening of the last well occurred with the achievement of the new steady state conditions.

A preliminary analysis of the variables trends recorded before, during and after the operational transient was performed in order to identify the initial and final steady state conditions and the main events that occurred during the shut-down and start-up operations. The interesting range for the OLGA simulation was identified between 00.00 of 28th of October (one day before the transient beginning) and 24.00 of 30th of October when all the recorded variables reached the new equilibrium state for sufficient time.

During this period three phases were identified:

PHASE I
It is characterised by the initial steady state conditions (except for the event related to the well testing of the T17 well). It goes from the beginning of the simulation to the closure of the first well.

PHASE II
It starts from the closure of the first well and it ends with the reopening of the first well (T20). It is characterised by the shut-down transient and by the cooling down of the network.

PHASE III
It is characterised by the start-up transient and by the achievement of the new steady state. It goes from the first well reopening to the end of the simulation.

The main events are listed in Tab. 4.

4. INPUT DECKS PREPARATION

4.1 Pipelines Discretization.

The current OLGA input deck has been prepared by updating an already existing one that was used to simulate a previous operational transient/1/. A few changes have been made in order to take into account the production increase (4 new wells) and a new best fit procedure to calculate the pipes roughness[1]. A detailed nodalization of the whole network composed of 17 branches divided into about 600 sections is shown in Tab. 3 and Fig. 3.

The flow lines from the wells to the satellites areas have been simulated as horizontal lines since their actual profile is very smooth and elevation changes occur only when streets or channels are crossed. They have been discretisated with one or two horizontal pipes divided into sections having a length of 200 - 300 meters.

The air coolers, located in the satellite areas, are controlled by keeping the outlet temperature in a narrow range around a selected set point value. The heat exchanger model in OLGA is very simplified since it does not perform the heat transfer calculations but simply does a heat balance of the unit, assuming that the heat exceeding the outlet heat (calculated on the base of a constant temperature) is transferred to the environment. For the simulation of the pressure drop across the coolers, normally it is convenient to use a local loss coefficient but due to mass flow oscillations (with period and amplitude related respectively to time step and loss coefficient values) an equivalent choke valve has been used (see Fig. 4).

Care was taken in the nodalization of the main flow line due to its changes in elevation. These do not exceed 30 meters but are homogeneously distributed along the pipelines and thus affecting the mass distribution. Since the real profile of the main flow line is relatively complex, an equivalent line has been used to simulate it, thereby respecting the following general criteria:

1. To maintain the total length of the flow line.
2. To maintain the inlet and outlet flow line elevations.
3. To maintain the predominant expected flow pattern/s through the flow line.

In the Oil Centre there are two valves on the main flow line. The former is an On/Off valve located at the inlet of the Oil Centre; the latter is a control valve located just before the slug catcher, manually actuated during the reopening phase in order to allow a slow

---

[1]For all the flow lines of the field the pressure drops were calculated with the OLGA code for different values of roughness and the value that minimised the mean square error was chosen for all the field/2/.

pressurisation of the slug catcher. Since its relative opening was not recorded, some simulations have been performed in order to have a rough idea of the flow area of the valve. Use was made of a very simplified nodalization obtained by adding the equivalent volume of the commingling flow lines to the real length of the main flow line in order to maintain the right transportation system inertia.

The slug catcher has been simulated by using the separator component with a unitary efficiency. The equipment is controlled by keeping the pressure and the liquid level constant by means of two actuated valves.

4.2 Boundary Conditions Determination.

4.2.1 Mass flow rates at the well flow lines inlet

The mass flow rates of oil and gas in steady state conditions were calculated for each well flow line inlet by converting the volumetric flow rates, measured during the last well testing, by means of the PVT Table. Then the total mass flow rate calculated at the test separator pressure and temperature assuming zero water flow rate has been used as inlet source leaving to the code the calculation of the oil and gas ratio at the down stream choke conditions. The temperature of the source was imposed constant during all the simulation.

The wells shut down started at 00.00 of 29th of October following a defined plan that envisaged the closure of one well every 15 - 30 minutes starting from the farthest from the Oil Centre. The consistence of the effective closing times with the planned ones has been verified for six wells where the well head pressures were recorded. The wells were stopped in a short time by closing the Wing valve and then the choke; for the simulation a closure time of 100 seconds was assumed.

The wells start-up began at 03.00 of 30th of October following a plan previously defined that envisaged the opening of the wells step by step reaching the pre-existent choke valve position in 3 - 12 hours, depending on the well. During this phase the mass flow rate increase was calculated for each well on the basis of the planned well head pressure decrease with respect to the time. Comparing them with the well head pressure trends recorded for the six wells mentioned above it has been found that the plan was not strictly followed and consequently, for these wells, the real well head pressure trends have been used. For the other ones no data were recorded and the mass flow rates were calculated using the planned values.

The following general correlation was applied to evaluate the mass flow rate supplied by every well:

$$\Gamma = \Gamma_N [\ [\mathbf{P}_0 - \mathbf{P}]\ /\ [\mathbf{P}_0 - \mathbf{P}_N]\ ]^{1/2}$$

in which:

$\mathbf{P}$ = W.H.F.P. measured during the start-up.
$\mathbf{P}_N$ = W.H.F.P. measured at full production before the shut-down.

P_0 = W.H.S.P. measured when the well is closed.
Γ_N = Oil mass flow rate measured at full production before the shut-down.
Γ = Oil mass flow rate calculated during the start-up.

The application of this simple formula is possible since the fluid is a single phase liquid from the reservoir to the choke. In Fig. 5 the calculated total mass flow is compared with the programmed one.

4.2.3 Pressure at the Slug Catcher outlet

The slug catcher has been simulated with the relative control system (valves on the gas and liquid lines actuated respectively by the pressure and level signals) and consequently the control valves back pressures were used as boundary conditions at the system outlet by assuming a constant value of 30 bar during all the simulation time.

5. RESULTS EVALUATION.

5.1. First phase:

This phase is related to the initial steady state conditions and it took one day of simulation time. As it can be seen from Tab. 5 and also from the several variable trends reported in the figures, a very good agreement among the measured and the simulated data was achieved during this phase.

The well testing of T17 (started at 10.00 a.m.) was the only significant event that occurred in this phase. Fig. 7 shows the effect of this operation on the T2 production manifold pressure. In fact this pressure was recorded by the pressure transducer located at the outlet of the T9 well flow line (manifold inlet). The pressure decrease, shown in this figure, during the well testing time (9 hours) is related to the reduction in the flow through the production line (see Fig. 2) due to the by pass of the T17 well from the production to the test manifold. This operation, that normally also brings about a lower temperature at the air cooler banks (production and test) outlets, has mainly a local effect on the satellite area where it occurs and leaves almost unchanged the conditions of the whole transportation system.

5.2 Second Phase.

This phase is related to the shut-down operation and it took about 27 hours of simulation time starting from 00.00 of 29th of October. The effective wells closure time was of about 3 hours starting from the farthest well (VF1). Fig. 6 and 7 show the pressure reduction at the manifold in the T2 and T4 satellite areas. During these operations the pressure and level set points at the slug catcher remained constant until the last well closure. Thereafter the pressure set point was reduced to 39 bar followed by the closure of the two valves upstream of the slug catcher, thus dividing the network into three separate systems:

1. All the flow lines from the wells to the On/Off valve.
2. The flow line held between the On/Off and the inlet slug catcher control valve.
3. The slug catcher with its controllers.

As shown in Fig. 16 and 17, which record the gas and liquid flow rates trends at the slug catcher outlets, the outlet control valves were closed three-four hours later than the inlet valves. After the slug catcher isolation the most important phenomena that occurred were the system cool down and depressurisation. The pressure and temperature trends after the system isolation are fairly different in the three insulated parts of the system due to their different heat capacity and heat exchanged.

Fig. 6 and 7 are representative of the main flow line, Fig. 8 of the line held between the two valves upstream the slug catcher and Fig. 9 of the slug catcher.

The comparison between the simulated trends and the experimental ones shows a very good agreement for the flow rates and for the pressures and a less good agreement for the temperatures. The good agreement of the pressure trends in the satellite areas and at the oil centre inlet shows that the cool down of the lines has been well simulated.

The discrepancies at the chokes and air coolers outlets (fig. 10-13) should be due to the simplified schematisation of the lines that does not take into account the higher elevations at the points where the temperature probes are located compared to the points of the simulated ones. In the pipe's upper part (in contact with the probes) this leads to lower gas temperature than the equilibrium temperatures simulated by the code, at zero flow when a full separation of gas and liquid phases occurs.

These non equilibrium real conditions are due to the low heat capacity of the gas connected to the low heat and mass transfer between the liquid bulk at the lower levels and the gas at the upper levels.

From Fig. 12 and 13 it is possible to identify the air coolers switch OFF and ON times (see respectively points A and B), that were not simulated with the code (given the OLGA limitations) and that can also be responsible for the disagreement between the two temperatures trends. In fact the switch OFF time (point A) strongly affects the initial conditions of the fluid into the unit during the shut-down phase.

Instead the disagreement of the temperatures at the slug catcher (Fig. 14) are due to a thermal non equilibrium condition connected with the fast depressurisation of the unit, when the pressure set point was decreased to 39 bar.

5.3 Third Phase.

The real start-up phase of the plant started just after the opening of the ON/OFF valve in the Oil Centre, at about the same time as the slug catcher inlet control valve reopening.

Fig. 8 shows the fast pressure increase in the network lines at the same pressure value (see Fig. 6) as the one in the flow line between the ON/OFF and the control valves.

The inlet slug catcher control valve was reopened step by step allowing a slow pressurisation of the separator (see Fig. 9) and, at the same time, the pressure set point was increased to 49 bar. All the simulated trends during the start-up phase show a little delay with respect to the recorded ones due to the uncertainty as to the actual time at which some wells were opened as explained before.

The pressure increase is well simulated in all the system (see Figs. 6 - 9) as well as the outlet slug catcher flow rates (see Figs. 16 - 17). At the beginning of the start-up the outlet liquid control valve was manually operated, thus bringing about sharp fluctuations in the outlet flow rate. The starting values of the temperatures simulated are higher than the measured ones for the reasons explained above but the fast temperature increases during the start-up phase are well simulated in all the system.

Fig. 11 shows that the down stream choke temperature measured during the start-up phase of the well is lower than the simulated ones. This is due to the cooling down of the fluid into the tubing in the shut-down phase and also to the higher heat exchanged through the tubing in the start-up phase. These phenomena, that affect the temperature increase in real conditions, were neglected in the simulation since it was assumed as a boundary condition, a source at a constant temperature, consistent with the normal operative conditions.

Fig. 10 shows a simulated temperature at the new equilibrium conditions lower than the real one since the well production was increased.

Part 2 - STEADY STATE CONDITIONS AND SEVERE SLUGGING PHENOMENON IN THE PREZIOSO FIELD

6. DESCRIPTION OF THE OFF-SHORE FIELD

Perla-Prezioso is an off-shore field located near the island of SICILY. It was developed with two platforms, but at present only the Prezioso platform is producing with 7 wells, while the Perla platform is shut-down for high water-cut problems. A 12 inch sealine, 29 km long and buried one meter below the sea bottom delivers the production to the Gela Oil Centre on-shore. The sealine cross the Perla platform with two risers (upward and downward).

Fig. 18 shows a schematic configuration of the transportation plant.

The fluid is a heavy oil with viscosity strongly dependent on the temperature. The fluid composition is reported in Tab. 6.

7. INPUT DECK PREPARATION

The simulation is only related to the sealine and the nodalization has been performed strictly by respecting the profile of the line. The resulting nodalization consists of a single branch composed of 8 pipes divided into 58 sections thickened near the elevation changes, as shown in Tab. 7, in order to permit a suitable discretization of the risers .

Given the fluid viscosity characteristics, particular care has been taken in the simulation of the heat exchange with the environment. For this reason a sensitivity study has been performed in order to evaluate its effect on the pressure drop of the line.

The measured temperature and the nominal mass flow rate, calculated from the single phases volumetric flow rates measured at the production separator, has been used as inlet boundary conditions while a constant pressure of 5.9 bar has been used as boundary conditions at the line outlet.

8. RESULTS EVALUATION

8.1 Sensitivity Analysis

A first group of calculations have been performed in order to evaluate the effect of the sea temperature on the code results. In fact, the high viscosity of the involved fluid and its strong dependence on the temperature could be the cause of a considerable dependence of the friction pressure gradient on the temperature itself. It should be observed that the pipeline is not insulated; in fact it has only a thin protective layer of polyurethane in contact with the pipe covered by a layer 5 cm thick of concrete. This involves a considerable dependence of the fluid temperature profile on the sea temperature.

The sea temperature values selected for these calculations cover the uncertainty range and they are: 13, 15 and 17°C. In Figs 19-21 the fluid temperatures, the liquid viscosity and the pressure profiles related to the three sea temperature values are shown. In Tab. 8 the values of pressure and temperature measured on Prezioso, Perla and Gela are compared with the calculated ones.

Fig. 20 shows the strong effect of sea temperature on the liquid viscosity along the pipe. From Tab. 8 it is possible to verify that the better agreement between the simulated and measured inlet pressures is obtained with a sea temperature of 17°C (CASE 3).

Fig. 13 shows that in all three cases the fluid temperature reaches the sea temperature after about 10 km from the inlet. Assuming that the temperature profile is constant after this equilibrium point it can be seen that also the pressure drop has a significant effect on the fluid viscosity.

8.2 Terrain Induced Slugging Analysis

Fig. 22 reports the pressure trends in some representative positions along the pipelines calculated by the OLGA (CASE 3 of Tab. 8). The pressure oscillations at the bottom of the upward and downward risers on the Perla platform (exactly in phase opposition) indicate a typical terrain induced slugging phenomena. The trends of the collapsed levels in the two risers have been plotted in Fig. 23. It is evident that, after a gradual refilling of the upward riser, the pressure of the stagnant gas at the bottom reaches a value high enough to enable the slug to be partially pushed out of the riser.

Fig. 22 shows that the inlet pressure to the sea line is not affected by the pressure oscillations in Perla.

At the moment the presence of the severe slugging phenomena predicted by the code in Perla cannot be verified because suitable instrumentation is not available. Pressure transducers and instrumentation capable of supplying information on the flow structure and composition are expected to be installed on Perla platform in order to collect more qualified data in steady state and transient conditions.

9. CONCLUSIONS

From an overall results analysis of the simulations performed the following conclusions can be drawn:

1. The OLGA 91 computer code is capable of performing simulations in steady state and in transient conditions of a complex network as well as of a transportation line containing risers (upward, downward) with good results.
2. The OLGA input deck preparation requires a tune-up activity that is very important to reproduce the field data as accurately as possible and must be performed roughly in real time with the field data collection in order to verify the engineering assumptions.
3. The heat exchanger model in OLGA is a very simplified model that may be sufficient to perform the steady state simulations but should be improved for the transient conditions.
4. For the simulation of the shut-down conditions the selection of a flowline schematisation guaranteeing the thermal equilibrium of the phases into the nodes is very important; this is required to be able to take into account the gas and liquid separation during the transient.
5. With high viscosity oils, the correct simulation of the thermal losses is very important since the strong variation of the oil viscosity, as a function of the fluid pressure and temperature, strongly affects the flowline pressure drops.
6. The OLGA simulation activity is very important to identify measurement points and suitable instrumentation to perform a representative field data collection.

REFERENCES

/1/ A. Mazzoni, M. Villa (Agip) - G. De Toma (Enea) - M. Bonuccelli (C.P.R.): "Performance of the OLGA computer code in the analysis of the multiphase transportation system of Trecate-Villafortuna AGIP's oil field", August 1992.

/2/ G.De Toma, D. Mazzei, T. Crescenzi (Enea): "Valutazione preliminare delle capacità di simulazione dinamica del trasporto multifase in sistemi complessi del codice OLGA, con l'ausilio di dati di campo", December 1992.

/3/ J. Haugen: "OLGA 91 Model and Numeric Guide" IFE/KR/F-91/149, December 1991.

/4/ J. Haugen: "OLGA 91 User Guide" IFE/KR/F-91/147, December 1991.

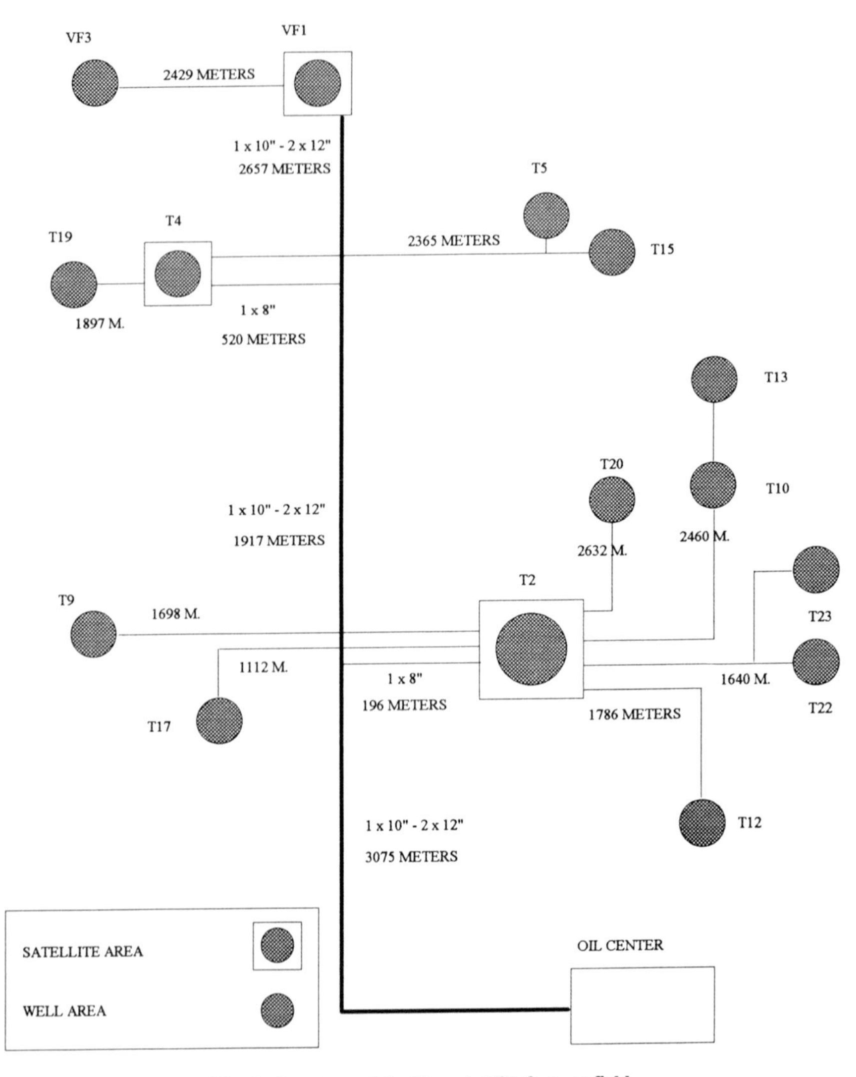

Fig. 1 - Structure of the Trecate-Villafortuna field

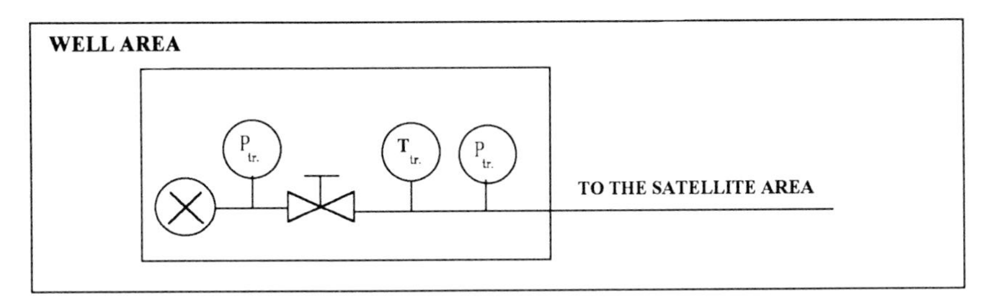

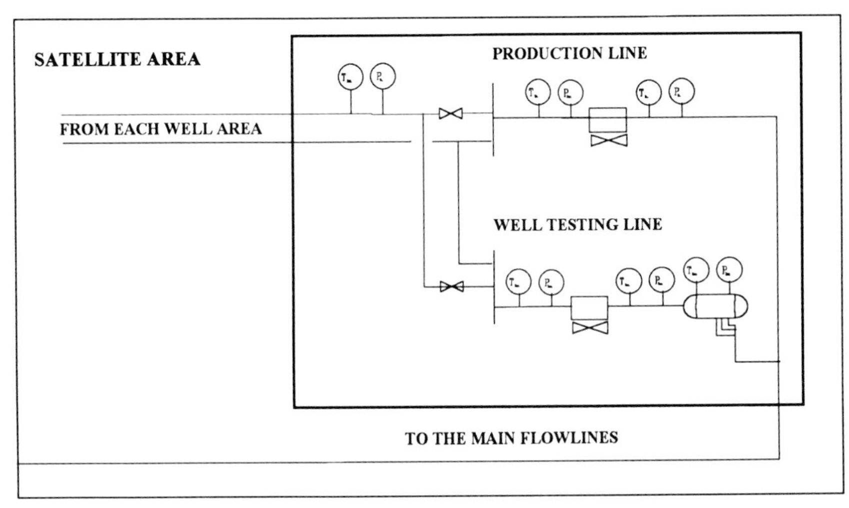

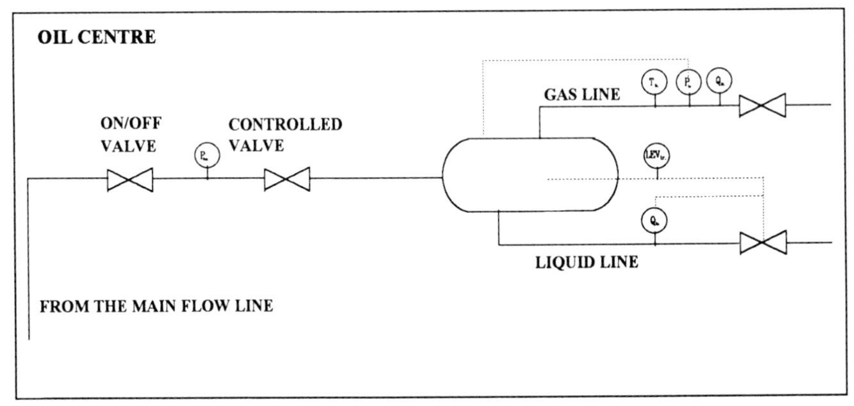

Fig. 2 Configurations of the typical areas

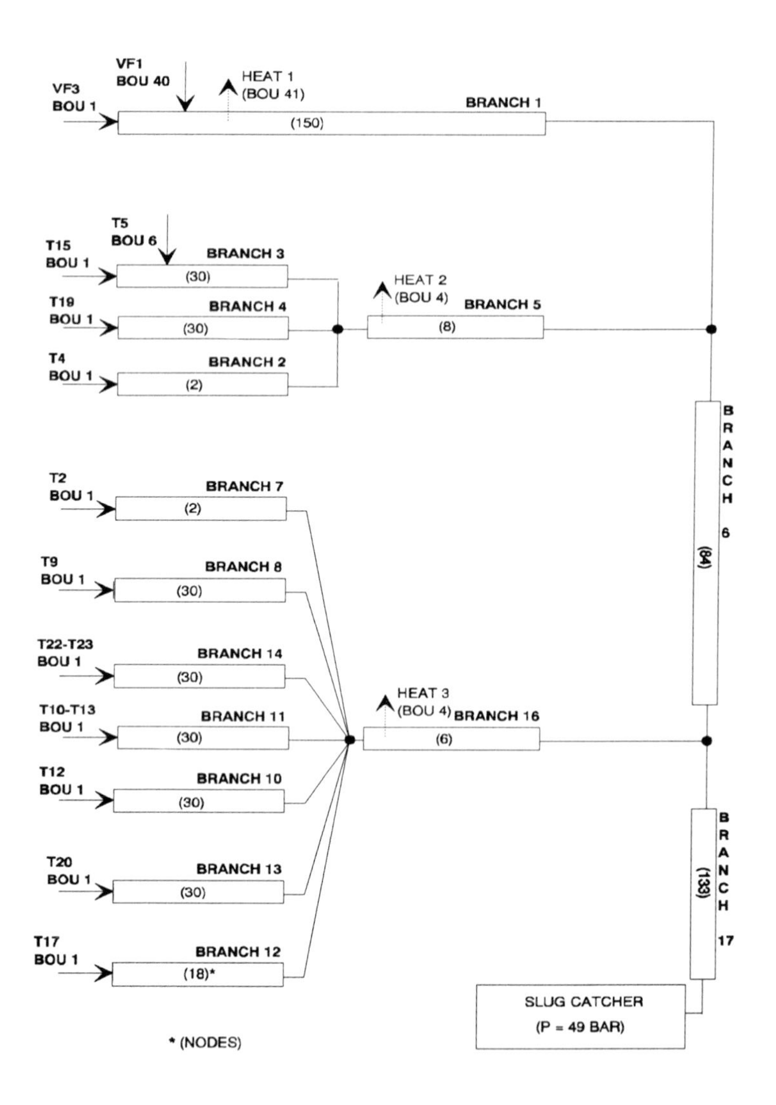

Fig. 3 Nodalization of the Trecate network for OLGA91

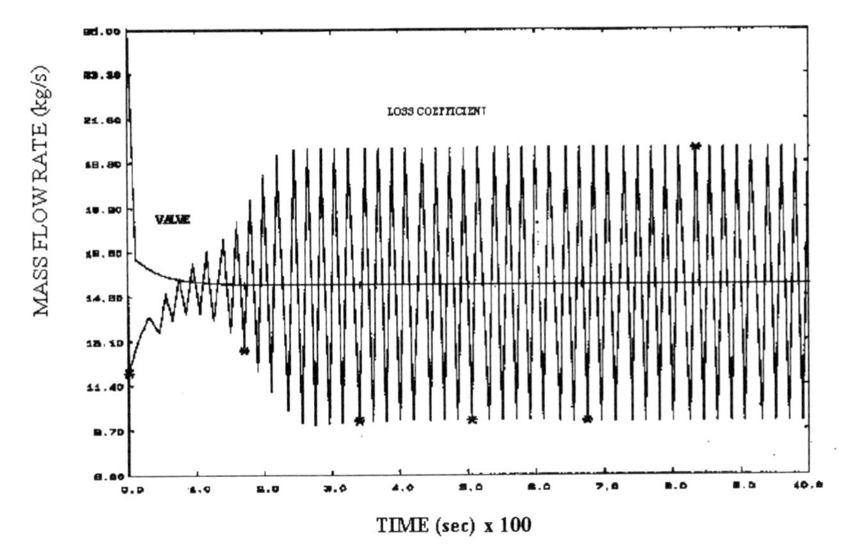

Fig. 4 - Mass flow rate at the air cooler outlet

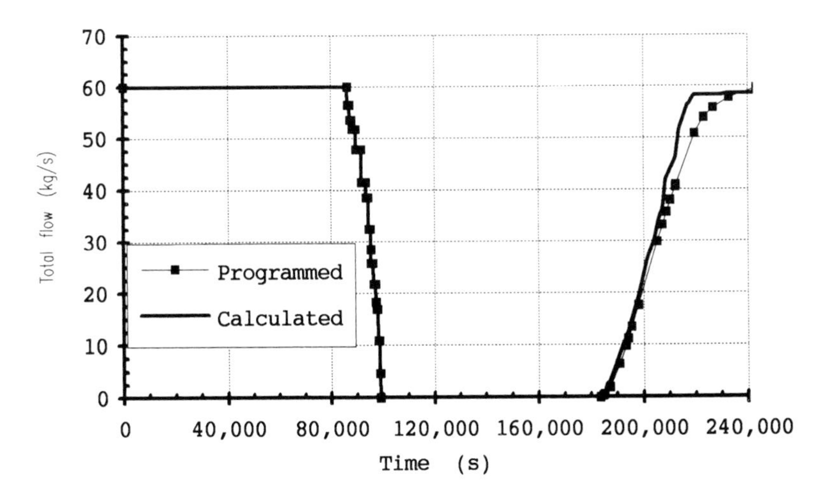

Fig. 5 - Calculated and planned total mass flow rate

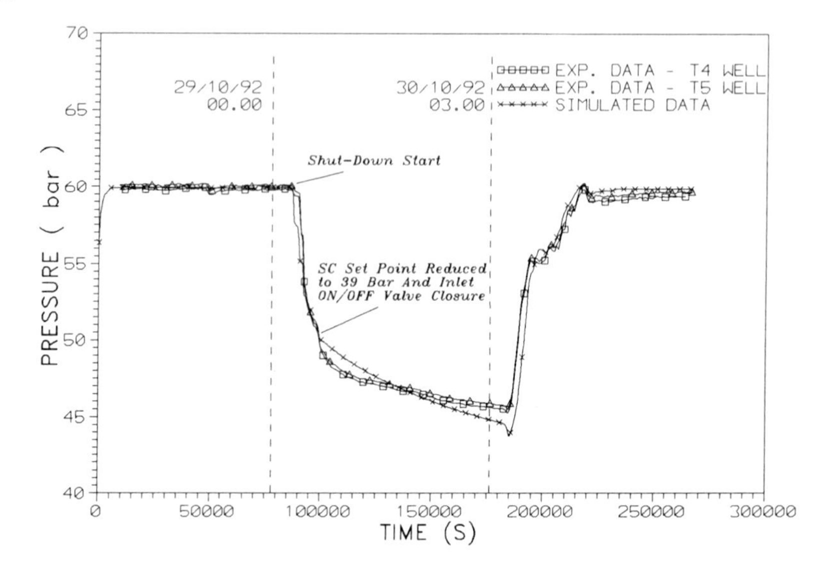

Fig. 6 - Manifold pressure in T4 area

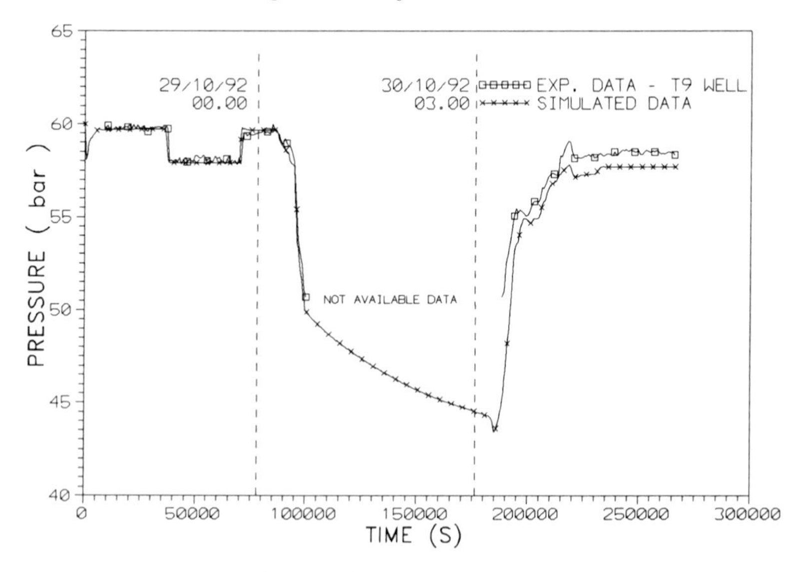

Fig. 7 - Manifold pressure in T2 area

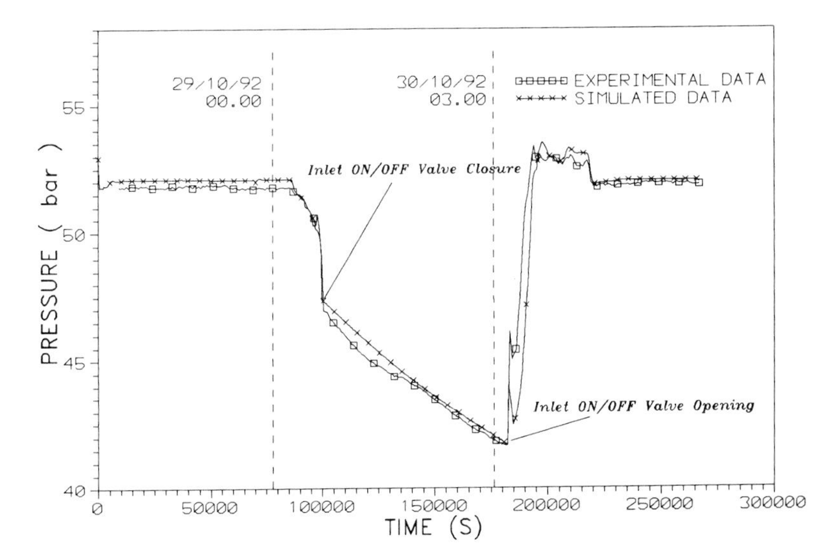

Fig. 8 - Manifold pressure at the Oil Centre inlet

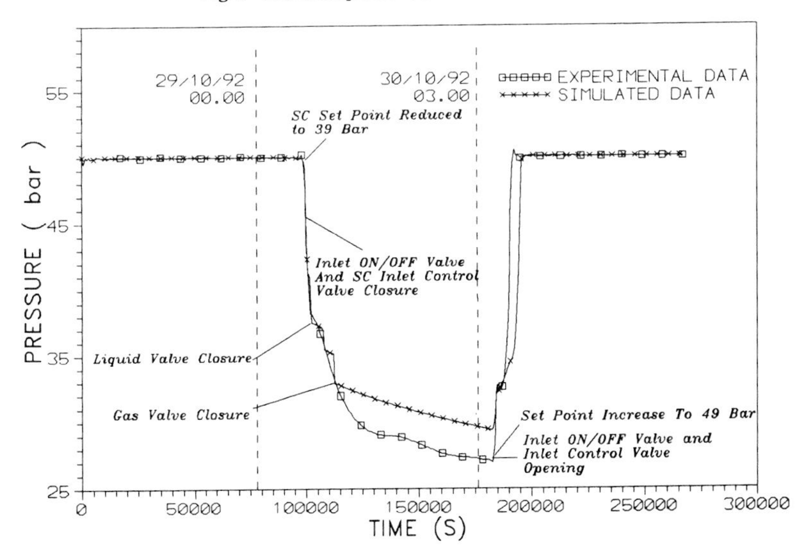

Fig. 9 - Slug Catcher pressure

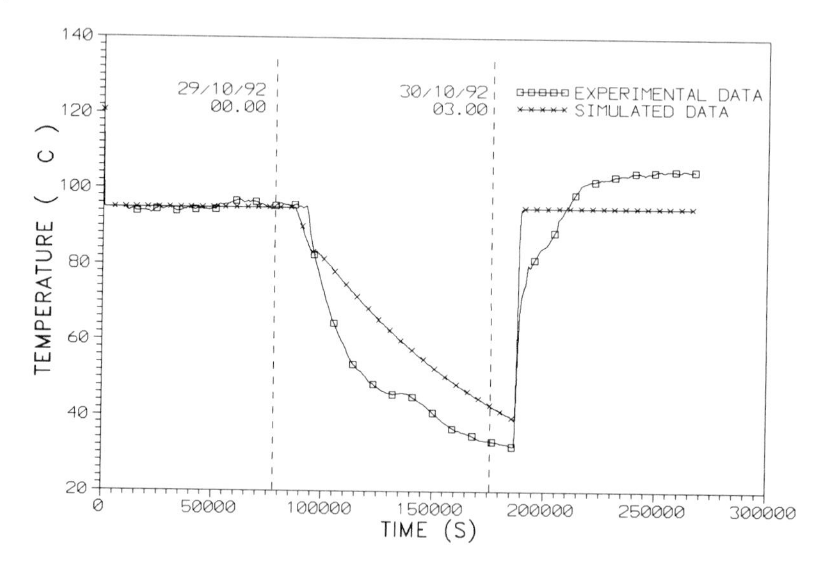

Fig. 10 - T4 Downstream choke temperature

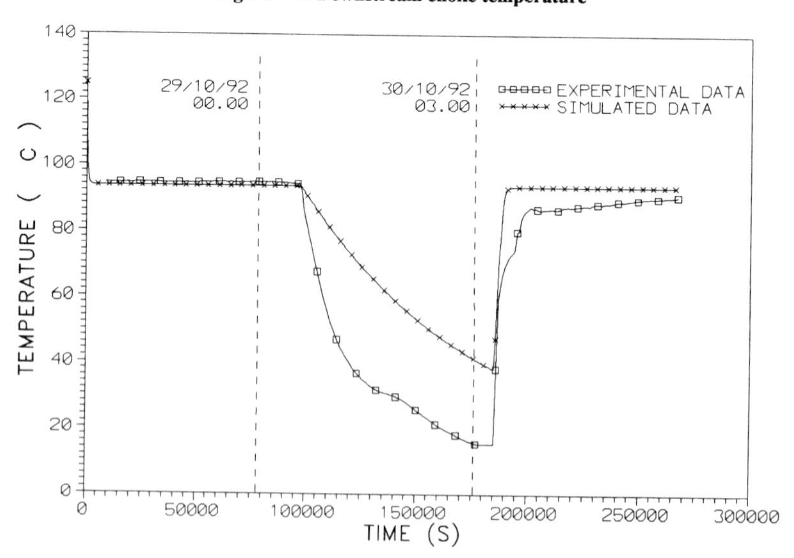

Fig. 11 - T9 Downstream choke temperature

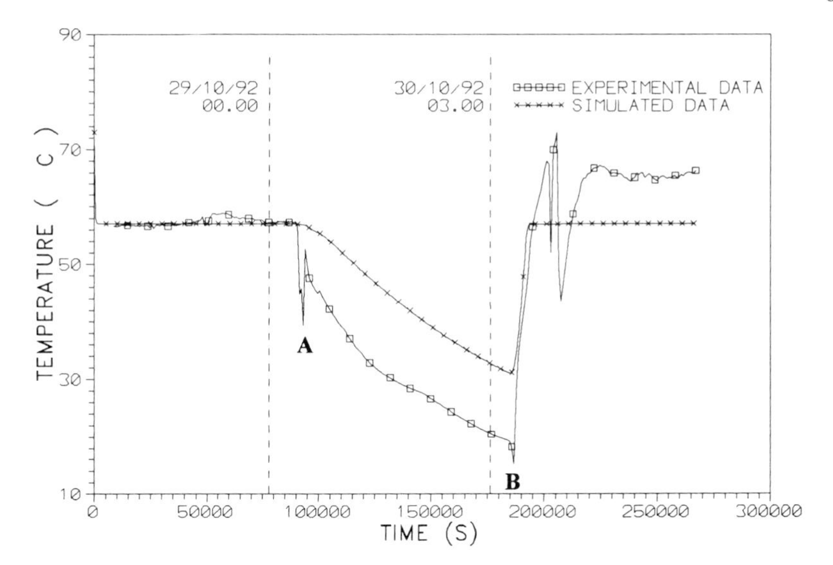

Fig. 12 - Downstream air cooler temperature in T4area

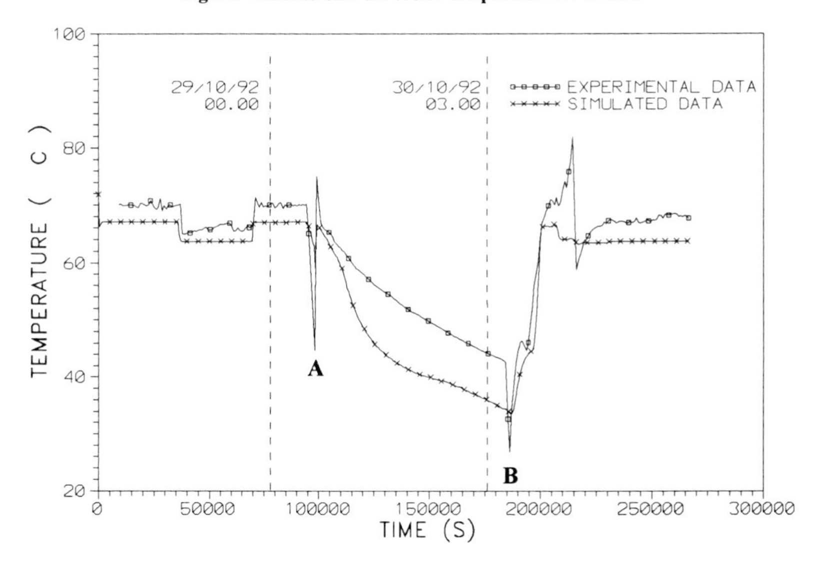

Fig. 13 - Downstream air cooler temperature in T2area

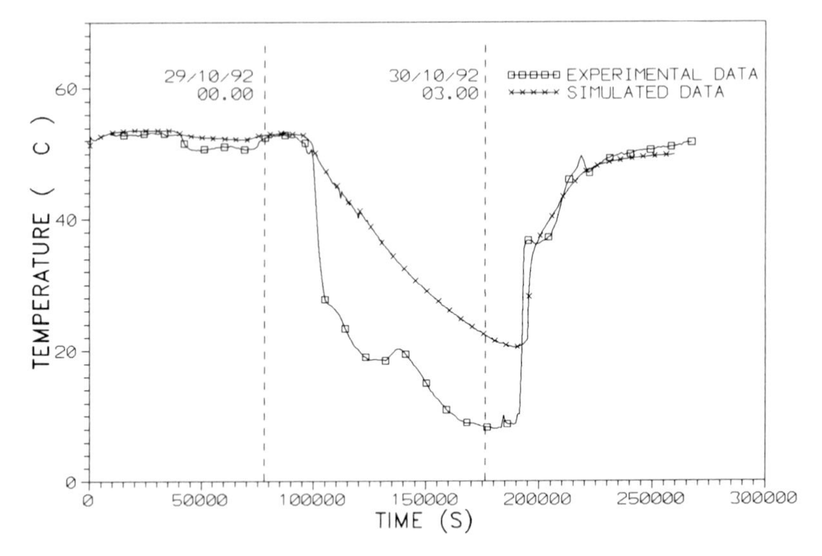

Fig. 14 - Slug Catcher temperature

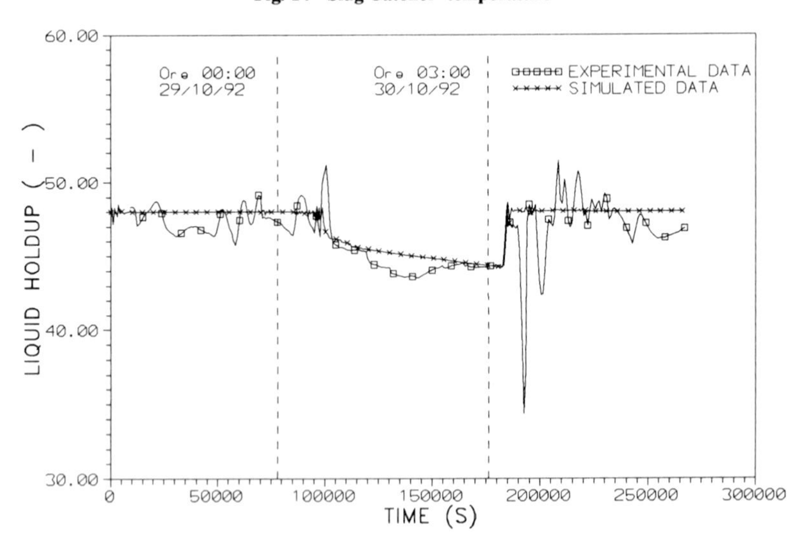

Fig. 15 - Slug Catcher liquid level

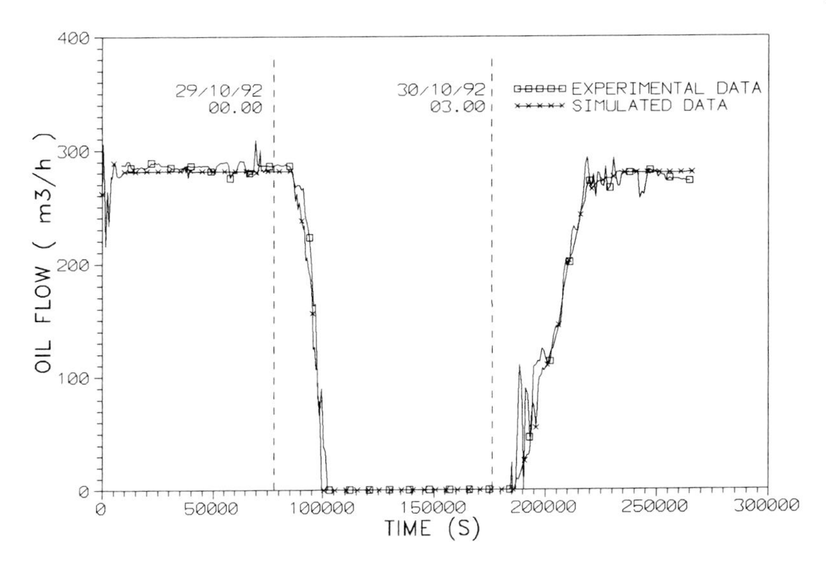

Fig. 16 - Slug Catcher oil flow rate

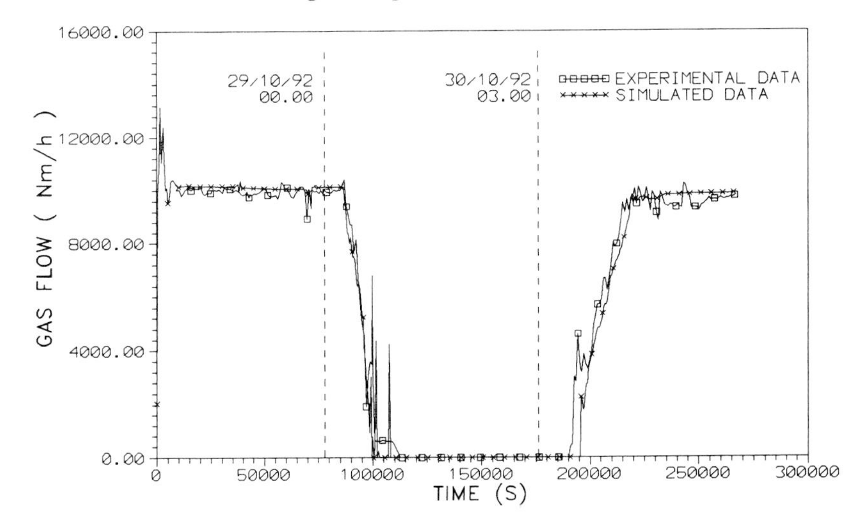

Fig. 17 - Slug Catcher gas flow rate

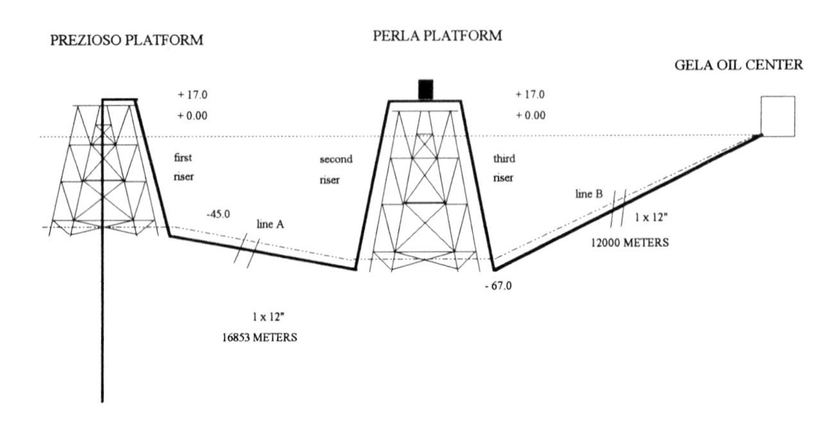

Fig. 18 - Prezioso-Perla-Gela sealine profile

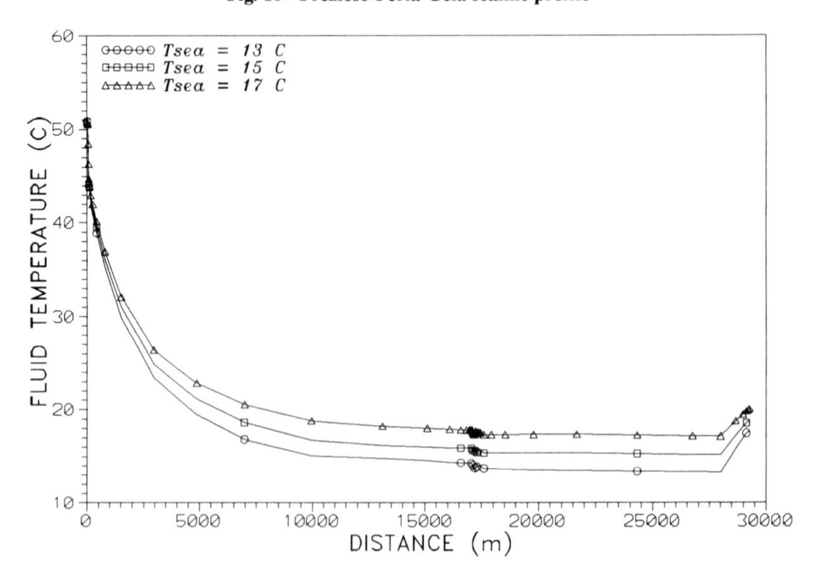

Fig. 19 - Fluid temperature profiles.

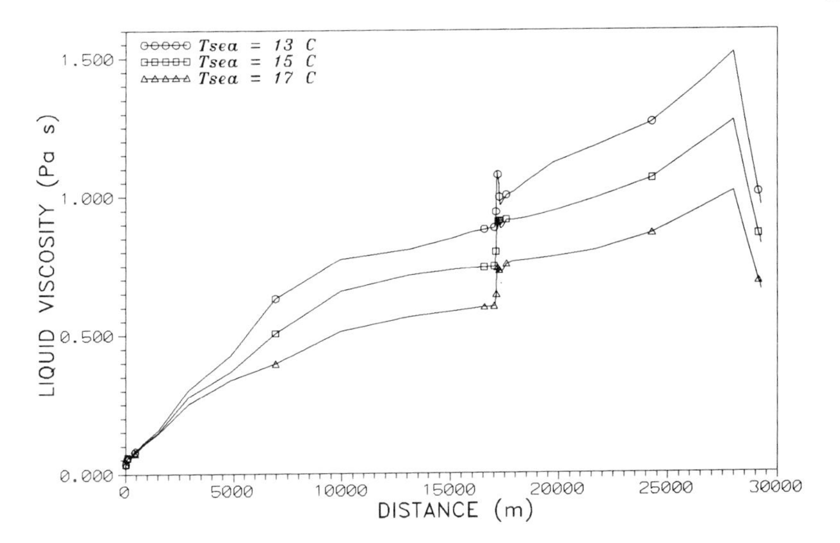

Fig. 20 - Liquid viscosity profiles.

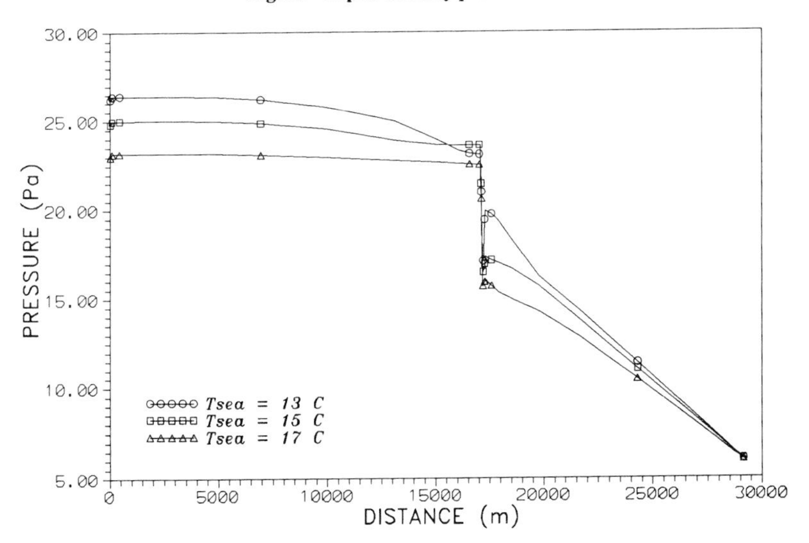

Fig. 21 - Pressure profiles.

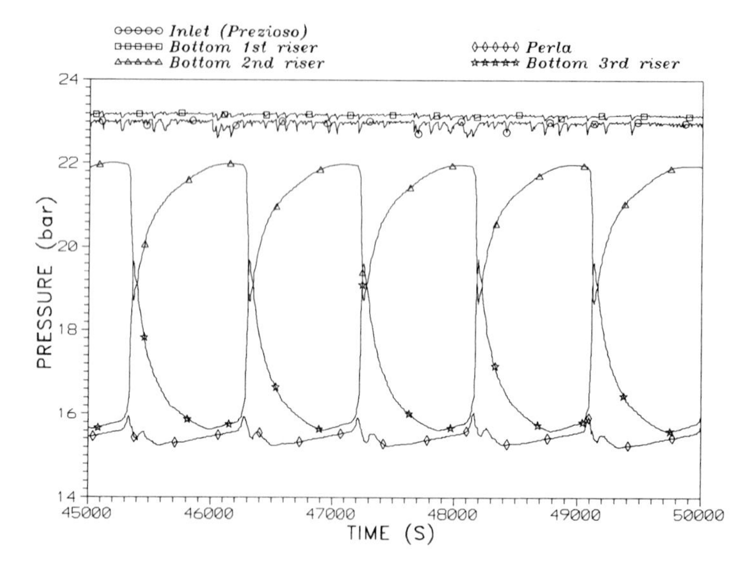

Fig. 22 - Calculated pressure trends in varius positions of the Prezioso sealine.

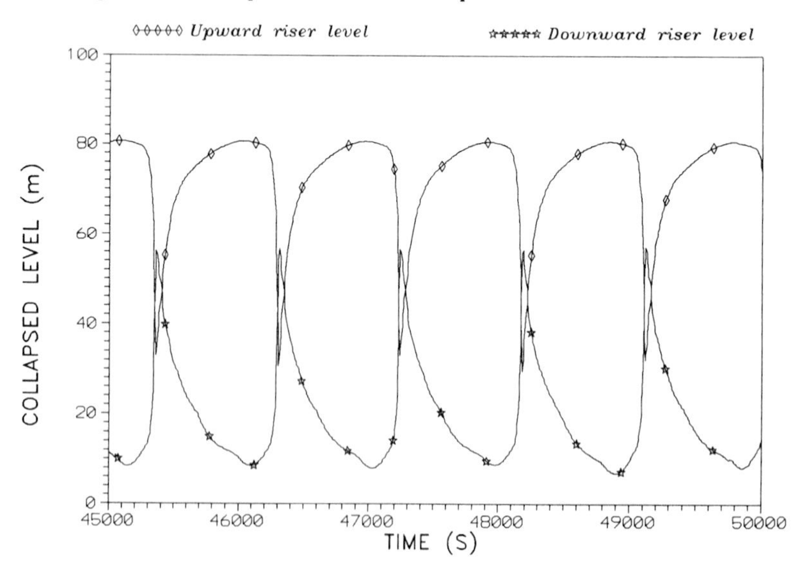

Fig. 23 - Collapsed level trend.

Number of wells	15	
Depth	5600 - 6300	m
Oil flow	6300	m3/d
Gas flow	440000	Nm3/d
G.O.R.	70	Nm3/m3
Boiling Point	170	bar
	160	°C
Water Cut	< 2	%
B.H.F.P.	1100	bar
W.H.F.P.	300 - 600	bar
B.H.T.	160 - 180	°C
W.H.T.	90 - 130	°C

Tab. 1 - Main characteristic of the Trecate-Villafortuna field

COMPONENT	% Mol.
CO2	1.100
N2	0.970
H2S	0.041
C1	35.150
C2	3.350
C3	2.730
i-C4	1.150
n-C4	1.710
i-C5	1.430
n-C5	1.630
C6	4.660
C7	5.030
C8	5.480
C9	4.790
C10	4.380
C11	2.070
C12	1.980
C13+	22.390

Tab. 2 - Average fluid composition

BRANCH NUMBER	LENGHT (m)	NUMBER OF SECTIONS	NOMINAL DIAMETER	WELLS		EQUIPMENTS	
				NAME	SECT. LOC.	TYPE	SECT. LOC.
1	2429	40	4"	VF3	1	Heat Exchanger	41
				VF1	40		
	2657	110	10"				
2	30	2	4"	T4	1		
3	2365	30	4"	T15	1		
				T5	6		
4	1897	30	4"	T19	6		
5	520	8	8"			Valve	4
						Heat Exchanger	4
6	1917	84	12"				
7	73	2	4"	T2	1		
8	1698	30	4"	T9	1		
10	695	12	4"	T12	1		
	1091	18	6"				
11	2460	30	6"	T13	1		
				T10	2		
12	1112	18	4"	T17	1		
13	200	3	3"	T20	1		
	2432	27	4"				
14	1640	30	6"	T22	1		
				T23	2		
16	196	6	8"			Valve	4
						Heat Exchanger	4
17	3180	132	12"			On/Off Valve	131
						Inlet. SC.	
	18	1	118"			Control. Valve	133
						Slug Catcher	133
						Valve	134

Tab. 3 - Main characteristics of the nodalization

TIME seconds	OCTOBER 92 dd	hh	mm	EVENTS
Steady State Phase				
36596	28	10	9	T17 WELL TRANSFERRED TO TEST MANIFOLD
36596	28	10	9	T2 WELL TRANSFERRED TO SBS MANIFOLD
70149	28	19	29	T17 TRANSFERRED TO PRODUCTION MANIFOLD
Shut Down Phase				
86400	29	0	0	SHUT-DOWN VF1
87312	29	0	15	SHUT-DOWN T4
88200	29	0	30	SHUT-DOWN VF3
89661	29	0	54	SHUT-DOWN T5
91833	29	1	30	SHUT-DOWN T15
93600	29	2	0	SHUT-DOWN T19
94500	29	2	15	SHUT-DOWN T17
95336	29	2	28	SHUT-DOWN T10
95400	29	2	30	SHUT-DOWN T13
96300	29	2	45	SHUT-DOWN T22
97204	29	3	0	SHUT-DOWN T12
97467	29	3	4	SHUT-DOWN T9
98100	29	3	15	SHUT-DOWN T20
98701	29	3	25	SHUT-DOWN T2
98779	29	3	26	SLUG CATCHER PRESSURE SET POINT REDUCED TO 39 bar
99000	29	3	30	SHUT-DOWN T23
99356	29	3	35	CLOSING OIL CENTRE INLET ON/OFF VALVE
99356	29	3	35	CLOSING SLUG CATCHER INLET CONTROL VALVE
107760	29	5	56	CLOSING SLUG CATCHER OUTLET LIQUID CONTROL VALVE
112280	29	7	11	CLOSING SLUG CATCHER OUTLET GAS CONTROL VALVE
Start Up Phase				
182320	30	2	38	OPENING OIL CENTRE INLET ON/OFF VALVE
182580	30	2	43	OPENING SLUG CATCHER INLET CONTROL VALVE
183210	30	2	53	SLUG CATCHER PRESSURE SET POINT INCREASED TO 49 bar
183600	30	3	0	START-UP T20
184200	30	3	10	START-UP T15
184620	30	3	17	START-UP T9
185130	30	3	25	START-UP T5
186380	30	3	46	OPENING SLUG CATCHER OUTLET LIQUID CONTROL VALVE
187200	30	4	0	START-UP T17
187200	30	4	0	START-UP T22
187200	30	4	0	START-UP T23
187222	30	4	0	START-UP T4
189630	30	4	40	OPENING SLUG CATCHER OUTLET GAS CONTROL VALVE
190800	30	5	0	START-UP T19
193280	30	5	41	START-UP T10
194150	30	5	55	START-UP T12
198000	30	7	0	START-UP VF3
207330	30	9	35	START-UP T2
212400	30	11	0	START-UP VF1

Tab. 4 - Main events during the operational transient

BRANCH SECTION		DESCRIPTION	STEADY STATE		TRANSIENT
			FIELD DATA	OLGA 91.	
		VF1 AREA AND CONNECTED WELLS:			
1	1	DOWNSTREAM CHOKE PRESSURE VF1	57	57.3	
1	40	DOWNSTREAM CHOKE PRESSURE VF3	58	56.5	
		DOWNSTREAM CHOKE TEMPERATURE VF3	97	118	
1	40	UPSTREAM AIR COOLER PRESSURE	57	56.5	
1	41	DOWNSTREAM AIR COOLER PRESSURE	55	55.2	
		DOWNSTREAM AIR COOLER TEMPERATURE	50	50	
		T4 AREA AND CONNECTED WELLS:			
2	1	DOWNSTREAM CHOKE PRESSURE TR4	60	60	
		DOWNSTREAM CHOKE TEMPERATURE TR4	96	95	
3	6	DOWNSTREAM CHOKE PRESSURE TR5	74	75	
		DOWNSTREAM CHOKE TEMPERATURE TR5	135	126	
3	1	DOWNSTREAM CHOKE PRESSURE TR15	77	76.8	
4	6	DOWNSTREAM CHOKE PRESSURE TR19	61	60.8	
5	3	MANIFOLD PRESSURE TR4			Fig. 6
		MANIFOLD PRESSURE TR5			Fig. 6
		UPSTREAM AIR COOLER TEMPERATURE	110	116	
5	4	DOWNSTREAM AIR COOLER PRESSURE	54	54.6	
		DOWNSTREAM AIR COOLER TEMPERATURE	57	57	Fig. 12
		T2 AREA AND CONNECTED WELLS:			
7	1	DOWNSTREAM CHOKE PRESSURE TR2	60	60	
		DOWNSTREAM CHOKE TEMPERATURE TR2	117	120	
8	1	DOWNSTREAM CHOKE PRESSURE TR9	59	60	
		DOWNSTREAM CHOKE TEMPERATURE TR9	95	93.8	
11	2	DOWNSTREAM CHOKE PRESSURE TR10	63	63.7	
10	1	DOWNSTREAM CHOKE PRESSURE TR12	61	61	
		DOWNSTREAM CHOKE TEMPERATURE TR12	129	130	
11	1	DOWNSTREAM CHOKE PRESSURE TR13	61.5	63.7	
12	1	DOWNSTREAM CHOKE PRESSURE TR17	61	59.7	
13	1	DOWNSTREAM CHOKE PRESSURE TR20	67	76.2	(*)
		DOWNSTREAM CHOKE TEMPERATURE TR20	110	109	
14	1	DOWNSTREAM CHOKE PRESSURE TR22	65	63.4	
		DOWNSTREAM CHOKE TEMPERATURE TR22	114	114	
14	2	DOWNSTREAM CHOKE PRESSURE TR23	65	63.4	
		DOWNSTREAM CHOKE TEMPERATURE TR23	112	114	
16	3	MANIFOLD PRESSURE TR9			Fig. 7
		UPSTREAM AIR COOLER TEMPERATURE	114	115	
16	4	DOWNSTREAM AIR COOLER PRESSURE	54	54.5	
		DOWNSTREAM AIR COOLER TEMPERATURE	64	67	Fig. 13
		OIL CENTRE:			
17	131	MANIFOLD PRESSURE			Fig. 8
	133	SLUG CATCHER PRESSURE			Fig. 9
	133	SLUG CATCHER LIQUID CONTENT			Fig. 15
	133	OIL FLOW			Fig. 16
	133	GAS FLOW			Fig. 17
	133	SLUG CATCHER TEMPERATURE			Fig. 14

(*) Local Pressure Metering not Calibrated

UNITS:		
	Pressure	**bar**
	Temperature	**°C**

Tab. 5 - Comparison between measured and calculated values in steady state

COMPONENT	% Mol.
CO_2	10.11
N_2	0.32
H_2S	0.44
C1	18.20
C2	1.12
C3	1.65
i-C4	0.65
n-C4	1.28
i-C5	1.08
n-C5	1.01
C6	3.01
C7	3.10
C8	3.41
C9	3.20
C10	3.19
C11	1.37
C12	1.41
C13+	45.45

Tab. 6 - Fluid composition of Prezioso field

ZONE	PIPE NUMB.	SECT. NUMB.	SECT. LENGHT	ZONE	PIPE NUMB.	SECT. NUMB.	SECT. LENGHT
Prezioso platform	1	1	15	second riser	4	30	16.4
		2	15			31	16.4
first riser	2	3	15			32	16.4
		4	15			33	16.4
		5	15			34	16.4
		6	15	Perla platform	5	35	15
line A	3	7	15			36	15
		8	30	third riser	6	37	16.4
		9	60			38	16.4
		10	120			39	16.4
		11	240			40	16.4
		12	480			41	16.4
		13	960	line B	7	42	15
		14	1920			43	30
		15	1932			44	60
		16	2237			45	104
		17	3735			46	208
		18	2624			47	416
		19	1312			48	832
		20	656			49	1664
		21	328			50	2120
		22	186			51	3145
		23	60			52	1716
		24	30			53	850
		25	15			54	425
		26	15			55	213
		27	15			56	106
		28	15	GelaOil	8	57	53
		29	15	Centre	8	58	53

Tab. 7 - Main characteristic of the Prezioso-Perla-Gela sealine nodalizzation.

	Prezioso		Perla		Gela	
	Pressure (bar)	Temperature (°C)	Pressure (bar)	Temperature (°C)	Pressure (bar)	Temperature (°C)
Measured	21.0	51.0	15.0	17.5	5.9	22.0
Calc. Case 1	26.2	51.0(*)	17.1	14.0	5.9(*)	17.7
Calc. Case 2	24.8	51.0(*)	16.4	15.7	5.9(*)	18.8
Calc. Case 3	23.0	51.0(*)	15.3	17.5	5.9(*)	20
(*) Imposed value.						

Tab. 8 - Comparison between measured and calculated values of pressure and temperature on Prezioso and Perla platforms and in the separator of the Oil Centre of Gela.

LUBRICATING FILM FLOW BEHAVIOR AND PRESSURE DROP MODEL FOR OIL/WATER CORE ANNULAR FLOW

JAUA, J.A., GUEVARA, E. AND ZAGUSTIN, K.

INTEVEP, S.A., S.A

ARNEY, M.S.

UNIVERSITY OF MINNESOTA

ABSTRACT

A theoretical and experimental study to investigate the flow behavior of the water lubricating film in oil/water core annular flow pattern in horizontal pipes, and a model to calculate pressure drop under the same conditions, are presented. Visual observations were made using polystyrene particles. As expected, a recirculation zone in the valleys between waves was observed. This, together with the velocity distribution measurements, supports the assumption that the main contribution to the frictional pressure drop in the core annular flow pattern is concentrated in a region between the pipe wall and a cylinder encompassing the core waves. The width of this region is the minimum distance between the pipe wall and the oil/water interface. The experimental results showed that this assumption holds for relatively high water velocities. That is, it was found that the motion of the water in the above mentioned region is insensitive to the recirculating zone.

Based on the above findings, a model to calculate the pressure drop in a horizontal pipe under core annular flow mode was developed. The model assumed that the water motion in the annulus between the pipe wall and a cylinder encompassing de core waves, is predominantly of the Couette type. Comparisons were made using experimental data obtained in the 1 km x 203 mm ID Pipe san tome test loop, using viscosities between 3.0-110 Pa.s and those of other sources. It was found that the model predicts most of the experimental data to within +/- 15% with a maximum error of 24%.

1. INTRODUCTION

The most convenient and usually the most economical method for transporting crude oil over land is through pipeline. In this mode of transportation the power required to move the oil is a direct function of the oil viscosity which, for different crude oils, may vary over a wide range. The flow of viscous crude oils under typical pipeline conditions tends to be laminar and in this regime the pressure gradients necessary to move the oil at a given through-put is proportional to the viscosity of the oil; therefore, the pressure gradient in a pipeline transporting a high viscous crude oil is considerable and the pumping facilities become much larger than the required for transporting a low viscosity crude oil. This increases initial investment, maintenance and operational costs, thereby increasing the cost per barrel transported.

There are significant reserves of heavy and extra heavy viscous crude oils and bitumens in Venezuela, Canada, The United States and Europe, these may have viscosities in the order of 1500 P at room temperature and consequently can not be transported by conventional methods . Instead they require the viscosity to be reduced for economical and technical reasons. There are several methods available for reducing the pressure gradient necessary to maintain a given through-put. Either through the addition of a hydrocarbon diluent or through the installation of heating equipments at short intervals along the pipeline. The former method can only be used in the unusual case in which there is an abundant supply of light oil in the same region as the heavy oil; heating is inconvenient and costly. Those two are the favored means used to transport high viscosity crude oil and bitumens in Venezuela (Parra, 1979 and Cabrera.1982). However, since fuel costs are becoming high and light crude oils used as diluent will not be available in the near future, there is a need to look for alternatives which will reduce both pumping and fuel consumption.

Therefore, much attention has been devoted at INTEVEP, S.A.,. To further develop the core annular flow technique among others, as a mean to transport heavy and extra heavy crude oils and bitumens through pipelines (Guevara et al, 1988 and 1991). In this method of transportation a less viscous immiscible fluid such as water is introduced into the flow to act as a lubricating layer for absorbing the shear stress existing between the wall of the pipe and the fluid being transported.

The prediction of the pressure losses under oil/water core annular flow is thus important for designing purposes. It is then the main objective of this article to present a simple model to determine the pressure losses through a pipeline when transporting viscous crude oils under core annular flow conditions.

2. PREVIOUS WORK

Even though the contact of two immiscible liquids frequently occurs in the chemical and petroleum industries, the methods for designing processes and equipment has not been satisfactorily developed from a fluid-dynamics point of view. The process variables are numerous and the analytical solutions are very complex and generally based on simple assumptions. Because of this fact, the research effort on the topic has been focused on particular solutions to specific fluids.

When two immiscible liquid phases - for example, oil and water - are injected into a pipe, different flow patterns might be formed. The characteristics of core-annular flow formation in liquid/liquid are still not very well understood.

Russell and Charles (1959) were the first authors to record laboratory experiments on the viscosity reduction by mixing water with oil, and analytically solved the problem through a mathematical analysis, considering the concentric flow in a circular pipe. These authors derived the interface position assuming equal density liquids, from a forced balance. They derived expressions for the water and crude flow rates respectively, from which an expression for the pressure gradient reduction factor was developed. Calculation were performed using viscosities from 0.1 to 1000 cP and pipe diameters from 12.7 to 508 mm. Similar works have been reported by Charles (1963) and Epstein (1963)

Bentwich et al (1970) studied theoretically the eccentric laminar core annular flow. The laminar velocity profiles calculated by Bentwich (1964) were integrated to give the volumetric flow rates in a pipe of two immiscible liquids with an eccentric circular interface.

Ooms et al (1984) presented a theoretical model, based on the lubrication theory as a possible mechanism for maintaining stable core annular flow. In the lubricating film model an eccentric oil core with an asymmetric wavy interface is assumed. The oil core, with a density lower than that of the water film, will rise to the top of the pipe until the buoyancy force is counterbalanced by the lubricating force. In the upper half of the pipe a very thin film of water will be present between the oil and the pipe wall; the remainder of the water will flow in the lower part of the pipe underneath the oil core. Using estimated wave parameters from photographs and pressure losses found in core flow tests with a viscous oil in a 50 mm ID pipe, the model allows pressure loss predictions to be made for different pipes sizes. Moreover, limitations to the core velocities for which the lubricating film model es valid, may be specified. The lower critical velocity, defined as the core velocity at which the interfacial waves touches the pipe wall, corresponds to the threshold velocity observed in practice for stable/unstable core flow transition. This model was latter extended by (Oliemans et al, 1987) to account for the turbulence in the water lubricating layer. However, these models are numerically complicated and in addition required as input the water film thickness and the waves characteristics at the oil/water interface. On the other hand, the application of the lubrication theory to explain the mechanism whereby core annular flow forms, has been recently challenged by Joseph (1993) by introducing the concept of flying core flow.

Brauner(1991) developed a model for liquid-liquid core annular flow in horizontal. A general development of the appropriated correlative parameters is obtained through a dimensional analysis of the momentum equations for each liquid phase. The model puts under a common framework all possible flow situations of laminar-laminar a, turbulent-turbulent and mixed flow regimes, for a wide range of viscosity and density ratios. Comparison with available experimental data was found satisfactory. The model does not check for the existence of core annular flow.

3. LUBRICATING FILM FLOW BEHAVIOR

From flow visualizations through a 203 mm ID transparent test section in the San Tome test loop (Zagustin et al, 1988) , it became clear that the oil-water interface is made up of waves of finite amplitude, whose characteristics are basically a function of the injected water fraction (Trallero, 1992). The presence of these waves of finite amplitude modifies the hydrodynamics of the water flow and therefore affects the magnitude of the wall shear stress which is directly related to the pressure gradient of the system.

Based on previous experiences related to finite amplitude waves Zagustin (1970) and Manson et al (1985), and by filming the motion of particles in the water phase, using a moving camera, which allowed to freeze the wave's shape in time, it was possible to observe that the water flow separates from the crests of the waves, forming a recirculating region in the valley between waves. The main water flow was thus restricted to the annulus formed between the pipe wall and a cylinder encompassing the core waves. The width of this region corresponds to the minimum distance between the pipe wall and the oil-water interface.

In order to further substantiate these findings, an experimental work was carried out (Jaua, 1989) to simulate the motion of the water layer between the pipe wall and the oil core. The test section was designed in such a way that the water flows between a stationary wavy wall and a moving boundary, corresponding to the case of an observer moving with the oil core. In this way, the effect of the wave characteristics such as the wave amplitude and length, could be investigated. Extensive flow visualization were made using polystyrene particles, for water velocities ranging from 0.3 to 0.5 m/s for different wave's amplitude and lengths. In addition, He measured the velocity profile across the water layer and the pressure distribution along the waves. It was found that Zagustin's (1988) observations were valid for relatively high water velocity and that the motion of the water in the above mentioned region is insensitive to the flow in the recirculating zone.

4. PRESSURE DROP MODELING

Having in mind the concepts underlined in the previous section, Zagustin (1988) develop a simplified model to compute the pressure drop for core annular flow in a pipe. The following assumptions were made:

(i) The frame of reference is chosen such that the oil core is at rest.

(ii) The main contribution to the frictional pressure drop in core annular flow is concentrated in a region between the pipe wall and a cylinder encompassing the core waves. The width of this region is the minimum distance between the pipe wall and the oil-water interface.

(iii) The flow in this region is predominantly of the Couette type.

Analogy is then made with the couette flow between parallel plates and the following expressions for the friction factor (El Telbany et al,1982) were assumed to apply:

$$C_f = \frac{2.0}{N_R} \qquad N_R < 1000 \qquad (1)$$

$$\sqrt{\frac{C_f}{2}} = \frac{0.19}{LOG(N_R)} \qquad N_R > 1000 \qquad (2)$$

Where C_f is the friction factor and N_R the reynolds number and is defined as

$$N_R = \frac{U_c x b x \rho}{\mu} \qquad (3)$$

Where U_c is the central line velocity, m/s
b is half the separation between the plates, m
ρ is the density of the water, kg/m^3
μ is the viscosity of the water, cP

The analogy required that in equation (3) MIN and V_w, be used instead of b and U_c respectevily. Here V_w is equal to the actual core velocity V_o (= $V_{so}/(1\text{-}C_w)$).

V_{so} is the superficial oil velocity. Thus,

$$NR = \frac{V_w x MIN x \rho_w}{\mu_w} \tag{4}$$

MIN corresponds to the minimum distance between the pipe wall and the oil-water interface. The density and the viscosity are as defined in equation (3).

The pressure gradient can then be calculated from:

$$\frac{\Delta P}{L} = \frac{4 x \tau_w}{D} = \frac{4}{D} x C_{fx} \frac{1}{2} x \rho_{wx} V_w \tag{5}$$

In order to be able to evaluate the the pressure gradient and then the pressure drop, the value of MIN has to be known. Trallero et al (1988) using an ultrasonic device, carried out measurements of MIN under differnt conditions and found that it can be correlated as a function of the theoretical value of the water layer thickness defined for a concentric configuration, as follows

$$MIN = \frac{\lambda}{2.5} \tag{6}$$

and

$$\lambda = \frac{D}{2}(1 - \sqrt{1 - C_w}) \tag{7}$$

Where D is the pipe internal diameter , m
C_W is the injected water fraction ($= Q_W/(Q_W + Q_O)$)
Q_W and Q_O are the water and oil volumetric flow rate, m^3/s

Knowing the pipe diameter, the oil and water volumetric or mass flow rates together with the fluid properties, the pressure drop can be calculated from (1), (2), (4), (5),(6) and (7).

5. COMPARISON WITH EXPERIMENTAL DATA

To determine the accuracy of the predictions of the proposed model to calculate pressure losses for oil-water core annular flow, four different independent sets of data were chosen from the limited available experimental works in this area. Specifically Charles et al (1961) in a horizontal 25 mm ID pipe, density ratio of 1.0 and the viscosity ratios of 0.14 and 0.053, Oliemans (1986) in a horizontal 50 and 203 mm ID pipe with an oil viscosity of 2300 cP and a density ratio of 1.03, Guevara et al (1988) in a horizontal 203 mm ID pipe with oil viscosity ranging from 5000 to 120000 cP and a density ratio of realy 1.0, and Arney et al (1993) with a 15.9 mm ID pipe, density ratio of 1.0 and a viscosity ratio of 0.0445. The results of the comparison are summarized in Table 1 and in Figures 1, 2, 3, and 4.

It can be seen that the model's predictions of the experimental data is good. It performed better with the dat of Charles and Oliemans with an absolute relative error no greater than 15%. In the other two cases, the error is of the order of 24%. In the case of the data from Guevara et al, the larger difference may be due to the fact that during those tests oil sticks on the wall of the pipe given larger pressure drops than those predicted by the model

6. CONCLUSIONS

The application of the proposed model to predict the pressure drop for oil water core annular flow using four different and independent sets of data, lead to the following conclusions:

(1) A simple model has been developed that allow to compute the pressure losses when transporting viscous oil through a pipeline operating under core annular flow conditions with nearly equal liquid densities and no-slip between the liquid phases.

(2) For two of the sets of data, the model predicted with an absolute relative error of 15%. For the other two the error is as high as 24% mainly due to the oil sticking on the pipe wall.

(3) The model does not take into account the effect of the oil sticking on the pipe wall.

(4) The model does not check for the existence of core annular flow .

REFERENCES

Arney, M. S. ; Bai, R. ; Guevara, E. ; Joseph, D. And Liu, K. Friction Factor And Holdup Studies Por Lubricated Pipelining. To Be Published In The Int. J. Multiphase Flow.

Bentwich, M. Two-Phase Viscous Axial Flow In Pipes. Trans. Asme (Sect. D) - J. Basic Eng., 86 (4): 669-672, 1964.

Bentwich, M.; Kelly, D.A.I.;Epstein, N. Two-Phase Eccentric Interface Laminar Pipeline Flow. Trans. Asme (Sect. D) - J. Basic Eng., 92 (1): 32-36,1970.

Brauner, N. Two-Phase Liquid-Liquid Annular Flow, Int. J. Multiphase Flow Vol. 17, No 1, Pp. 59-76, 1991.

Cabrera, H. 1982, Production And Transport Of Heavy Oil By Blending With Lighter Crude: 2nd Int. Conf. On Heavy Crude And Tar Sands, Unitar, Caracas, Venezuela, February.

Charles, M. E. ; Govier, G. W. ; Hodgson, G. W. The Horizontal Pipeline Flow To Equal-Density Oil-Water Mixtures. Can. J. Chem. Eng., 39 (1): 27-36, 1961.

Charles, M. E. The Pipeline Flow Of Capsules. Part 2. Theoretical Analysis Of The Flow Of Cylindrical Forms. Can. J. Chem. Eng., 41 : 46-51,1963.

Epstein, N. Concentric Laminar Flows. Can. J. Chem. Eng. , 41 (4) : 181, 1963.

El Telbany, M. M., Reynolds, A.J. The Structure Of Turbulent Plane Couette Flow, Journal Of Fluid Engineering, Vol. 104 : 367, Sept. 1982.

Glass, W. Water Addition Aids Pumping Viscous Oils. Chem. Eng. Prog., 57 (3) : 116, 1961.

Guevara, E. ; Zagustin, K. ; Trallero, J. L. ; Diaz, T. And Zamora, G. 1988, Estudio De Las Caracteristicas Del Flujo Anular: Medidas De La Caida De Presion En El Circuito De 8 Pulgadas (Set) En San Tome, Parte Ii. Intevep, S.A., Los Teques.

Guevara, E. ; Zagustin, K. ; Zubillaga, V. And Trallero, J. L. 1988, Core Annular Flow: The Most Economical Method For The Transportation Of Viscous Hydrocarbons : 4th Int. Conf. On Heavy Crude And Tar Sands, Unitar, Edmonton, Canada, August.

Guevara, E. ; Zubillaga, V. ; Zagustin, K. ; Zamora, G. And Duijvestijn, A. , 1991 . Core Annular Flow : Preliminary Operational Experience And Measurements In A 152 Mm X 55 Km Pipeline. 5th Int. Conf. On Heavy Crude And Tar Sands, Unitar, Caracas, Venezuela, Vol. 3, 237 - 243.

Hasson, D. ; Mann, U. ; Nir, A. Annular Flow Of Two Immiscible Liquids. Can. J. Chem. Eng., 48 (10) : 514 - 525, 1970.

Jaua, J. A. Estudio Del Comportamiento De La Pelicula De Agua En Flujo Anular. Tesis De Grado, Universidad Metropolitana, Departamento De Ingenieria Mecanica, 1989.

Joseph, D. Privet Communication, 1993.

Manson, B. R. ; Bangwalla, A.A. And Mann. J.A. Low Reynolds Circular Couette Flow Past A Wavy Wall. Physics Of Fluids Vol. 28, No 9, Sept. 1985.

Ooms, G. ; Segal, A. ; Van Der Wees, A.J. ; Meerhoff, R. ; Oliemans, R. V. A. A Theoretical Model For Core-Annular Flow Of A Very Viscous Oil Core In A Water Annulus Through A Horizontal Pipe. Int. J. Multiphase Flow, 10 (!) : 41 - 60, 1984.

Oliemans, R. V. A. The Lubricating Film Model For Core-Annular Flow, Ph.D. Thesis, Delft University Press, Technische Hogeschool Delft, The Netherlands, (1986)

Oliemans, R.V.A. ; Ooms, G. ; Wu, H.L. And Duijvestijn, A. Core-Annular Oil/Water Flow : The Turbulent - Lubricating - Film Model And Measurements In A 5 Cm Pipe Loop. Int. J. Multiphase Flow Vol. 13, No 1, Pp. 23 - 31, 1987.

Parra, D. 1979, Transport Of Heavy Crude : 1st Int. Unitar Aostra - U.S. Dept. Energy Conf. On Future Of Heavy Crude Oil And Tar Sands, Alberta, Canada, June

Russell, T. W. F. ; Charles, M. E. The Effect Of The Less Viscous Liquid In The Laminar Flow Of Two Immiscible Liquids. Can. J. Chem. Eng. , 37 (1)) : 18 - 24, 1959.

Trallero, J.L. ; Zagustin, K. And Guevara, E. 1988, Estudio De Las Caracteristicas Del Flujo Anular : Ondas Interfaciales, Intevep, S.A., S.A., Los Teques.

Trallero, J.L. 1992 Estudio De Las Ondas Interfaciales En Flujo Anular, Tesis De Maestria, Universidad Central De Venezuela, Facultad De Ingenieria, Departamento De Hidraulica, Julio.

Zagustin, K. Flow Characteristis Over A Finite Amplitude Waves With Moving Boundary Conditions. Xii Congress Of Coastal Ing. Washington,1970.

Zagustin, K. ; Guevara, E. ; Trallero, J. L. And Nunes, G. 1988. Core Annular Flow Characteristics Studies : Core Annular Flow Modeling. Intevep, S.A., S.A. , Los Teques.

TABLE 1 : EVALUATION OF THE PROPOSED MODEL

Source	No. of data points	Average relative error (%)	Average absolute relative error (%)	Standard deviation (%)	Average error (Pa) $x10^4$	Average absolute error (Pa) $x10^4$	Standard deviation (PA) $x10^4$
Oliemans (1986)	36	5.2	12.4	17.4	0.0194	0.0394	0.0565
Charles et al (1961)	28	12.0	15.2	15.1	0.02	0.027	0.041
Arney et al (1993)	101	21.7	23.2	15.5	0.014	0.014	0.010
Guevara et al (1988)	141	10.2	24.0	27.5	0.23	1.63	2.6

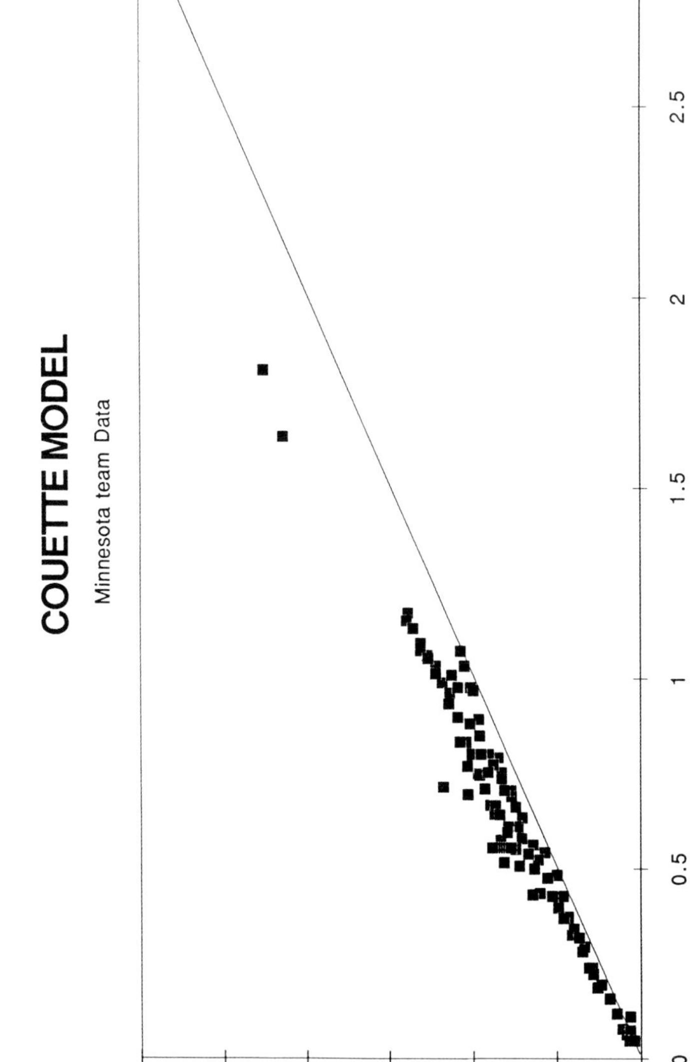
COUETTE MODEL
Minnesota team Data
Calculated pressure drop (KPa/m)
Measured pressure drop (KPa/m)
0
0.5
1
1.5
2
2.5
3

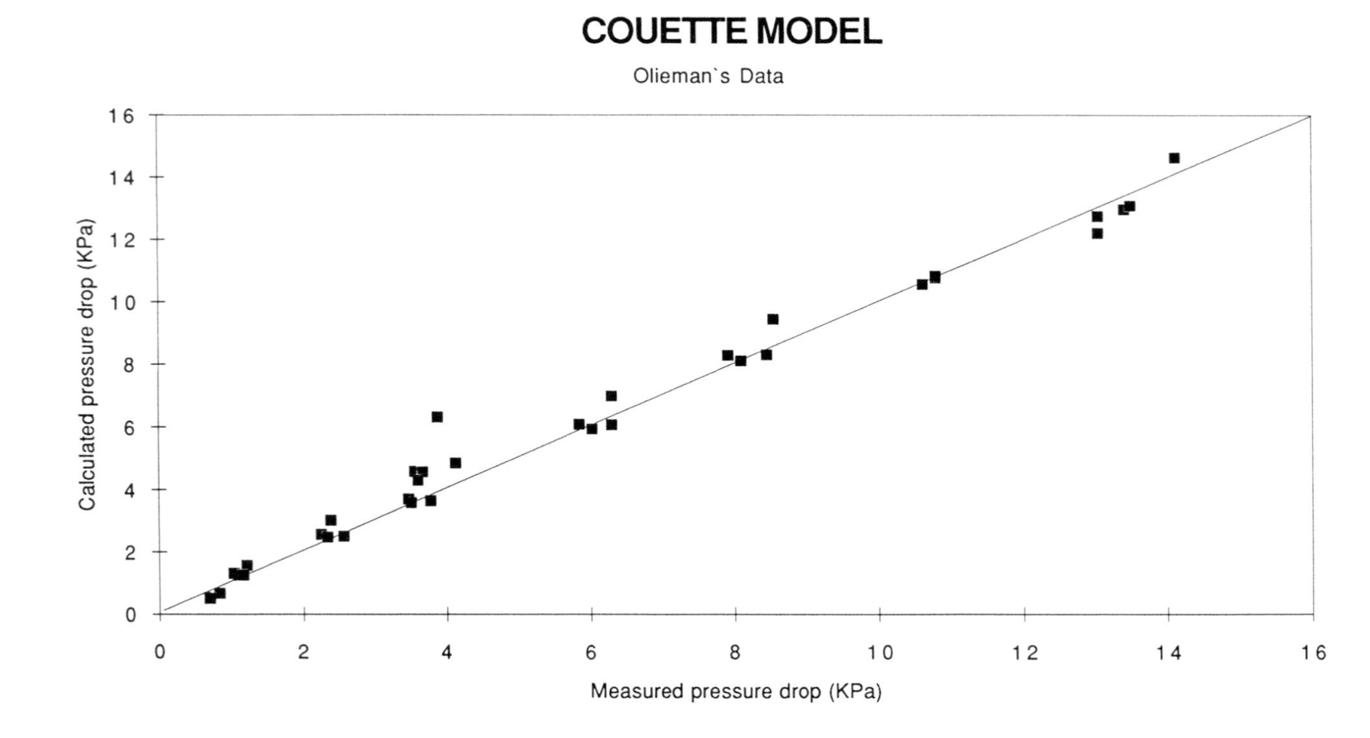

COUETTE MODEL
Olieman`s Data
Calculated pressure drop (KPa)
Measured pressure drop (KPa)
0
2
4
6
8
10
12
14
16

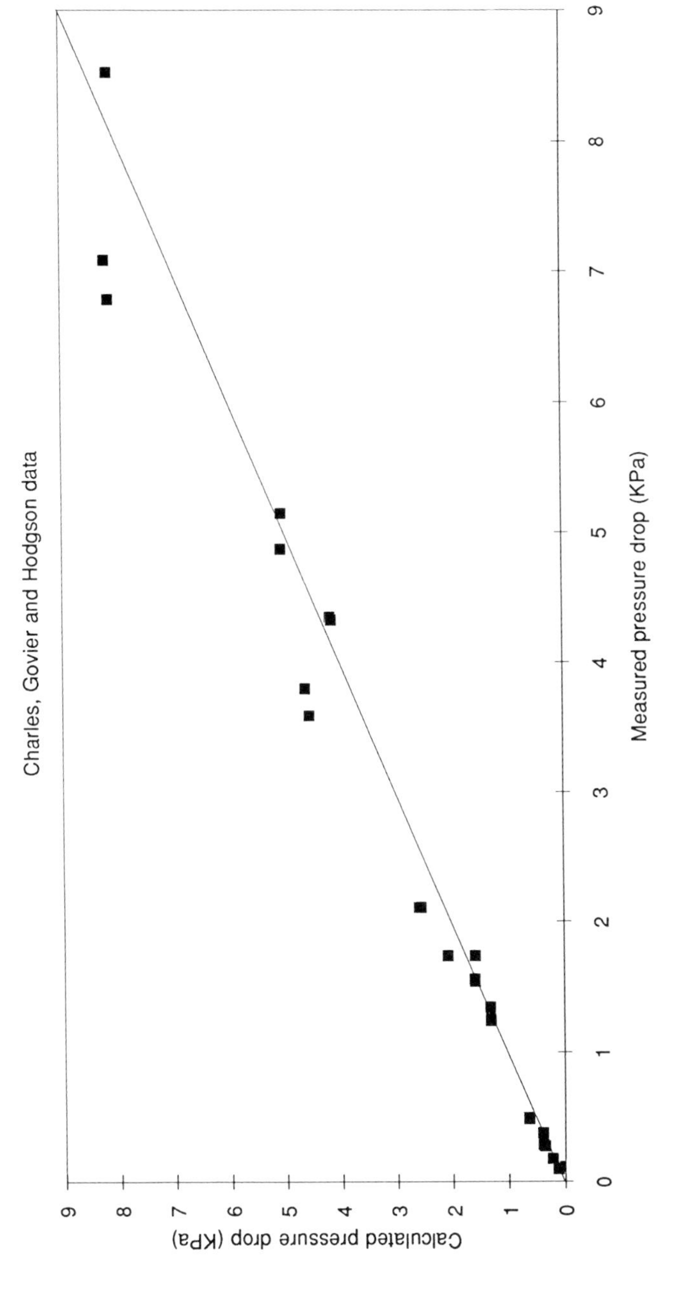
COUETTE MODEL
Charles, Govier and Hodgson data
Calculated pressure drop (KPa)
Measured pressure drop (KPa)
0
1
2
3
4
5
6
7
8
9

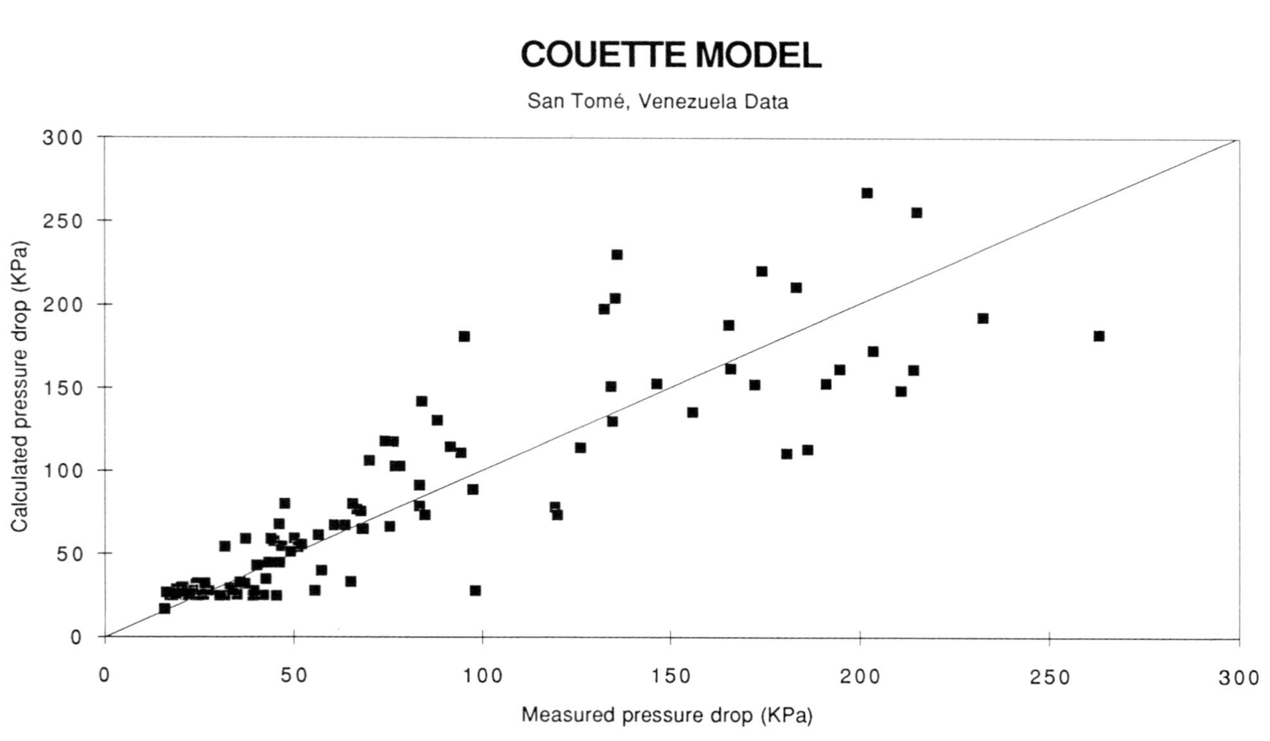
COUETTE MODEL
San Tomé, Venezuela Data
Calculated pressure drop (KPa)
300
250
200
150
100
50
0
0
50
100
150
200
250
300
Measured pressure drop (KPa)

ON FRACTAL DIMENSION AND MODELLING OF SLUG AND BUBBLE FLOW PROCESSES

M. Bernicot, H. Dhulesia (TOTAL, FRANCE)
and
P. Deheuvels (Université Paris VI, FRANCE)

ABSTRACT

A multiphase flow in a pipeline with stationary input often generates a dichotomous stochastic pattern, characterized by an alternance of slugs and gas bubbles of random sizes within the pipeline. In order to design the appropriate equipment and facilities to cope with such a situation, it is essential to develop a suitable *mathematical model*, enabling to describe properly the random fluctuations of the slug and bubble length sequences outflowing at the pipeline outlet. The marginal distributions of individual slug lengths have been proved empirically to be well-fitted by *Inverse Gaussian distributions* by Dhulesia et al. (1991). We first describe a physical model which explains for this phenomenon and discuss some of its implications. We then consider the global stochastic behaviour of slug flow time series, and investigate through statistical analysis performed on several experimental sequences their long-range dependence properties. In particular, we estimate the *self-similarity coefficients* H pertaining to each of these data sets, following the approach of Saether et al. (1991). We finally conclude by the description of a probabilistic model which combines the requirements of having preassigned fractal dimension expressed in terms of the index H, together with specified Inverse Gaussian marginal distributions. Such a model turns out to be a reasonable candidate for analysis and simulation of slug flows in the pipeline.

1. INTRODUCTION

Operation of multiphase production systems is now becoming a more or less routine operation based on *experimentally adjusted procedures.* On the other hand, we are still lacking of universally approved and validated *operating philosophy and operating methods* associated with *design methods and simulation models* to be used for such installations (ex. pigging philosophy, slug-catcher design method, etc.).

Field developments and operations based on the use of two-phase gas-oil and three-phase gas-oil-water flow *mechanistic models* have received considerable attention in the recent literature. Comparisons of these model predictions to available experimental data banks confirm a significant improvment of their accuracy in the last years. We refer to Dukler and Fabre (1992), Fabre and Liné (1992) and the references therein for recent surveys of this modelling problem. In this paper, we are mainly concerned with the derivation of appropriate mathematical models to describe the *stochastic structure of slug and bubble flows.*

Two-phase horizontal flow patterns can be classified into three main types, *separated* (stratified / annular), *intermittent* and *dispersed.* Here, we will consider intermittent flow patterns which can be described by an alternance of slugs and bubbles within the pipeline. Moreover, we will consider non horizontal intermittent flow patterns as well.

2. MODELLING OF THE SLUG FLOW PATTERN

2.1. A GENERAL MODEL FOR SLUG AND BUBBLE LENGTH

A rough description of a slug flow is as in Figure 1 below. Neglecting dispersed bubbles within the slug and stratified flows within the bubbles, the flow pattern is characterized by a stochastic alternance of gas bubbles and liquid slugs. In the sequel, we denote by $X_1, Y_1, \ldots, X_n, Y_n, \ldots$, the respective lengths of the alternating slugs and bubbles which can be observed at a fixed position in the pipe and focus our interest in the modelling of this double time series.

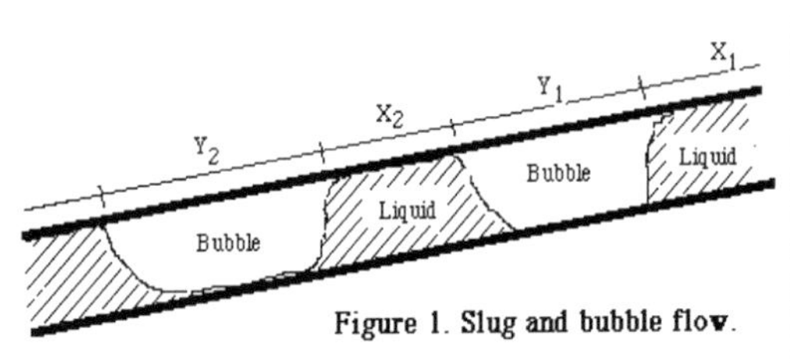

Figure 1. Slug and bubble flow.

The probability distribution of an individual slug length in a stationary flow has been found by Dhulesia et al. (1991) to be closely fitted by inverse Gaussian distributions which we briefly discuss. A random variable X following the common distribution of $X_1, X_2, \ldots$, is said to follow an *Inverse Gaussian* [IG] distribution $IG(\alpha, \nu, \lambda)$ of parameters α, $\nu > 0$ and $\lambda > 0$ if the probability density f of X is of the form

$$(2.1) \quad f(x) = \frac{\lambda^{1/2}}{(x-\alpha)^3\sqrt{2\pi}} \exp\left\{-\frac{\lambda(x-\alpha-\nu)^2}{2\nu^2(x-\alpha)}\right\} \text{ for } x > \alpha,\ f(x) = 0 \text{ for } x \leq \alpha.$$

In other words, the frequency of slug lengths less than or equal to x is equal to

$$(2.2) \quad P(X \leq x) = \int_0^x f(t)dt,$$

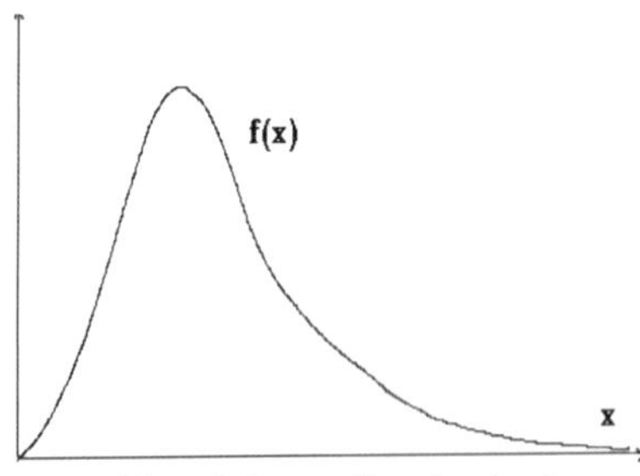

Figure 2. Inverse Gaussian density

with f as in (2.1). The expressions of α, ν and λ in terms of the moments of X are simple. Let $m=E(X)$, $M_k=E((X-m)^k)$, $\sigma^2=var(X)$ and $\beta_1=M_3^2/M_2^3$. We have

$$(2.3) \quad m = \nu+\alpha,\ \sigma^2=\nu^3/\lambda \text{ and } \beta_1 = 9\nu/\lambda.$$

Brill et al. (1981) proposed a *lognormal model* [LN] for the distribution of X. As shown by Dhulesia et al. (1991), a statistical discrimination between the IG and LN models is difficult to achieve on small samples. By extensive statistical analysis performed over a large number of data sets, the latter authors have reached nevertheless the conclusion that IG distributions provide in general a better fit than either of the LN and Gamma distributions.

We now provide a simple physical interpretation of this fact by considering the slug formation in a dip. Denote by Z(t) the volume of liquid flowing into the dip from the inlet and cumulated in the time-interval (0,t] (the *dip-input)*. The following assumptions on Z(t) are natural in a steady state flow.

(I1) The expectation E(Z(t)) and the variance var(Z(t)) of Z(t) are linear functions of t.
(I2) For every n disjoint intervals $(t_1,t_2), \dots ,(t_{n-1},t_n)$, the increments $Z(t_2)-Z(t_1), \dots ,Z(t_n)-Z(t_{n-1})$ are independent.
(I3) Each increment Z(t+h)-Z(h) is normally distributed.
(I4) Z(t) is a continuous function of $t \geq 0$ with Z(0)=0.

The conditions (I1-I2-I3-I4) characterize the fact that $\{Z(t),t \geq 0\}$ is a *Brownian motion with drift* (see e.g.Ch.7 of Karlin and Taylor (1975)). We will set accordingly for $t \geq 0$

(2.4) $Z(t) = \mu t + \sigma W(t)$,

where $\{W(t),t \geq 0\}$ is a standard *Brownian motion* (or *Wiener process*). We next assume that at time t = 0 there is no liquid accumulation in the dip, and denote by T_0 the first time following 0 where a slug is generated. The input Z(t) may be decomposed into

(2.5) $Z(t) = U(t) + V(t)$,

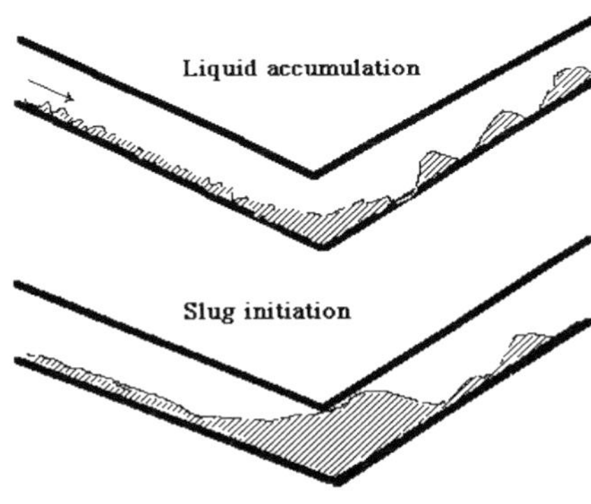

Figure 3.

where U(t) is the *dip cumulated liquid hold-up* at time t, and V(t) is the *cumulated dip output* in the time-interval (0,t]. The heuristic arguments used previously to model $\{Z(t),t \geq 0\}$ lead to the assumption that a suitable model for each of the processes $\{U(t),t \geq 0\}$, and $\{V(t),t \geq 0\}$ may be also assumed to be a Brownian motion with drift within the time-interval $[0,T_0]$. We will limit ourselves for the time being to the construction of stochastic models for these processes.

We assume that (see Figures 3 and 4)

(2.6) $U(t) = Z_1(t) - Z_2(t)$, $V(t) = Z_2(t) + Z_3(t)$, $Z(t) = Z_1(t) + Z_3(t)$,

and

(2.7) $Z_1(t) = \mu_1 t + \sigma_1 W_1(t)$, $Z_2(t) = \mu_2 t + \sigma_2 W_2(t)$, $Z_3(t) = \mu_3 t + \sigma_3 W_3(t)$,

for $0 \leq t \leq T_0$, where $\{W_i(t),t \geq 0\}$ are independent Wiener processes for i = 1,2,3. The physical interpretation of these quantities is as follows. Recall first that, for $0 \leq t \leq T_0$,

Z(t) *is the cumulated liquid volume inflowing into to the dip in the time-interval* (0,t];
U(t) *is the cumulated liquid volume accumulated within the dip in the time-interval* (0,t];
V(t) *is the cumulated liquid volume outflowing from the dip in the time-interval* (0,t].

We assume that part of the liquid in the dip hold-up is lost by being forwarded into the pipeline. This gives $Z_2(t)$. Likewise, part of the liquid inflowing into the dip remains there. This gives $Z_1(t)$, which can be interpreted as the virtual quantity of liquid which would have accumulated within the dip had there been no loss. Finally, part of the liquid which flows into the dip is immediately forwarded (or lost) in the pipeline. This gives $Z_3(t)$. The cumulated outflow from the dip is then equal to $V(t) = Z_2(t) + Z_3(t)$ in the time-interval $(0,t]$. With this notation, the cumulated liquid volumes in the time-interval $(0,t]$ with $0 \leq t \leq T_0$ are denoted as follows.

$Z_3(t)$ *is the cumulated liquid volume flowing non-stop through the dip in* $(0,t]$;
$Z_2(t)$ *is the cumulated liquid volume lost from the dip hold-up in* $(0,t]$;
$Z_1(t)$ *is the virtual dip liquid hold-up without loss in the time-interval* $(0,t]$.

It will be convenient to set in the sequel

(2.8) $\mu_U = \mu_1 - \mu_2$ and $\sigma_U^2 = \sigma_1^2 + \sigma_2^2$.

The constants μ, μ_U, μ_i, σ, σ_i, $i = 1,2,3$ are related as follows. We have

(2.9) $\mu = \mu_1 + \mu_3$, $\mu_U = \mu_1 - \mu_2 \geq 0$, $\mu_i \geq 0$ for $i = 2,3$, $\sigma^2 = \sigma_1^2 + \sigma_3^2$.

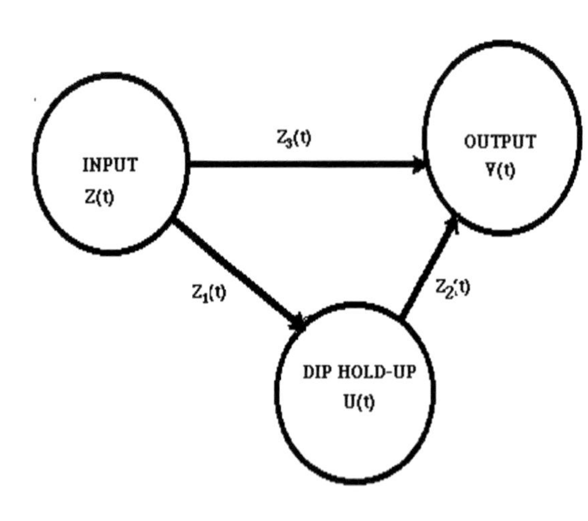

Figure 4. The dip accumulation and transfer model.

We now discuss the physical meaning of our assumptions. Since the flows may be inverted at times, there is no contradiction to assume that either $Z(t)$, $Z_1(t)$, $Z_2(t)$ or $Z_3(t)$ becomes eventually negative. On the other hand, the dip hold-up $U(t) = Z_1(t) - Z_2(t)$ must be non negative to have a meaning. If $\mu_U = \mu_1 - \mu_2 > 0$ in (2.9), the latter condition holds for all $t \geq t_0$, with t_0 denoting the largest value of $t \geq 0$ for which $\mu_U t + \sigma_U W(t) = 0$. If such is the case, we may reasonably neglect the negative fluctuations of the process $\mu_U t + \sigma_U W(t)$ in the time-interval $(0, t_0)$. In the remaining case where $\mu_U = \mu_1 - \mu_2 = 0$, some other interpretation has to be given for $U(t)$. This point will be considered later on.

We now turn to the definition of T_0. We assume that a slug is generated (see Figure 3) when the liquid hold-up U(t) within the dip reaches a critical level C.
The time $t = T_0$ of the first slug initiation after 0 is therefore given by :

(2.10) $T_0=\inf\{t\geq 0:U(t)\geq C\}$.

It follows immediatly from (2.6), (2.7), (2.9) and (2.10) that the distribution of T_0 is of the following two possible types, according to whether $\mu_U > 0$ or $\mu_U = 0$.

Case 1. $\mu_U = \mu_1-\mu_2 > 0$ (see Figure 3). T_0 follows an *Inverse Gaussian distribution* $IG(0,C/\mu_U,C^2/\sigma_U^2)$, with density given by (see e.g. (2.1), (2.2), (2.3), Karlin and Taylor (1975), p.363, and Deheuvels and Steinebach (1990))

$$(2.11)\quad f_T(x) = \frac{C}{\sigma_U\, x^{3/2}\sqrt{2\pi}}\exp\left(-\frac{(x\mu_U - C)^2}{2\,\sigma_U^2\, x}\right)\quad \text{for } x > 0.$$

By applying (2.3) with $\alpha = 0$, $\nu = C/\mu_U$ and $\lambda = C^2/\sigma_U^2$, we have

$$(2.12)\quad E(T_0) = C / \mu_U \text{ and } \mathrm{var}(T_0) = C\sigma_U^2 / \mu_U^3.$$

This case (Figure 3) corresponds to *a pronounced dip in which liquid accumulates more or the less linearly with respect to time and with a positive rate* before the first slug occurs.

Case 2. $\mu_U = \mu_1-\mu_2 = 0$ (see Figure 5). In this case, T_0 has distribution with density given by (see e.g. Karlin and Taylor (1975), p.347, Basu and Wasan (1975)).

$$(2.13)\quad f_T(t) = \frac{C}{\sigma_U x^{3/2}\sqrt{2\pi}}\exp\left(-\frac{C^2}{2\sigma_U^2 x}\right)\quad \text{for } x > 0.$$

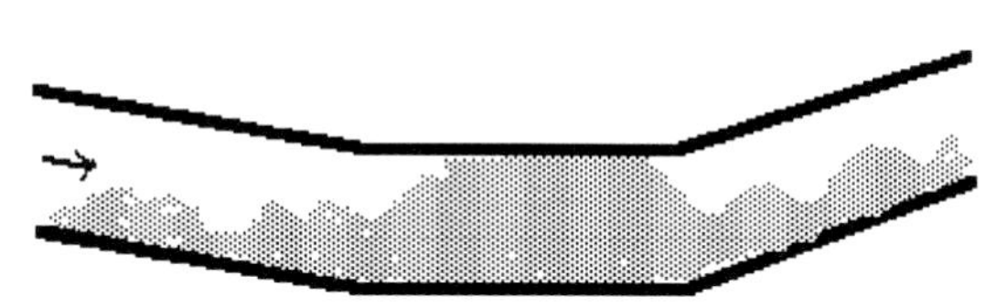

Figure 5. Slug initiation in a moderately inclined pipe.

This case corresponds to moderately inclined pipes in which no significative dip-type accumulation occurs, and where slugs are generated when the random fluctuations of the *stratified* (or *stratified wavy*) flow in the vicinity of an appropriate generating zone reach a critical level. Following our previous remarks, U(t) may become negative in the latter case and cannot any more bear the physical meaning of a dip liquid hold-up. To account for this observation, we need to modify the above model as follows.

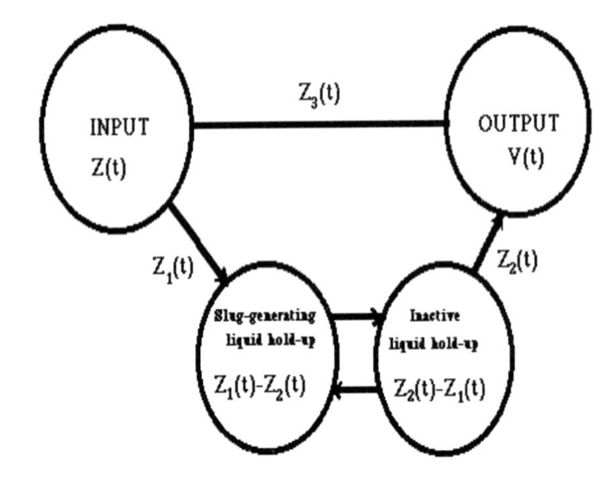

Figure 6. Moderately inclined pipe accumulation and transfer model.

We interpret (Figure 6) $U(t) = Z_1(t)-Z_2(t)$ as the liquid hold-up of a *slug-generating zone* when $U(t)>0$. On the other hand, when $U(t)<0$, we consider that no liquid is present in this *active zone*, and that $-U(t)$ corresponds to the liquid hold-up of an *inactive* part of the pipe (with respect to slug generation). This allows $U(t)$ to be negative. Under these assumptions, the liquid hold-up of the slug-generating zone is given by

$$(2.14) \quad U^*(t) = \max\{U(t),0\} = \max\{Z_1(t)-Z_2(t),0\}.$$

There is no need to modify the definition (2.10) of T_0 in this case since then, by (2.14), we have $T_0 = \inf\{t \geq 0 : U^*(t) \geq C\}$ for $C \geq 0$.

Remark 2.1. In spite of the fact that the distributions characterized by (2.11) and (2.13) may look similar at first sight, they are of very different natures. In Case 1, T_0 has moments of all orders, whereas in Case 2, the expectation of T_0 is infinite and the distribution (2.13) exhibits a *Paretian heavy tail behavior.* We have namely if (2.13) holds

$$(2.15) \quad P(T_0 > x) = \frac{2C(1+o(1))}{\sigma_A\sqrt{2\pi}} x^{-1/2} \quad \text{as } x \to \infty.$$

A random variable T is said to follow a *Pareto-type distribution of index* α if $P(T > x)$ is regularly varying of order $-\alpha$ as $x\to\infty$, that is (see e.g. Bingham et al. (1987)), for any fixed $\lambda > 0$, we have $P(T > \lambda x) / P(T > x) \to \lambda^{-\alpha}$ as $x\to\infty$. These distributions are named after the late Vilfredo Pareto (1848-1923) and occur frequently in econometry (see e.g. Johnson and Kotz (1970), p.233-249). They exhibit typically a *heavy tail behavior* in the following sense. Since $P(T > x)$ decreases to 0 roughly as $x^{-\alpha}$ as $x\to\infty$, it can be shown that the maximum M_n of n independent observations following this law increases roughly as $n^{1/\alpha}$ as $n\to\infty$. Especially for small values of α, one may expect to observe very large values of M_n. As far as slugs are concerned, there is empirical evidence that phenomenon of this type occur when the inlet or the outlet of the pipeline is submitted to a rapid variation of its operating conditions. We will consider elsewhere this problem of simulation of fast transient phenomena.

We will not discuss the differences between these two cases further and turn our attention to the derivation of the *length X_0 of the slug generated at time T_0*. Towards this aim, we make the assumption that the slug, once created, collects all the liquid which has flowed through the dip (or through the *active slug-generating zone)* during the time-interval $[0,T_0]$. By (2.10), the slug has initial volume equal to $C = U(T_0) = Z_1(T_0) - Z_2(T_0)$. Since the *virtual liquid volume* $V(T_0) = Z_2(T_0) + Z_3(T_0)$ which needs to be added to C may eventually be negative, we need truncate the sum as follows. We let

$$(2.16)\quad X_0 = \max\{Z(T_0),0\} = \max\{U(T_0)+V(T_0),0\} = \max\{Z_2(T_0) + Z_3(T_0),0\}.$$

The case where $X_0 = 0$ may be interpreted to correspond to the situation when the slug has disappeared shortly after it was created.

We now make use of the following two properties of Wiener processes. In the first place, if $\{W(t),t \geq 0\}$ is a Wiener process, then it is also the case for $\{\lambda^{-1/2}W(\lambda t),t \geq 0\}$ for an arbitrary constant $\lambda > 0$. This property entails that if T is a positive random variable independent of $\{W(t),t \geq 0\}$, then $W(T) = T^{1/2}Y$, where Y is a standard normal N(0,1) random variable independent of T. In the second place, whenever two standard Wiener processes $\{W'(t),t \geq 0\}$ and $\{W''(t),t \geq 0\}$ are independent, then, for any constants a and b with $a^2+b^2 \neq 0$, the processes

$$(2.17)\quad \{(a^2+b^2)^{-1/2}(aW'(t)+bW''(t)),t \geq 0\} \text{ and } \{(a^2+b^2)^{-1/2}(bW'(t)-aW''(t)),t \geq 0\}$$

are independent standard Wiener process. Let now $W_4(t)$ be defined by

$$(2.18)\quad (Z_2(t) - \mu_2 t) + \frac{\sigma_2^2}{\sigma_1^2+\sigma_2^2}(U(t) - \mu_U t) = \sigma_2 W_2(t) + \frac{\sigma_2^2}{\sigma_1^2+\sigma_2^2}\{\sigma_1 W_1(t) - \sigma_2 W_2(t)\}$$

$$= \frac{\sigma_1\sigma_2}{\sigma_1^2+\sigma_2^2}(\sigma_2 W_1(t) + \sigma_1 W_2(t)) = \left(\frac{\sigma_1^2\sigma_2^2}{\sigma_1^2+\sigma_2^2}\right)^{1/2} W_4(t).$$

Since, by (2.6), (2.7) and (2.8), $U(t) - \mu_U t = \sigma_1 W_1(t) - \sigma_2 W_2(t)$, it follows from (2.17), (2.18) and the independence of W_1 and W_2 that $\{W_4(t),t \geq 0\}$ is a standard Wiener process independent of $\{U(t),t \geq 0\}$. Since, by (2.10), T_0 depends only upon $\{U(t),t \geq 0\}$, that is of $\{\sigma_1 W_1(t)-\sigma_2 W_2(t),t \geq 0\}$, it follows that $\{W_4(t),t \geq 0\}$ and T_0 are independent. Therefore, $W_4(T_0) = T_0^{1/2}Y_1$, where Y_1 is a random variable following a standard normal N(0,1) distribution independent of T_0 and $\{W_3(t),t \geq 0\}$. This, in combination with (2.18) taken with $t = T_0$, (2.9), and the fact that $U(T_0) = C$, yields

$$(2.19)\quad Z_2(T_0) = \mu_2 T_0 - \frac{\sigma_2^2}{\sigma_1^2+\sigma_2^2}(C - \mu_U T_0) + \left(\frac{\sigma_1^2\sigma_2^2}{\sigma_1^2+\sigma_2^2}\right)^{1/2} T_0^{1/2}Y_1$$

$$= -C\frac{\sigma_2^2}{\sigma_1^2+\sigma_2^2} + \left((\mu_1-\mu_2)\frac{\sigma_2^2}{\sigma_1^2+\sigma_2^2} + \mu_2\right)T_0 + \left(\frac{\sigma_1^2\sigma_2^2}{\sigma_1^2+\sigma_2^2}\right)^{1/2} T_0^{1/2}Y_1.$$

By (2.7), $Z_3(T_0) = \mu_3 T_0 + \sigma_3 W_3(T_0)$. Since T_0 is independent of $\{W_3(t),t \geq 0\}$, we may set $W_3(T_0) = T_0^{1/2}Y_2$, where Y_2 is a random variable following a standard normal N(0,1) distribution and independent of T_0 and Y_1. By all this, it follows from (2.16) and (2.19) that there exists a random variable Y_0 following a standard normal N(0,1) distribution, independent of T_0 and such that :

$$(2.20)\quad X_0 = \max\left\{-C\frac{\sigma_2^2}{\sigma_1^2+\sigma_2^2} + \left(\mu_1\frac{\sigma_2^2}{\sigma_1^2+\sigma_2^2} + \mu_2\frac{\sigma_1^2}{\sigma_1^2+\sigma_2^2} + \mu_3\right)T_0 + \left(\left(\frac{\sigma_1^2\sigma_2^2}{\sigma_1^2+\sigma_2^2}\right)^{1/2} Y_1 + \sigma_3 Y_2\right)T_0^{1/2},\ 0\right\}$$

$$= \max\{C_1 + C_2 T_0 + C_3 T_0^{1/2} Y_0,\ 0\},$$

where

$$(2.21)\quad C_1 = -C\frac{\sigma_2^2}{\sigma_1^2+\sigma_2^2},\ C_2 = \mu_1\frac{\sigma_2^2}{\sigma_1^2+\sigma_2^2} + \mu_2\frac{\sigma_1^2}{\sigma_1^2+\sigma_2^2} + \mu_3 \text{ and } C_3 = \left\{\frac{\sigma_1^2\sigma_2^2}{\sigma_1^2+\sigma_2^2} + \sigma_3^2\right\}^{1/2}.$$

The general form of distribution given in (2.20) gives the following particular cases.

Case 1.1. The Inverse Gaussian case. When $\sigma_2 = \sigma_3 = 0$, $\sigma_1 > 0$ and $\mu_1 - \mu_2 > 0$, (2.6), (2.7) and (2.20) imply that

$$(2.22)\quad X_0 = (\mu_2+\mu_3)T_0,\ U(t) = (\mu_1-\mu_2)t + \sigma_1 W_1(t),\ V(t) = (\mu_2+\mu_3)t \text{ for } 0 \le t \le T_0,$$

$$(2.23)\quad E(X_0) = C\frac{\mu_2+\mu_3}{\mu_1-\mu_2},\quad \mathrm{var}(X_0) = C^2\left(\frac{\mu_2+\mu_3}{\mu_1-\mu_2}\right)^3(\sigma_1^2+\sigma_2^2).$$

We so obtain that X_0 follows an *Inverse Gaussian distribution*. This exact situation is also approximatively obtained when the dip output V(t) is either negligible or very regular, with respect to the input Z(t). It is *typical of a dip located far from the pipeline inlet,* since otherwise the input Z(t) would be nearly linear. Another case which yields approximatively an Inverse Gaussian distribution is when T_0 is large in which case $T_0^{1/2}$ is negligible with respect to T_0.

Case 1.2. The truncated-normal case. When $\sigma_1 = \sigma_2 = 0$ and $\mu_1 - \mu_2 > 0$, (2.10) implies that $T_0 = C / (\mu_1-\mu_2)$. Therefore, (2.6), (2.7) and (2.20) imply that

$$(2.24)\quad X_0 = \max\left\{\frac{C\mu_3}{\mu_1-\mu_2} + \sigma_3\left(\frac{C}{\mu_1-\mu_2}\right)^{1/2} Y_2,\ 0\right\},$$

$$U(t) = (\mu_1-\mu_2)t,\ V(t) = (\mu_2+\mu_3)t + \sigma_3 W_3(t) \text{ for } 0 \le t \le T_0.$$

If, in addition, $\mu_3 = 0$, it follows from (2.21) and (2.24) that the distribution of X_0 is *half-normal* and given by

$$(2.25)\quad X_0 = C_4 \max\{Y_4, 0\},$$

where Y_4 follows a standard normal N(0,1) distribution, and

$$(2.26)\quad C_4 = \left(\frac{C}{\mu_1-\mu_2}\sigma_3^2\right)^{1/2}.$$

This case is in agreement with the *half-normal* and *truncated-normal* models of Bernicot et al. (1988). It occurs when the dip input U(t) is very regular and is *typical of a dip located close to the pipeline inlet.*

The preceding arguments hint that one should observe all possible intermediate situations between these two extreme cases.

In practice, and as will be shown later on, models close to the *inverse Gaussian distribution* tend to be more frequent.

In particular, our experiments hint that *truncated normal distributions* occur more or less as a transient model in the vicinity of the slug generating zone. After some time, the slug distribution seems to evolve systematically towards an Inverse Gaussian form.

Pareto-type distributions such as that given in (2.15) are a step towards the understanding of heavy slugging phenomenon which specifically occur if the inlet or outlet conditions undergo a rapid variation. This problem will be considered elsewhere.

We conclude the description of our model by considering the behaviour of the system after the time T_0 where the first slug is generated.

Since T_0 is a stopping time with respect to each of the Wiener processes $\{W_i(t), t \geq 0\}$, the processes $\{W_i(T_0+t)-W_i(t), t \geq 0\}$ are again independent Wiener processes. It is therefore logical to assume that the slug generating process we have described above starts afresh each time that a new slug occurs.

This leads to assume that the sequence $X_0, X_1, \ldots$ of slugs successively generated is composed of independent and identically distributed [i.i.d.] random variables with distribution given by (2.20).

This *independence assumption* may hold as a local approximation, but could eventually be untrue in the case of *long range dependence.*

Whether long range dependence phenomenon occur or not is the object of the discussion in Section 3. The same holds for the successive slug inter-arrival times $T_0, T_1, \ldots$ with distribution given by (2.11) or (2.13). Obviously, the sequences $\{X_n, n \geq 0\}$ and $\{T_n, n \geq 0\}$ are dependent.

For the sake of conciseness, we will not discuss the related problem of bubble lengths. The results are essentially similar as above, but depend further on the additional factor of the dynamics of surrounding slugs.

Apparently, a bubble length distribution law may be a more regular inverse Gaussian law than its corresponding slug length distribution law which may be eventually perturbated by *slug coalescence phenomenon.*

2.2 ANALYSIS OF SOME EXPERIMENTAL DATA SETS

A large set of experimental data was obtained through the Multi-phase Pipelines and Equipment research project (MPE) of the British Hydromechanics Research Group (BHRG). These experiments were designed to be used for the evaluation of the accuracy of existing correlations as well as for the possible development of new correlations with respect to pressure drop, liquid hold-up and slug characteristics (length, velocity, frequency, void fractions, etc.).

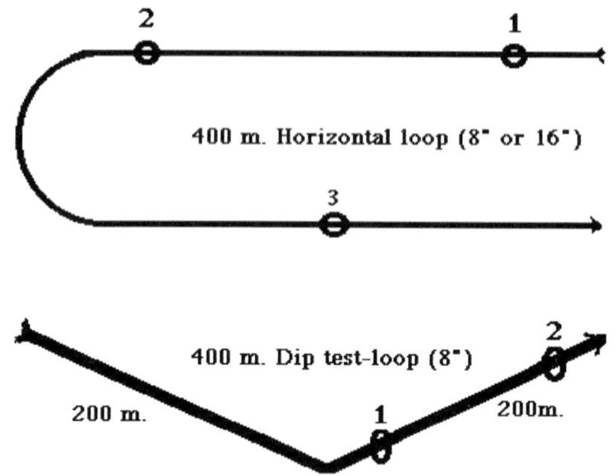

Figure 7. Test loop configurations.

Since these data sets were not initially intended to be used for an accurate analysis of slug sequences, no special care was taken for quality control with respect to the corresponding measurements.

Selected BHRG data obtained on air-water mixtures have been gathered for the following three different configurations, corresponding to the diameter of the pipe (16" or 8") and its profile (horizontal or dip) (Figure 7).

The following histograms illustrate the different types of distributions which can be observed in either case.

A. Horizontal configuration.

We present histograms of slug lenghts and corresponding slug velocities in Figure 8 for a typical case of 16 ¨ loop in horizontal configuration. Similar histograms for a typical case of 8¨ test loop in horizontal configuration is shown in Figure 9.

As shown in Figure 8, the variations of slug length distributions along the pipeline length is not significant but the corresponding slug velocities distributions show a considerable variation. In contrast, Figure 9 demonstrates an opposite behaviour i.e. a large variation in slug length distributions and almost no variation in slug velocity distributions.

These preliminary observations seem to confirm the recent studies by Grenier (1992) that slug lengths and slug velocities are apparently not highly correlated. However a systematic and detailed analysis of a large number of data will be necessary to confirm the above conclusion. It is of interest here to observe that slug velocities appear to follow distributions belonging to the same families of laws predicted by our mathematical model. We do not offer presently any logical explanation for this phenomenon which will be investigated elsewhere.

Interestingly, besides the cases which obviously belong to the skew-type distributions among which *Inverse Gaussian* laws can be considered as the most appropriate family, several data sets are heavily truncated near the origin and can be analyzed by *truncated normal* laws just as well.

B. Dip configuration

Such a typical case is shown in Figure 10. This figure corresponds to the 8¨ test loop in a dip configuration.A truncated Gaussian distribution may be observed at a point located not far from the dip point (slug formation point). This distribution becomes an Inverse Gaussian distribution as it approaches the pipeline outlet.

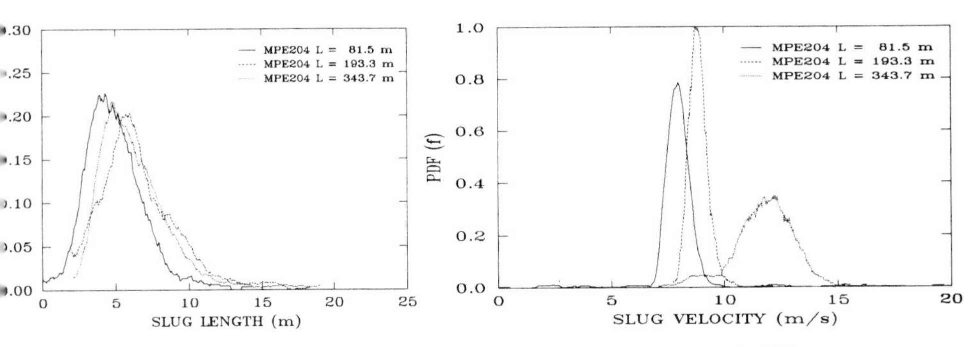

FIGURE 8– 16" LOOP IN HORIZONTAL CONFIGURATION

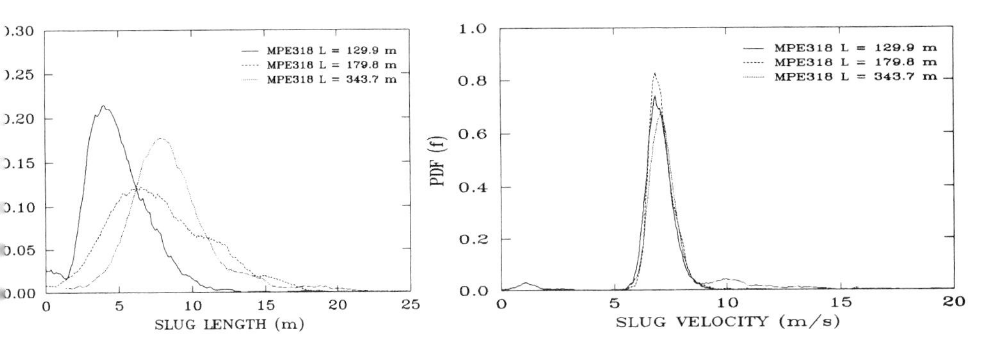

FIGURE 9 – 8" LOOP IN HORIZONTAL CONFIGURATION

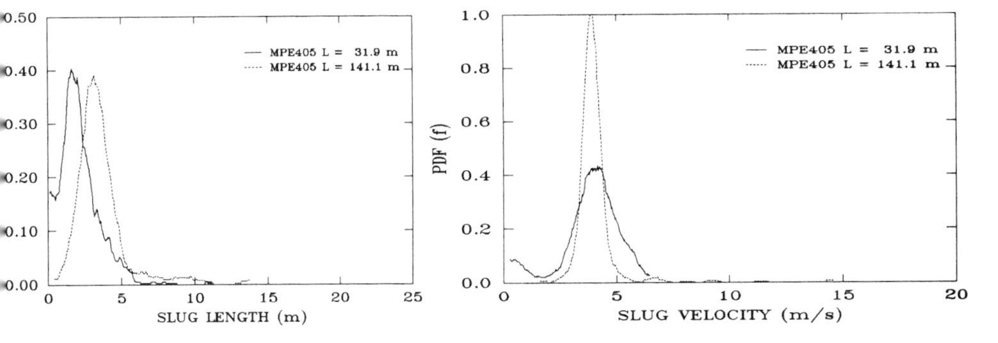

FIGURE 10 – 8" LOOP IN DIP CONFIGURATION

3 THE FRACTAL DIMENSION OF SLUG FLOWS

3.1. INTRODUCTION

Saether et al. (1990) were first to suggest that slug flow patterns could exhibit long-range dependence properties. Their analysis was based on experimental results obtained on Institute for Energy Technology (IFE) air-oil-water small-scale atmospheric test loop and on Norwegian Technical Research Center (SINTEF) Two Phase Flow Lab. Their estimates, based on the R/S statistical analysis initially proposed by H.E.Hurst and discussed below reached the conclusion that the corresponding series had a *self-similarity* (or *Hurst) coefficient* H generally above 0.5, and ranging between 0.53 and 0.76. Moreover, these estimates increase linearly with the total superficial velocity of the two-phase or three-phase fluid.

Our first attempt of a fractal analysis based on experimental results originating from the BHRG test loop was fully negative in the sense that the values of the self-similarity coefficients H estimated from this data set by classical methods was found not significatively different of 0.5 under stable flow conditions. Under the relatively low flow-rates experimental conditions, this led to the conclusion that no long range dependence properties could be detected on such data sets by the preceding methods (Manneville (1991)).
A new and very efficient tool has become available for the estimation of H. The program FRAGTEST performs an analysis partly based on the maximum-likelihood type estimator due to Fox and Taqqu (1986). For reasons which will be given later on, such estimators are considerably more accurate than that given by the R/S analysis. In Section 3.3 below, we present the results of our experimental study. Section 3.2 collects some useful facts concerning the physical meaning of self-similarity and fractional dimension for slug flow processes.

3.2. FRACTAL DIMENSION AND LONG RANGE DEPENDENCE

Consider the *cumulated slug-length sequence* based on $X_1, X_2, \ldots$, and defined by

$$(3.1) \quad S(t) = \sum_{i=1}^{\lfloor t \rfloor} X_i, \text{ with } \lfloor t \rfloor \text{ denoting integer part of t, and } \sum_{i=1}^{0} X_i = 0.$$

The *fractal analysis* of $\{S(t), t \geq 0\}$ consists to assess that this process is approximatively self-similar with parameter H. In general, a *self-similar process* $\{Z(t), t \geq 0\}$ with self-similarity parameter $H \in [0,1]$ is (see e.g. Taqqu (1977)) as a separable process with stationary increments, satisfying $Z(0) = 0$, and invariant (in distribution) under changes of scale in the following sense. For any $c > 0$, we have

$$(3.2) \quad \{c^{-H} Z(ct), t \geq 0\} =_d \{Z(t), t \geq 0\},$$

where "$=_d$" denotes equality in distribution. Up to a scale factor, the only Gaussian self-similar process is the *fractional Brownian motion* introduced by Mandelbrot and Van Ness (1968), denoted by $\{W_H(t), t \geq 0\}$. This process has continuous trajectories and satisfies, for $H \in (0,1)$,

$$(3.3) \quad W_H(0) = 0, \; E(W_H(t)) = 0, \; E(W_H(s)W_H(t)) = \tfrac{1}{2}\{s^{2H} + t^{2H} - |s-t|^{2H}\} \text{ for } s,t \geq 0.$$

When H = 1/2, (3.3) yields the *standard Wiener process* (or *Brownian motion*, denoted in the sequel by $W(t) = W_{1/2}(t)$. This process is characterized, among the family of fractional Brownian motions by the fact that it has independent increments, which implies in particular that the best mean-square prediction of W(t+h) given W(t) is W(t) itself for any h>0. Such is not the case for fractional Brownian motions with $H \neq 1/2$ which exhibit properties of *long-range dependence*. The *Hausdorff dimension* of the trajectory of a fractional Brownian motion with index H is equal to 2-H (see e.g. Theorem 16.7 of Falconer (1990)), which explains for the name of *fractal* or *fractional Brownian motion*. We refer to Taqqu (1985) for general references on *self-similar processes* and stationary processes with long-range dependence properties.

Let μ denotes the mean of X_n, σ^2 the variance of X_n and r_k, the correlation coefficient between X_n and X_{n+k}. The sequence is weakly stationary when these quantities are independent of n.

A *weakly stationary sequence* $\{X_n, n\geq 1\}$ has its moments related to its *spectral density* $\{h(x), -\pi\leq x\leq\pi\}$ via the relations

(3.4) $$E(X_n) = \mu,\ var(X_n) = \sigma^2,\ cov(X_n, X_{n+k}) = \sigma^2 r_k = \sigma^2 \int_{-\pi}^{\pi} e^{ikx} h(x)dx.$$

Such a sequence is said to exhibit a *long-range dependence property* whenever for, some constant $C_1 > 0$ and self-similarity parameter $0 < H \neq 0.5 < 1$,

(3.5) $$r_k = (1+o(1))C_1 H(2H-1)k^{2H-2} \text{ as } k\to\infty,$$

(where "o(1)" denotes a quantity tending to 0) or equivalently if the spectral density h satisfies, for some constant $C_2 > 0$,

(3.6) $$h(x) = (1+o(1))C_2 |x|^{1-2H} \text{ as } x\to 0.$$

It follows in particular from (3.6) that, contrarily as in the *standard case* where H = 0.5, the spectral density h(x) of a process with long-range dependence tends either to zero or to infinity as $x\to 0$. Of primary importance here, when $0.5 < H < 1$, $h(x)\to\infty$ as $x\to\infty$, which corresponds to the so-called *persistent phenomenon*, where high (resp. low) values of the sequence repeat more than often on a long-range basis. We refer to Mandelbrot and Van Ness (1968) and the references therein for a description of this problem. Under general assumptions which we will not discuss (see e.g. Taqqu (1975)), the cumulated long-range dependent process S(t) defined, via (3.1), from a sequence $\{X_n, n\geq 1\}$ satisfying (3.4), (3.5) and (3.6), can be suitably approximated, up to scale and shift factors, by a *fractional Brownian motion of index H*.

Given a finite sequence $\{X_1,\ldots,X_n\}$ of observations, it is therefore of crucial interest for statistical modelling of the process $\{X_n, n\geq 1\}$ to estimate from the data the corresponding self-similarity index H *assuming that it exists*.

Of particular importance is to check whether the possible value of H=1/2 is acceptable or not.

In the latter case, and as far as long-range limiting properties of the process are concerned,

S(t) behaves very much like a partial sum of independent and identically distributed random variables. On the other hand, one needs more sophisticated models to cope with a value of $H \neq 1/2$,. The first estimator of H to be historically used is based on the Hurst R/S analysis whose limiting properties are discussed by Mandelbrot (1975) and Mandelbrot and Taqqu (1979). This estimator was used by Saether et al. (1990). Unfortunately, as shown in the following Remark 3.1, its convergence is very slow and requires enormous data sizes to allow to decide on sound statistical bases whether $H = 0.5$ or not.

Remark 3.1. Assume that the X_i are independent and identically distributed random variables such that $E|X_i|^{2+\delta} < \infty$ for some $\delta > 0$. Let $\mu = E(X_i)$ and $\sigma^2 = \mathrm{var}(X_i)$. By the strong invariance principles of Komlós, Major and Tusnady (1975, 1976) and Major (1976) that there exists a standard Wiener process $\{W(t), t \geq 0\}$ such that

$$(3.7)\quad \lim_{T\to\infty}\left\{T^{-1/(2+\delta)} \sup_{0\leq t\leq T} |S(t) - t\mu - \sigma W(t)|\right\} = 0 \text{ a.s.}$$

The *R/S statistic* is defined by

$$(3.8)\quad R = r(T) = \sup_{0\leq t\leq T}\{S(t) - \tfrac{t}{T}S(T)\} - \inf_{0\leq t\leq T}\{S(t) - \tfrac{t}{T}S(T)\},$$

and

$$(3.9)\quad S = s(T) = \left\{\frac{1}{\lfloor T\rfloor}\sum_{i=1}^{\lfloor t\rfloor}(X_i - \overline{X})^2\right\}^{1/2}, \text{ with } \overline{X} = \frac{1}{\lfloor T\rfloor}\sum_{i=1}^{\lfloor t\rfloor} X_i.$$

It follows from (3.7), (3.8) and (3.9) that

$$(3.10)\quad |r(T) - \sigma(\sup_{0\leq s\leq 1}\{W(Ts) - sW(T)\} - \inf_{0\leq s\leq 1}\{W(Ts) - sW(T)\})| = o(T^{1/(2+\delta)}) \text{ a.s.,}$$

whereas

$$(3.11)\quad s(T) \to \sigma \text{ as } T\to\infty.$$

The Wiener process being self-similar with index $H = 1/2$, $B_T(s) = T^{-1/2}\{W(Ts)-sW(T)\}$, follows the distribution of the *Brownian bridge* $W(s)-sW(1)$. We have therefore

$$(3.12)\quad \left|\frac{r(T)}{s(T)} - T^{1/2}(\sup_{0\leq s\leq 1} B_T(s) - \inf_{0\leq s\leq 1} B_T(s))\right| = o(T^{1/(2+\delta)}).$$

Let

$$(3.13)\quad R_T = \sup_{0\leq s\leq 1} B_T(s) - \inf_{0\leq s\leq 1} B_T(s).$$

By (3.12), setting $\rho = \frac{1}{2} - \frac{1}{2+\delta}$, we have

$$(3.14)\quad \hat{H}_T = \frac{1}{\log T}\log\left(\frac{r(T)}{s(T)}\right) = \frac{1}{2} + \frac{\log R_T}{\log T} + o(T^{-\rho}) \text{ as } T\to\infty.$$

In this case, we see that the Hurst (or R/S) estimator $\hat{H}_T$ of the self-similarity index H of the cumulated process $\{S(t), t\geq 0\}$ introduced in (3.14) is therefore converging to the exact

S(t) behaves very much like a partial sum of independent and identically distributed randomvariables. On the other hand, one needs more sophisticated models to cope with a value of $H \neq 1/2$,. The first estimator of H to be historically used is based on the Hurst R/S analysis whose limiting properties are discussed by Mandelbrot (1975) and Mandelbrot and Taqqu (1979). This estimator was used by Saether et al. (1990). Unfortunately, as shown in the following Remark 3.1, its convergence is very slow and requires enormous data sizes to allow to decide on sound statistical bases whether H = 0.5 or not.

Remark 3.1. Assume that the X_i are independent and identically distributed random variables such that $E|X_i|^{2+\delta}<\infty$ for some $\delta > 0$. Let $\mu = E(X_i)$ and $\sigma^2 = \mathrm{var}(X_i)$. By the strong invariance principles of Komlós, Major and Tusnady (1975, 1976) and Major (1976) that there exists a standard Wiener process $\{W(t),t\geq 0\}$ such that

$$(3.7)\quad \lim_{T\to\infty}\left\{T^{-1/(2+\delta)}\sup_{0\leq t\leq T}|S(t) - t\mu - \sigma W(t)|\right\} = 0 \text{ a.s.}$$

The *R/S statistic* is defined by

$$(3.8)\quad R = r(T) = \sup_{0\leq t\leq T}\{S(t) - \tfrac{t}{T}S(T)\} - \inf_{0\leq t\leq T}\{S(t) - \tfrac{t}{T}S(T)\},$$

and

$$(3.9)\quad S = s(T) = \left\{\frac{1}{\lfloor T\rfloor}\sum_{i=1}^{\lfloor t\rfloor}(X_i - \overline{X})^2\right\}^{1/2}, \text{ with } \overline{X} = \frac{1}{\lfloor T\rfloor}\sum_{i=1}^{\lfloor t\rfloor}X_i.$$

It follows from (3.7), (3.8) and (3.9) that

$$(3.10)\quad |r(T) - \sigma(\sup_{0\leq s\leq 1}\{W(Ts) - sW(T)\} - \inf_{0\leq s\leq 1}\{W(Ts) - sW(T)\})| = o(T^{1/(2+\delta)}) \text{ a.s.,}$$

whereas

$$(3.11)\quad s(T) \to \sigma \text{ as } T\to\infty.$$

The Wiener process being self-similar with index H = 1/2, $B_T(s) = T^{-1/2}\{W(Ts)-sW(T)\}$, follows the distribution of the *Brownian bridge* W(s)-sW(1). We have therefore

$$(3.12)\quad \left|\frac{r(T)}{s(T)} - T^{1/2}(\sup_{0\leq s\leq 1}B_T(s) - \inf_{0\leq s\leq 1}B_T(s))\right| = o(T^{1/(2+\delta)}).$$

Let

$$(3.13)\quad R_T = \sup_{0\leq s\leq 1}B_T(s) - \inf_{0\leq s\leq 1}B_T(s).$$

By (3.12), setting $\rho = \frac{1}{2} - \frac{1}{2+\delta}$, we have

$$(3.14)\quad \hat{H}_T = \frac{1}{\log T}\log\left(\frac{r(T)}{s(T)}\right) = \frac{1}{2} + \frac{\log R_T}{\log T} + o(T^{-\rho}) \text{ as } T\to\infty.$$

In this case, we see that the Hurst (or R/S) estimator $\hat{H}_T$ of the self-similarity index H of the cumulated process $\{S(t),t\geq 0\}$ introduced in (3.14) is therefore converging to the exact value H = 1/2 with a rate $O_P(1/\log T)$, which is very unaccurate unless one disposes of very large samples.

In spite of the fact that R/S estimators of H provide a good starting point in estimating the fractal dimension of a time series $\{X_n\}$, they are in general unreliable unless huge data sets are made available. We have therefore considered a more sophisticated analysis through the program FRAGTEST based in part on the estimators introduced by Fox and Taqqu (1986), which converge with an asymptotically optimal rate of $O_P(T^{-1/2})$ naturally much better than the $O_P(1/\log T)$ rate of the R/S analysis. These estimators are appropriate only for $1/2 \leq H < 1$. This, however, is not a restriction with respect to the results of Saether et al. (1990) which have obtained estimates of H varying between 0.59 and 0.76.

3.3. ANALYSIS OF SOME EXPERIMENTAL DATA SETS

The present study is based on experimental data measured on the BHRG test loops. Three configurations have been used: 8" dip, 8" horizontal and 16" horizontal. In each of the following tables of results, we give the sample size, the mixture velocity and the estimation of H. For each data set, the first number records the experiment, and the second the location of the slug-length measurement. For instance the sets 318-1, 318-2 and 318-3 correspond to the experiment #318, achieved on an 8" diameter pipe and on a horizontal 400 m long loop. The numbers 1, 2 and 3 correspond to the different locations as given in Figure 7.

Configuration: 8" dip

Set	Sample size	Velocity m/s	H
404-1	823	1.70	0.57
404-2	823	1.70	0.63
405-1	427	2.95	0.51
405-2	427	2.95	0.51
406-1	808	2.59	0.51
407-2	808	2.59	0.61
409-1	178	2.13	0.76
409-2	178	2.13	0.61
410-1	998	6.50	0.60
410-2	998	6.50	0.51
412-1	755	6.26	0.51
412-2	755	6.26	0.51

Configuration: 8" horizontal

Set	Sample size	Velocity m/s	H
318-1	887	5.35	0.57
318-2	887	5.35	0.55
318-3	887	5.35	0.51
321-1	621	6.68	0.58
321-2	621	6.68	0.51
321-3	621	6.68	0.51
322-1	712	5.85	0.54
322-2	712	5.85	0.53
322-3	712	5.85	0.63
328-1	600	7.28	0.57
328-2	600	7.28	0.64
328-3	600	7.28	0.51

Configuration: 16" horizontal

Set	Sample size	Velocity	H
201-1	465	4.15	0.51
201-1	465	4.15	0.53
201-3	465	4.15	0.55
204-1	475	9.70	0.51
204-2	475	9.70	0.54
204-3	475	9.70	0.59
206-1	407	17.0	0.65
206-2	407	17.0	0.51
206-3	407	17.0	0.58
207-1	374	4.30	0.51
207-2	374	4.30	0.51
207-3	374	4.30	0.53
211-1	224	3.59	0.51
211-2	224	3.59	0.51
211-3	224	3.59	0.51
212.1	1000	4.16	0.56
212-2	1000	4.16	0.51
212-3	1000	4.16	0.51

The main conclusion which can be infered from these measurements is that they exhibit some instability. High values of H are present for high mixture velocities in horizontal pipes (such as, for instance, for sets #328-2 (H=0.64), #206-1 (H=0.65)). on the other hand, for the dip configuration, high mixture velocities may be associated with values of H close to 0.5 (such as, for instance, for the sets #410-1, #412-1 and #412-2). The data is not sufficient to enable to conclude about this property.

4 CONCLUSION

A purely *statistical approach* of the slug length evolution within a pipeline has already demonstrated the interest of the Inverse Gaussian laws. The presently proposed *mathematical model of the slug generation phenomenon*, which is based on a set of simplified hypotheses, confirms the applicability of Inverse Gaussian laws and gives some insight on the reasons for which they generally provides a good fit to experimental data.

The other types of distributions predicted by the model may be observed at times in exceptional locations, such as near the slug-generating zone. The truncated-normal, or half-normal distributions seems to occur as a transient phenomenon, which evolves towards Inverse Gaussian distributions far from the pipe inlet. This observation needs however to be validated by additional data sets. Paretian distributions, which are predicted by the model in special cases, could account for the slugs of exceptional lengths which are experimentally observed under transient conditions.This type of behaviour should be described through refinements of our model taking into consideration the influence of transient phenomena on the slug flow process. This will be considered elsewhere.

Our analysis of the long-range dependence properties of experimental slug sequences obtained from the BHRG test-loops cannot be considered as fully conclusive. In spite of the fact that it is based on a theorically optimal method by an efficient program FRAGTEST, the quality of the data which has been used may influence in part the significance of our analysis. In particular, the stability of inlet, the outlet control conditions, the size of the data sets should need to be monitored more closely for future studies of this type.

We have obtained conclusions which come close to that of Saether et al. (1990) only for two data sets corresponding to the *highest mixture velocities*. In the latter case, it would seem that :

- The value of the self-similarity coefficient H is greater than 0.5 and increases with the gas velocity

- The slug flow process appears to exhibit a *fractal type behaviour* with long-range dependency.

For the data sets with lower gas velocities and/or less stable inlet / outlet conditions, no other conclusion could be made than that:

- The value of the self similarity coefficient H is usually *not significatively different from 0.5*

- The slug sequence exhibits a *purely random behavior*, similar as that of a sequence of independent and identically distributed random variables.

It is tempting to conclude from these results that a fractal behavior of the slug flow process should appear only for very high mixture velocities. Such a conclusion would be in agreement with the results with the results of Saether et al. (1990). The fact that the discrepancy of H to 0.5 cannot be detected in the other cases does not imply that it does not exist, but would lead to the conclusion that *a purely random model can be used* in these cases without a significant error.

To validate or to disprove this conclusion as well as any other practical conclusions which may be inferred from our model, it is necessary to improve our basic understanding of the slug generation, slug evolution and slug coalescence phenomena. This implies, among other requirements :

- The availability of very large data sets measured under high quality experimental conditions, and obtained under closely controlled inlet and outlet conditions

- Refinements of the presently available mathematical models taking care in particular of :

    - the superposition of several slug generation processes within the same pipeline, together with the associated slug evolution laws

    - the variations of inlet / outlet operating conditions of the pipeline

    - the still more complex slugging phenomenon in gas / oil / water three-phase flow

In conclusion it is important to precise that :

- this study is being performed within the scope of the EvE project and is parallel to the develoment of the transient pipeline simulation code TACITE

- one of the ultimate goals of our work is the integration of these stochastic models in the deterministic code TACITE in order to improve its prediction capability in matter of slug flow.

ACKNOWLEDGMENT

The authors wish to thank TOTAL EXPLORATION PRODUCTION for the permission to publish this paper. They also wish to acknowledge the British Hydrodynamics Research Group for providing slug characteristic data measured under Multi-phase Pipelines and Equipment Project.

5. REFERENCES

Basu, A.K. and Wasan, M.T. (1975). Scand. Actuarial J. 99-108.
Bendiksen, K. (1984). An experimental investigation of the motion of long bubbles in inclined tubes. Int. J. Multiphase Flow **10** 467-483.
Bendiksen, K.H. (1985). On the motion of long bubbles in vertical tubes. Int. J. Multiphase Flow **11** 792-812
Bernicot, M. and Drouffe, J.M. (1988). A slug length distribution law for polyphasic transportation systems. 20th. Ann. OTC, Paper 5649, Houston.
Bingham, N.H., Goldie, C.M. and Teugels, J.L. (1987). *Regular Variation.* Cambridge University Press, Cambridge.
Brill, J.P., Schmidt, Z., Coberly, W.A. Herring, J.D. and Moore, D.W. (1981). Analysis of two-phase tests in large diameter flow lines in Prudhoe Bay field. Soc. Pet. Eng. J. **271** 363-378.
Deheuvels, P. and Steinebach, J. (1990). On the sample path behavior of the first passage time of a Brownian motion with drift. Ann. Inst. Henri Poincaré, Probab. Statist. **26** 145-179.
Dhulesia, H., Bernicot, M. and Deheuvels, P. (1991). Statistical analysis and modelling of slug lengths. In: *Multi-Phase Production* (A.P.Burns, Edit.). Elsevier, London, 80-112.
Dukler, A.E. and Hubbard, M.G. (1975). A model for gas-liquid flow in horizontal and near horizontal tubes. Ind. Eng. Chem. Fundam. **14** 337-345.
Fabre, J. and Liné, A. (1992). Modelling of two-phase slug flow. Annu. Rev. Fluid. Mech. **24** 24-46.
Falconer, K. (1990). *Fractal Geometry, Mathematical Foundations and Applications.* Wiley, New York.
Fox, R. and Taqqu, M.S. (1986). Large-sample properties of parameter estimates for strongly dependent stationary Gaussian time series. Ann. Statist. **14** 517-532.
Grenier, P. (1992). Caractérisation des grandeurs statistiques de l'écoulement diphasique intermittent en conduite horizontale. Rapport de DEA, INP Toulouse.
Johnson, N.L. and Kotz, S. (1970). *Continuous Univariate Distributions 1.* Houghton Mifflin, Boston.
Karlin, S. and Taylor, H.M. (1975). A First Course in Stochastic Processes. (2nd. Ed.). Academic Press, New York.
Komlós, J., Major, P. and Tusnády, G. (1975). An approximation of partial sums of independent r.v.'s and the sample d.f. I. Z. Wahrsch. verw. Gebiete **32** 111-131.
Komlós, J., Major, P. and Tusnády, G. (1976). An approximation of partial sums of independent r.v.'s and the sample d.f. II. Z. Wahrsch. verw. Gebiete **34** 33-58.
Major, P. (1976). The approximation of partial sums of independent r.v.'s. Z. Wahrsch. verw. Gebiete **35** 213-220.
Mandelbrot, B.B. (1975). Limit theorems for the self-normalized range. Z. Wahrsch. verw. Gebiete **31** 271-285.
Mandelbrot, B.B. and Van Ness, J.W. (1968). Fractional Brownian motion, fractional noises and applications. S.I.A.M. Rev. **10** 422-437.
Mandelbrot, B.B. and Taqqu, M.S. (1979). Robust R/S analysis of long run serial correlation. *42nd. Session of the International Statistical Institute*, Manila, Invited Paper.
Saether, G., Bendiksen, K., Müller, J. and Frøland, E. (1990). Int. J. Multiphase Flow **16** 1117-1126.

Taitel, Y. and Dukler, A.E. (1977). A model for slug frequency during gas-liquid flow in horizontal and near horizontal pipes. Int. J. Multiphase Flow **3** 358-396.

Taqqu, M.S. (1975). Weak convergence to fractional Brownian motion and to the Rosenblatt process. Z. Wahrscheinlichkeit. verw. Gebiete **31** 287-302.

Taqqu, M.S. (1977). Laws of the iterated logarithm for sums of non-linear functions of Gaussian variables that exhibit a long range dependence. Z. Wahrscheinlichkeit. verw. Gebiete **40** 203-238.

Taqqu, M.S. (1985). A bibliographical guide to self-similar processes and long-range dependence. In: *Dependence in Probability and Statistics, A survey of Recent Results.* Eberlein, E. and Taqqu, M.S. eds. Birkhäuser, Boston 137-164.

Numerical Simulation of Fluid Flow Behaviour Inside, and Redesign of a Field Separator.

E.W.M. Hansen, H.Heitmann, B.Lakså*and M. Løes†

Abstract

Separation equipment is needed in the production of oil and gas, to meet the requirements of pure fluids. The design of such equipment is not optimal for processing of multiphase fluids as it has been shown in the past. The operating conditions for separators will change as the production in the field is changing. The amount of oil/gas and water will change relatively.

Separator sizing must satisfy several criteria for good operation during the lifetime of the producing field;

- Provide sufficient time to allow the immiscible gas, oil and water phases to separate by gravity
- Provide sufficient time to allow for the coalescence and breaking of emulsion droplets at the oil-water interface
- Provide sufficient volume in the gas space to accommodate rises in liquid level that result from a surge in liquid flow rate
- Provide for the removal of solids that settle to the bottom of the separator
- Allow for variation in the flow rates of gas, oil and water into the separator without adversely affecting separation efficiency

A computer code (FLOSS; FLOw Simulator for Separators) is being developed to improve separator effeciency and performance. The code can be used to analyze the fluid flow behaviour and the separator performance of oil, gas and water in two or three-phase gravity separators. It may also be used to identify improved separator configuration for both steady-state and transient operation.

This paper presents results from numerical simulations of the flow in the Gullfaks A 1st stage separtor which is 16 m long and 3.5 m in diameter. Fluid flow analysis is performed in the inlet zone and the liquid bulk flow zone inside the separator. The results of three dimensional numerical simulation show that the internal equipment has to be rearranged and modified to ensure good separation conditions when the watercut in production is raised in the future. The paper also presents the operational experience and the redesign of the 1st stage separator in the field.

*All from SINTEF, NORWAY
†STATOIL, NORWAY

1 Introduction

Separators used on platforms, in the North Sea, separating bulk flows of water/oil/gas are vessels in which gravity settling due to density differance of the fluids takes place. The simple design methods like those found in Ref. [2] and [3], do not match the complicated multi-phase fluid flow behaviour through such equipment in any great detail. This means that the design of the separtors is not accurate or the design is too conservative, which frequenly gives oversized vessels, both in volume and weight, and low separation efficiency, Ref. [1].

More elaborate design procedures has been developed and published recently, Ref. [4],[5],[6] and [7]. The new procedures take more into consideration the distributed fluid flow inside the separator vessels and the effects like the breakup/coalescence of drops/bubbles are important mechanisms. The need for new methodes has been emphasized in Ref.[8].

The production rates of the fluids through the separators are frequenly changing both in total volume and in the relative amount of water/oil/gas. To set the design basis for such a production system is an insecure business, so evaluation of factors affecting the separation and upgrading of the separators will be part of the process optimization efforts.

This paper discusses the modelling and numerical simulation of the distributed flowfield through the 1st stage three-phase separator on the Gullfaks A platform in the North Sea. The numerical fluid flow behaviour inside the separator was performed by FLOSS (FLOw Simulator for Separators). Operational experience and the numerical flow simulations led to redesign of the separator.

2 Gullfaks A Platform

The Gullfaks oil field is a large oil field in the Norwegian sector of the North Sea and the production on the Gullfaks A Platform was started in 1986/87.

The Gullfaks A production train is a three-stage, dual-train process. Figure 1 shows the process flow diagram in which the 1st stage horizontal three-phase separator is under investigation. Figure 2 gives the details inside the separator as it was designed and installed at the platform.

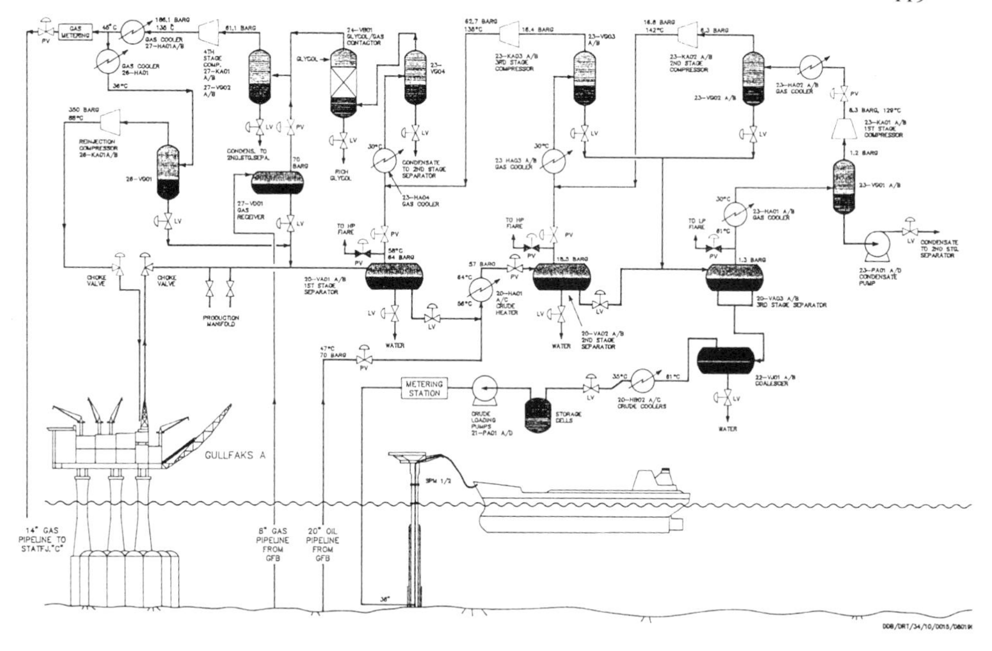

Figure 1: Gullfaks A process flow diagram. Production facilities consist of a three-stage, dual-train process.

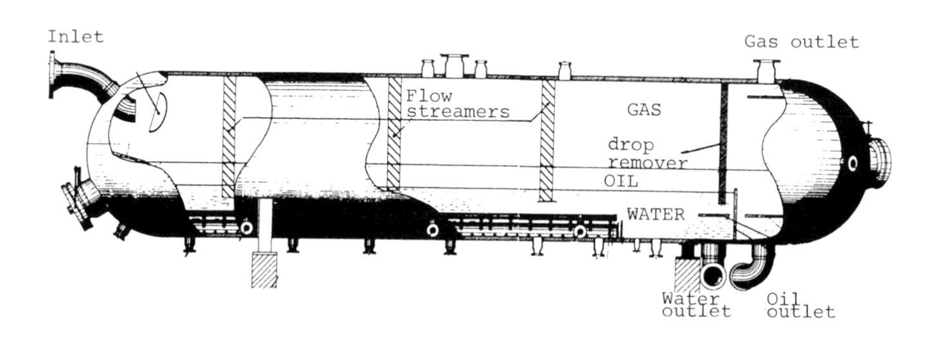

Figure 2: Gullfaks A 1st stage separator as designed.

The operational experience of the production process at Gullfaks A during the first four years in work was mainly, Ref. [9]:

- increased liquid capacity through separator
- water-level control failed with increased production of water

- emulsion problems inside separator
- sand accumelation and water-level control failure
- rising watercut in oil out of the separator

3 Flow Modelling

The operating conditions for the separator are; the pressure 68.7 bar and the temperature 55.4 ^{o}C. A GasOilRatio (GOR) of 100 Sm$^3/Sm^3$ was applied for the fluids. The densities and the viscosities of the fluids are given in table 1.

Table 1: Physical properties for the fluids in the 1st stage separator at 68.7 bar and 55.4 ^{o}C.

Fluid	Density kg/m^3	Viscosity Pas
Gas	49.7	0.000013
Oil	831.5	0.00525
Water	1030	0.00043

No dropsize distribution data of the liquid dispersion was available for the modelling. The design production rates and the future production rates are presented in table 2.

Table 2: Production rates in Gullfaks A 1st stage separator.

	Gas m^3/h	Oil m^3/h	Water m^3/h
1988 production	1640	1840	287
Future production	1640	1381	1244

Dropsize measurements in the inlet fluid dispersions to 1st stage separator were not available.

The separator is working half-filled with liquids in normal operation.

Figure 3 shows a sketch of the 1st stage separator with internals. The internals are designed as flowstreamers and dropremovals for the gas filled part of the separator. The internals will be extended into the liquid part of the separator when the vessel is half filled with liquid.

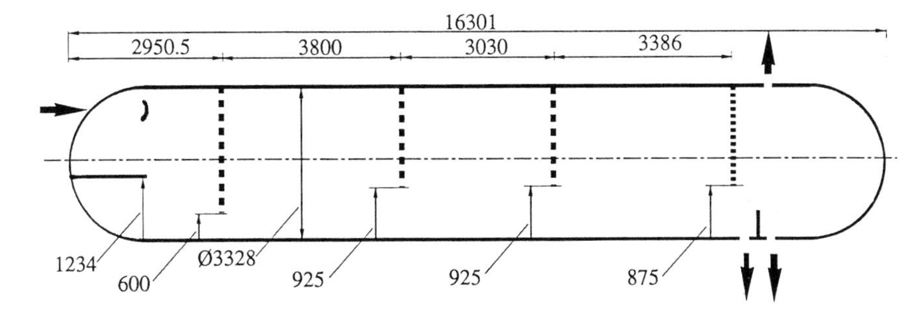

Figure 3: Sketch of 1st stage separator with internals. All measurements in mm.

3.1 FLOw Simulator for Separators (FLOSS)

FLOSS - FLOW Simulator for Separators, is a software package of "state-of-the-art" in computer codes which meets the requirements in modelling and simulation of distributed multiphase flow systems encountered in separation devices.

FLOSS was collected and developed in the ISEP-project, Ref. [10] and [11], to simulate the fluid flow behaviour inside 3-dimensional gas/oil/water separators by means of finite-volume numerical methods.

In general, for the use in offshore oil process, the separators are huge pressurized vessels. The separation system in an offshore oil process is going on in separators in a serial line system with separators working at different pressure stages. The multi-phase fluid flow, from the oil wells, run into the separators and the stream is separated into its different components to some extent in each separation stage. In each separator working at a fixed pressure, the inlet multiphase fluid flow enters the vessel as a high momentum jet, hitting some sort of a momentum breaker and the liquid splashes into the liquid pool in lower part of the vessel and the gas together with liquid drops flows in the upper part of the vessel.

Inside the liquid pool, near the inlet, the multiphase fluid flows as a dispersion with low horizontal velocity. The outcoming gas rises, and the oil and water will separate on the way to the outlet. In the upper part of the vessel, the oil and water particles together with condensed gas will fall down to the liquid interface as the multiphase fluid flows to the outlet. Both the upper and the lower part of the vessel may be equipped with internals to enhance the separation process.

With all the complex process which has been described above, the only fruitful way to understand the overall fluid flow system inside a separator may be to simulate the whole process or parts of it in a fully two-/multi-phase flow model in two or three dimensions. For the simulation of fluid flow and phase separation behaviour inside a gravity separator, it is helpful to characterize the different flow regimes or zones, see Figure 4. The modelling effort in the 1st stage separator has been concerned with zones 2 and 3, which is the inlet and momentum breaker zone and the bulk liquid flow zone.

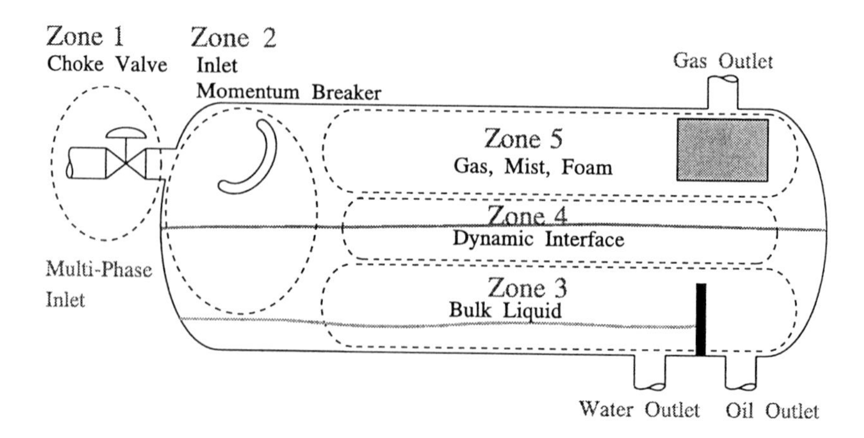

Figure 4: Typical Flow Zones in a Horizontal Gravity Separator.

3.2 Two-Phase Flow Modelling - Inlet Zone.

The inlet section of the separator in which all three phases are present has been modelled as a two-phase, gas/liquid, flow system. A gas/liquid jet is modelled to flow against a cup-shaped momentum breaker in a 3D geometry. The splash plate below the inlet jet inside the liquid pool has also been taken into consideration. The calculation of the two-phase flow behaviour in this zone will give the details of the distributed velocity field down in the liquid pool. The grid system used in the numerical flow simulations for the inlet zone was 11x8x15.

3.3 Homogenous Flow Modelling - Bulk Liquid Zone.

The velocities inside the liquid pool in a horizontal gravity separator are rather low, so for many reasons a homogeneous two-phase flow field may be assumed. However, the flow field behaviour in space is not homogeneous.

In the Gullfaks A 1st stage separator no information about the liquid/liquid dispersion characteristics was available so the simulations were performed with the assumption of homogeneous fluid properties with no slip between the phases inside the bulk liquid zone. An average of the liquid velocity distribution, developed in the inlet zone, was modelled to flow through the separator influenced by the internals in the liquid part. The porosity of the flowstreamers and the drop removal was taken into consideration. The grid system used in the numerical flow simulations for the bulk liquid zone was 23x4x5.

In both zones under investigation the flow was considered to be symetrical about a verical plane in the middle of the separator, thus only half of the separator volume was taken into account in the numerical simulations performed.

4 Results of Numerical Simulations

The numerical simulations were performed on a VAX 8600 and more details are found in Ref. [12].

4.1 Inlet Zone Flow

The results from the numerical simulation of the two-phase flow behaviour in the inlet part of the separator are presented in Figure 5 and 6.

The result for the year 1988 flowrate through the separator is shown in Figure 5. The computed velocity vectors and the liquid volume fraction in a vertical longitudinal plane midway between the centerline and the side wall are shown in Figure 5 a). The velocity vectors indicates a flow which has been turned after hitting the momentum breaker in the upper part of the separator and the flow in the liquid pool wil be turned by the splash plate to form a jetflow further downstream. The fluid volume belowe the splash plate is "dead-volume" inside the separator.

Figure 5 b) gives the velocity speed on the end of the splash plate in a crosswise plane. The speed is not homogeneous over the plate, and along the separator walls there is a backflow. Thus no plug flow situation is established in the inlet zone of the separator.

Figure 6 gives the results with an increased flowrate through the separator and the fluid flow behaviour is similar to what has been discussed in the previous simulation. Although the velocity distribution was more homogeneous over the splash plate in the crosswise direction, no backflow was simulated along the walls.

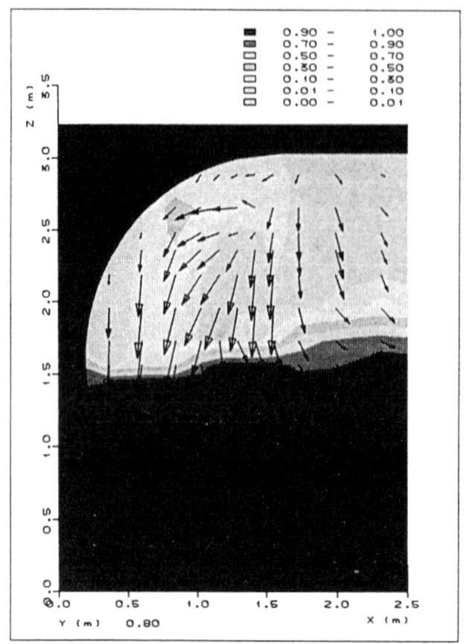

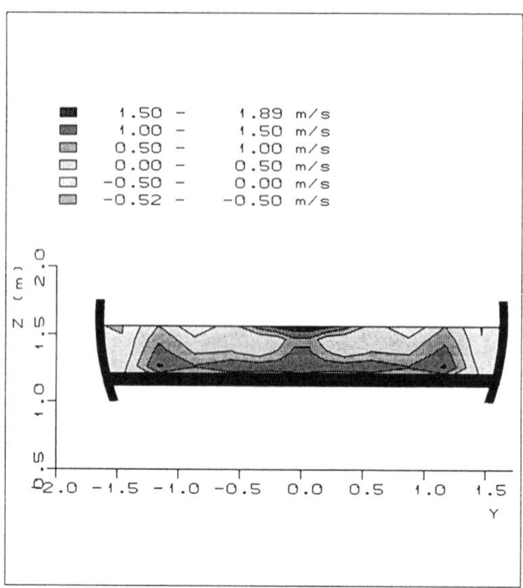

Figure 5: Computed velocity field behaviour in inlet zone of the separator with the year 1988 production rate. a) Velocity vectors and liquid volume fraction in a x-z plane, b) the horizontal velocity distribution in the liquid above the splash-plate.

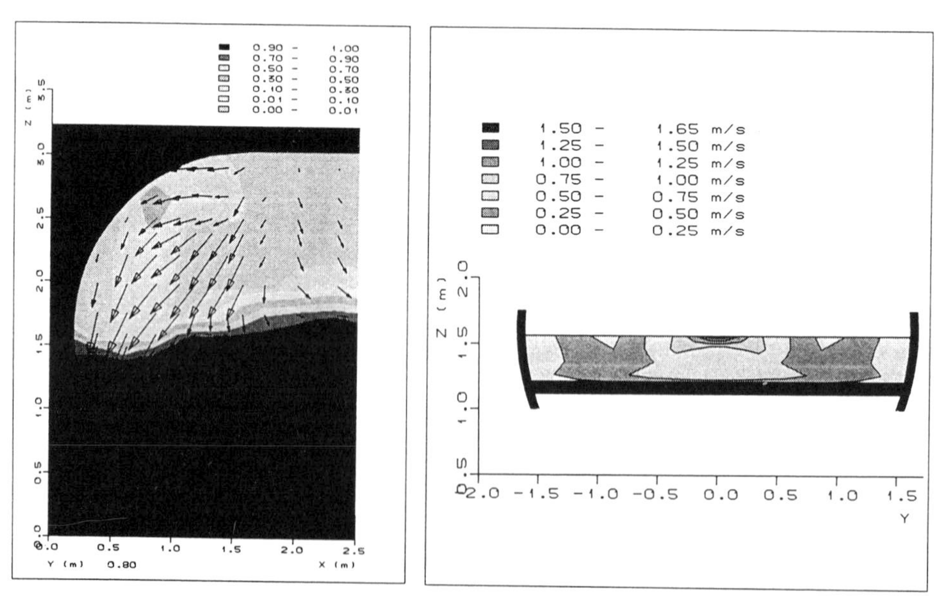

Figure 6: Computed velocity field behaviour in inlet zone of the separator with future production rate. a) Velocity vectors and liquid volume fraction in a x-z plane, b) the horizontal velocity distribution in the liquid above the splash-plate.

4.2 Bulk Liquid Flow

The numerical fluid flow simulations performed for the bulk liquid flow inside the separator are shown in Figure 7 and 10.

The computed velocity field behaviour with the year 1988 production rate through the separator is presented in Figures 7 and 8. The distributed velocity field in different planes, both horizontal and vertical , for half of the separator is shown. The placing of the internals is indicated in the figures and the flow field is highly influenced by the internals. Mixing and reduced separation efficiency will be caused by the internals extended down into the liquid pool. Backflows along the separator wall is shown due to three dimensional rotational systems caused by the internals.

In Figure 9 and 10 the numerical simulation with increased flowrate through the separator is shown. The overall flow field behaviour is similar to the previous one. Although the rotational systems created are more intense and the magnitude of the velocities are higher, which will give poorer separation efficiency with increased watercut.

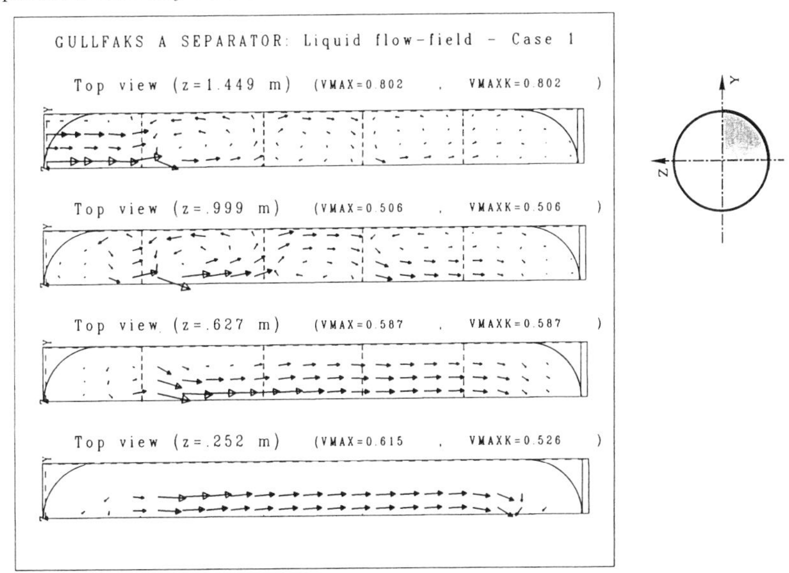

Figure 7: Computed velocity field behaviour in bulk liquid zone of the separator with year 1988 production rate. Velocity vectors in different horizontal planes, seen from above, are presented.

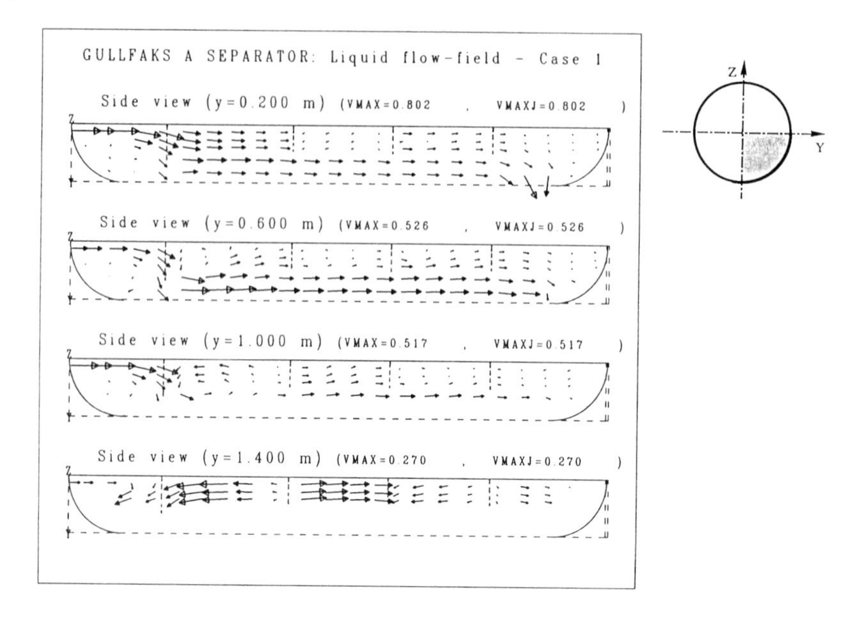

Figure 8: Computed velocity field behaviour in bulk liquid zone of the separator with year 1988 production rate. Velocity vectors in different vertical planes are presented.

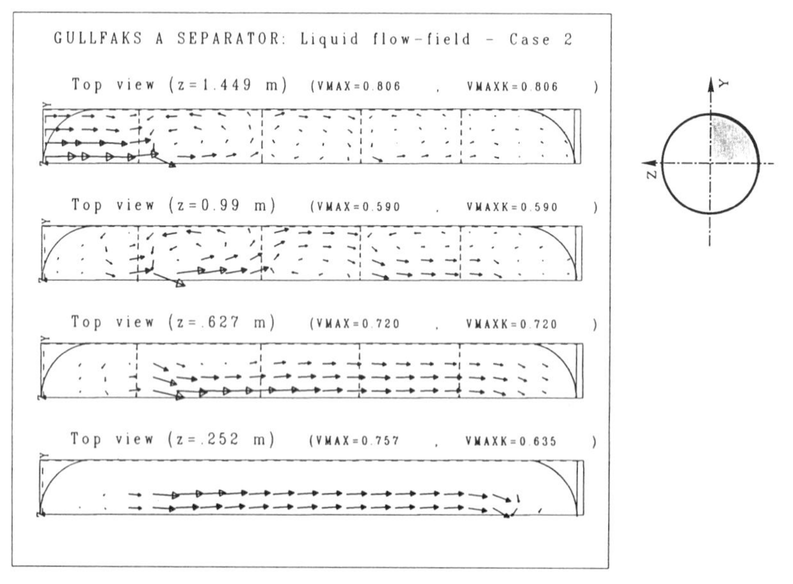

Figure 9: Computed velocity field behaviour in bulk liquid zone of the separator with future production rate. Velocity vectors in different horizontal planes, seen from above, are presented.

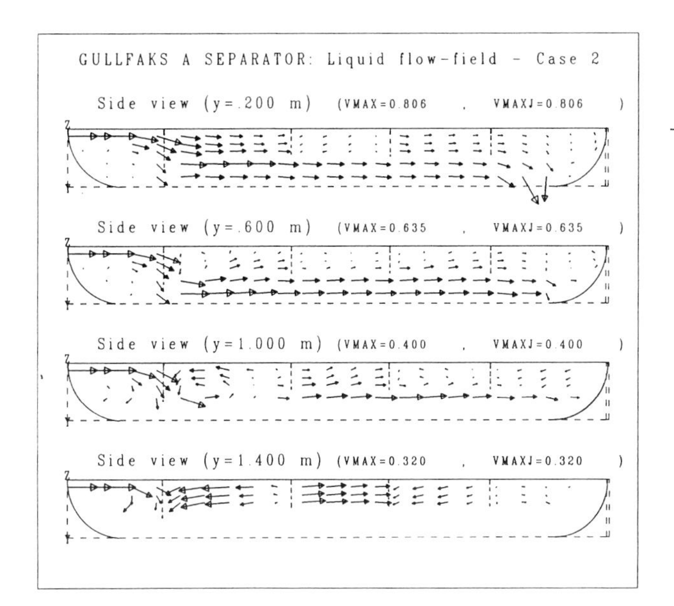

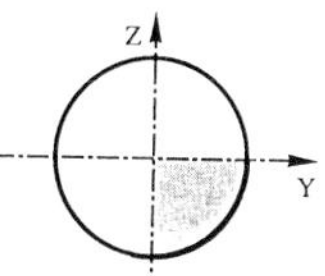

Figure 10: Computed velocity field behaviour in bulk liquid zone of the separator with future production rate. Velocity vectors in different vertical planes are presented.

5 Separator Modifications

Based on the operational experience and the research done the 1st stage separator was redesigned. The following changes were introduced, Ref. [9]

- The sand removal system was redesigned.
- The measurment devices for the liquid level control was rebuild to avoid sand to flow into the tubes.
- The flowstreamers and the drop removal were reduced not extending down into the liquid pool. Sand would foul the internals and the internals would also act like mixers which would decrease the separation efficiency of the water from oil.

The operational experience with the separator after the redesign was very good. Field tests were performed in 1990 to compare the flow through the redesigned and the original designed separators, the two 1st stage separators on the Gullfaks A platform. The tests gave favorable data for the redesigned separtor, both the water level control and the water in oil content out of the separator were improved.

6 Conclusion

The opertional experience gave an indication the design of internal equipment for operation and control of the Gullfaks A 1st stage separator was not optimized.

The research by numerical simulation of the distributed flowfield behaviour inside the separator presented both dead volumes, backflow and mixing in the flowfield.

Many separator vessels are already installed offshore and the relative amount of the flow composition will change with time in the different locations. The vessel itself may be reused but the internals may be redesigned to meet the requierements of better separation due to increased water content in the flow.

Knowledge of the fluid flow dispersion characteristics are needed to improve the distributed fluid flow simulation. Also more research about the breakup/coalescence processes inside the multi-phase flowfield in all zones in a separator are needed, in order to increase the separation effeciency in the bulk flow zones.

References

[1] Hancock W.P. and Hagen D.A.: Fine Tuning Increases Statfjord B Output 39 %, World Oil,1986.

[2] Arnold K. and Stewart M.: Surface Production Operations - Vol. 1, Gulf Publishing Co., Houston, 1986.

[3] Gordon I.C. and Fairhurst C.P.: Multi-Phase Pipeline and Equipment Design for Marginal and Deep Water Field Development,3rd International Conference on Multi-Phase Flow,Hague,1987.

[4] Swanborn R.A.: A New Approach to the Design of Gas-Liquid Separators for the Oil Industry,Ph.D.Thesis, Tech. Univ. Delft, 1988.

[5] Hafskjold B. and Dodge F.: An Improved Design Method for Horizontal Gas/Oil and Oil/Water Separators, SPE 19704, 64h Annual Technical Conference and Exibition of the Socity of Petroleum Engineers, San Antonio, 1989.

[6] Verlaan C.: Performance of Noval Mist Eliminators, Ph.D.Thesis, Tech. Univ. Delft, 1990.

[7] Hansen E.W.M., Heitmann H., Lakså B., Ellingsen A. Østby O., Morrow T.B. and Dodge F.T.: Fluid Flow Modelling of Gravity Separators., 5th International Conference on Multi-Phase Production, Cannes, 1991.

[8] Spence S.: The Importance of Proess Intensification in Offshore Separation Equipment., Conference on Multiphase Operations Offshore, London,1991.

[9] Solemslie K.: Prosessutstyr for petroleumsutvinning til havs. Driftserfaring Gullfaks og Veslefrikk.,NIF-møte, Gol, 1991.

[10] Morrow T.B., Hansen E.W.M. and Dodge F.T.: ISEP Program - Volume III: Separator Fluid Dynamics Studies, IKU Report No.: 09.3009.00/14/88, June 1988.

[11] Hansen E.W.M., Heitmann H., Lakså B., Ellingsen A. and Østby O.: ISEP Program - Volume VI: Computer Model Documentation - FLOSS, IKU Report No.: 09.3009.00/14/88, July 1988.

[12] Hansen, E.W.M., Heitmann H. and Lakså B.: Numerisk simulering av strømning i 1. trinns separator for Gullfaks-A plattformen, STF15 F88054, 1988.

MULTIPHASE PRODUCTION FLOWLINE MONITORING USING ULTRASONICS EMISSIONS

D G Wood, BP Exploration

SUMMARY

Field data collection on existing multiphase production flowlines is essential in order to develop and verify improved methods for the hydraulic design of future multiphase systems, and also to identify the cause of any operational difficulties with the existing systems. The recognised technique by which flow regime and slug flow data can be obtained on such systems is gamma ray densitometry. However, gamma ray densitometers are heavy, require calibration for individual flowlines, and have safety problems associated with the use of the sealed radioactive source. BP have now developed a compact, and safer, alternative approach for monitoring multiphase flowlines, based on the detection of the ultrasonic noise emitted by the production fluids.

This paper describes the patented ultrasonic emissions monitoring system and outlines the technique by which the acoustic data is analysed to provide information on the multiphase flow regime and slug flow characteristics. Data obtained on various field flowlines is also presented. Finally the acoustic monitoring and densitometry techniques are compared, and the respective merits of the two methods discussed.

1. INTRODUCTION

Multiphase schemes are now increasingly employed for the development of oil and gas fields. Methods for predicting pressure drop, flow regime and slug characteristics (1) are continuously being improved, leading to greater confidence in the design of flowlines and downstream processing facilities for multiphase systems.

Further development and verification of these design tools needs to be based on field operating data. BP has therefore conducted an extensive field data collection programme since 1985. The main aim of this programme has been to identify multiphase flow regime and measure slug characteristics in multiphase flowlines. Field trial work has been conducted in many of BP's fields including Magnus (2) and South East Forties (3).

The main instrumentation techique employed during this field trial work has been gamma ray densitometry. This technique, based on the attenuation of a gamma ray beam as it passes through the production fluids, provides an approximate measurement of the liquid hold-up in the flowline. The time dependent variations in this hold-up trace allow the flow regime to be determined. Also, cross-correlation of the signal between two sensors separated by a short axial distance, leads to the calculation of the velocity and length of individual events.

Gamma densitometry has proved to be a highly successful and reliable method of monitoring multiphase flowlines. However, the equipment is bulky, with each densitometer weighing between 40 and 80 Kg. The use of a radioactive source also leads to additional safety implications, such as the necessity to obtain licenses for transportation and operation. Finally, as a substantial proportion of the beam attenuation is due to the steel walls of the pipe, careful calibration is required for each pipe on which the equipment is used.

The development of an alternative non-intrusive instrumentation technique for monitoring multiphase flow systems was necessary to provide a more portable system that did not suffer from the safety drawbacks of densitometry.

One possible technique, based on the detection of ultrasonic emissions, was identified during initial testing of a prototype system in BP's Prudhoe Bay field. Following this initial testing BASEEFA certified equipment and data analysis techniques were developed in combination with extensive testing by BP at several field locations.

2. MEASUREMENT TECHNIQUE

Ultrasonic measurement systems fall into two categories - active and passive. With active techniques, the measurement system produces a standard signal, and it is the measured changes to this signal that provide information. Such techniques could be developed for the determination of liquid hold-up in homogeneneous two-phase flows through the measurement of the attenuation of an ultrasonic signal, or for the determination of the liquid level in a stratified flow through measurements of the reflections of an ultrasonic signal

With passive techniques no transmitted signal is employed. Instead the measurements are made by detecting the ultrasonic noise produced intrinsically by the flow of the fluids. This is the technique which will be discussed within this paper. The source of the ultrasonic emissions is the dissipation of the energy through fluid turbulence. Thus, the principle of the ultrasonic system is to identify events in the flow by measuring and characterising the noise emitted by the turbulence of the fluids.

The noise emissions from multiphase flow systems cover a wide range of frequencies. Spectral analysis indicates that intensity levels are highest at the upper limit of the audible frequency range, between 5 and 20 kHz. As the measurement frequency is increased above 20 kHz, the intensity steadily decreases. However, the choice of a suitable operating frequency for the measurement system must be based on the best signal to noise ratio. This occurs at a frequency of approximately 70 kHz. At this frequency the attenuation of ultrasonic noise through air is high, and thus external noise from process equipment cannot be detected.

Although monitoring at ultrasonic frequencies ensures the system only detects a signal via the pipe wall, some extraneous noise can be detected if the pipe vibrates. This noise can be minimised by locating the transducers close to a pipe support.

The measured signal intensity is dependent on several factors. These include flow dependent parameters such as the local two-phase density, the local fluid velocities and the intensity of mixing. Location dependent parameters, such as the transmission characteristics of the pipe wall, the quality of the acoustic coupling between the wall and the transducer and the pipe diameter, are also important. However, at a given location, the intensity of the signal detected by the transducer is dependent on the level of the turbulence energy and the rate of signal attenuation with distance, integrated over the transducer's detection range.

3. HARDWARE

Figure 1 shows a schematic of the ultrasonic emissions monitoring system. A photograph of a 3-channel system is shown on Figure 2. The equipment consists of a piezo-electric transducer and associated pre-amplifier for each channel, combined with a remote multi-channel signal processing unit.

The transducers are located in direct contact with the external wall of a metal pipe. Pipe insulation attenuates the noise and must therefore be removed before the transducers can be installed. Good acoustic coupling between the sensor and the metal is ensured by the use of magnetic clamps and silicon vacuum grease. To date the best results have been obtained using transducers that are sensitive to frequencies of around 70 kHz.

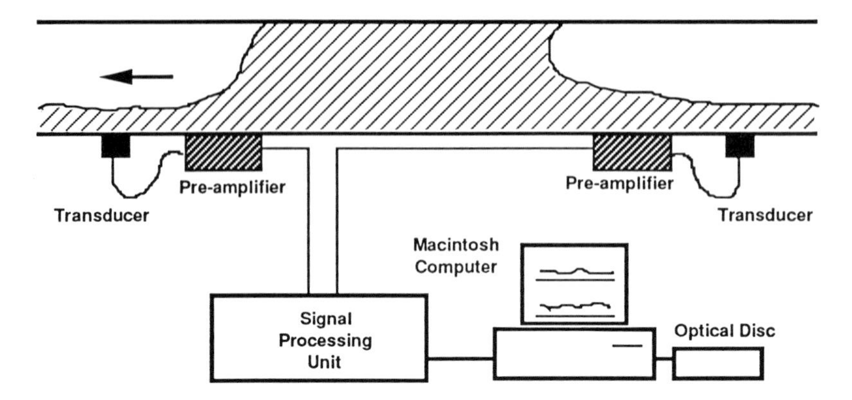

Figure 1 - Schematic of Ultrasonic Emissions Monitoring System

The pre-amplifier has two tasks. Firstly, amplification of the signal allows the main processing unit to be sited up to 500 metres away. Secondly, a high pass filter ensures that low frequency signals (below 50 kHz) are removed before transmission.

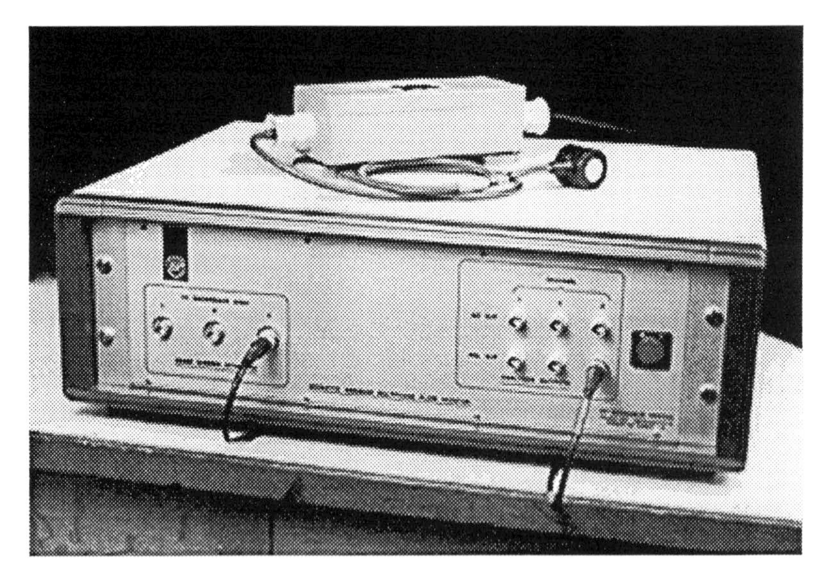

Figure 2 - 3-Channel Ultrasonic Emissions Measurement Systems

Although the piezo-electric crystal requires no external power supply, power must be transmitted to the pre-amplifier to operate the amplification and filtering circuits. A single co-axial cable is used both to supply the power to the pre-amplifier and to transmit the signal back to the processing unit.

The signal processing unit has a variety of tasks. It initially separates the AC instrument signal from the DC power signal. It then further amplifies and filters the signal, and averages the intensity over the sample period set by a time constant control. The signal amplification and time constant are adjustable. Finally the average signal intensity is converted to a DC voltage for output. A zener barrier is also contained within the signal processing unit to ensure that the equipment can be used within an electrically zoned area.

4. FLOW REGIME IDENTIFICATION

The ultrasonic emissions techniques can be used to identify a multiphase flow regime using the characteristic time dependent variations in turbulence energy that are associated with each type of flow. These variations lead to the expected signal intensity time trace for each horizontal flow regime shown on Figure 3.

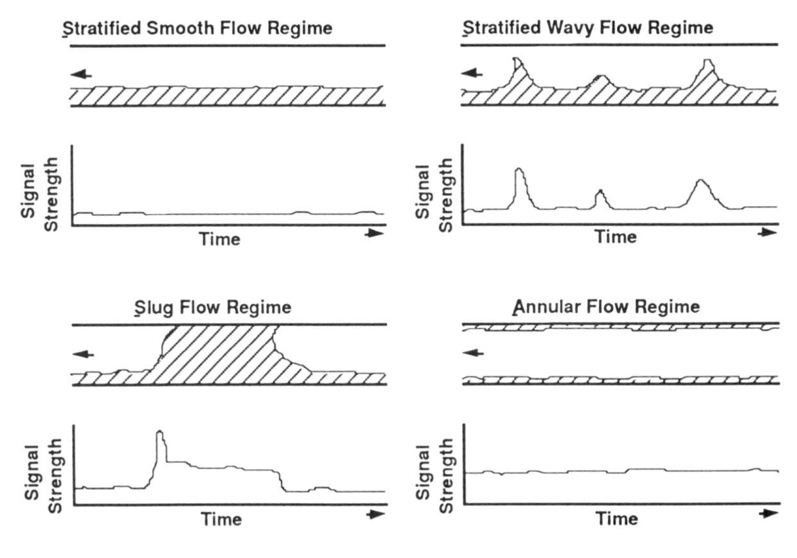

Figure 3 - Characteristic Emissions For Horizontal Flow Regimes

In stratified-smooth flow the signal level is relatively low with few fluctuations. In wavy flow sharp peaks in signal level occur as the wave front passes the sensor.

The trace for slug flow is more complicated. The signal initially rises rapidly from the background noise level to a peak as the slug front, where maximum turbulence occurs, passes the transducer. The signal level then decreases to a new value (still above the background level) as the slug body passes and finally, at the rear of the slug, there is a sharp decrease to the original background level. If the gas content of the slug body is relatively high rapid fluctuations in signal level may be observed.

At any given instrument location, the maximum signal level attained as a slug or wave passes the sensor is dependent on the gas content of the event, the velocity of the event and the intensity of mixing.

5. SLUG FLOW MEASURMENTS

To use the ultrasonic emissions equipment to detect slugs, and also to measure their characteristics, slugs must be distinguished from the other events which can occur within a multiphase pipeline. Any events, in addition to slugs, which consist of a variation in the local gas or liquid flowrate are detected by the equipment as a change in the signal level.

For horizontal flowlines these changes may be of two forms -

(1) Rapid changes associated with the propagation of other, non-slug events through the flowline, in particular large amplitude waves.

(2) Slow changes in the background level associated with variations in the pipeline inlet and outlet conditions, i.e. line depressurisation following the discharge of a slug into the separator.

Although slug traces can be identified visually, data logging and analysis software based on pattern recognition has been developed, for use on an Apple Macintosh, to analyse the traces automatically. This analysis software distinguishes slugs from waves using the duration and velocity of the event, and from changes in the background level by the rate of change of the signal level. This software also normalises the signal automatically, thus removing the requirement to re-calibrate the equipment between pipelines.

Although the flow regime can be determined using only one transducer and an estimate of the multiphase mixture velocity in the flowline, cross-correlation between the signals detected by two sensors, separated by a short axial distance, is required to provide accurate measurements of the slug characteristics.

The data analysis process consists of four main stages. The first stage involves digital filtering which removes any fluctuating noise due to pipe vibrations. The slug and wave events are then distinguished from changes in the background noise level. The time at which the front and rear of each event passes each sensor is recorded during this second stage of the analysis.

Cross-correlation between the two sensors is conducted in the next stage leading to the calculation of the velocity and length for each event (both slugs and waves). The final stage determines whether each event is a slug or a wave based on the duration and velocity of the event. In this process the event velocity is compared to the calculated mixture velocity. As waves are interfacial phenomena they have a translational velocity that is, in general, less than the mixture velocity, whereas slugs have translational velocities in excess of the mixture velocity.

6. FIELD TESTING

To date the equipment has been used in five field locations. Over 5000 hours of monitoring have been conducted on flowlines ranging from 4 to 28 inches in diameter, over a wide range of operating conditions. This has led to the optimisation both of the hardware and data logging settings, and of the techniques used for data analysis. This section describes some typical traces obtained with the equipment under different conditions, highlighting the capabilities and limitations of the system.

The majority of the development work with the equipment has been based on work conducted on low pressure, large diameter flowlines in Alaska. Figures 4-7 show typical traces obtained on four separate horizontal flowlines, highlighting the response of the system when installed on lines operating in common flow regimes. Each figure consists of traces obtained using two sensors located approximately 15 metres apart.

Figure 4 shows traces obtained on a flowline operating in the stratified-smooth flow regime. The signal level is relatively constant for both the upstream and downstream sensors with no significant variations in signal level.

Figure 5 shows traces obtained on a flowline operating in wavy flow, a regime typical of many of the Alaskan large diameter flowlines. It consists of a series of sharp peaks corresponding to the passage of waves past the sensor. The variation in the peak signal level from wave to wave corresponds to the variations in wave velocity. The rates of rise and decline of the signal level before and after the signal peak are similar. The significant differences in the shape of some of the wave traces between the two sensors are indicative of interfacial events that are changing rapidly as they pass through the flowline.

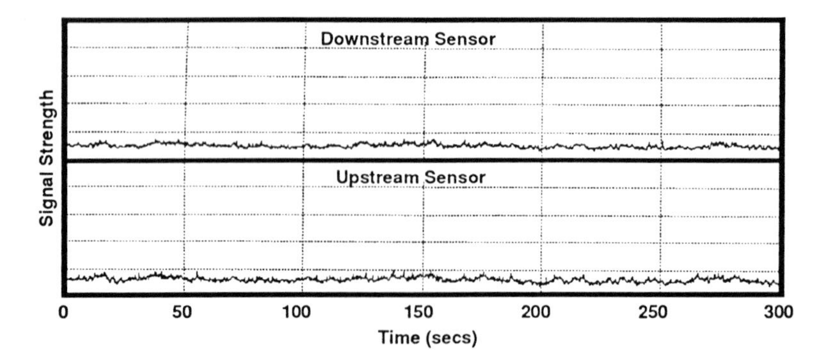

Figure 4 - Stratified Smooth Flow Regime

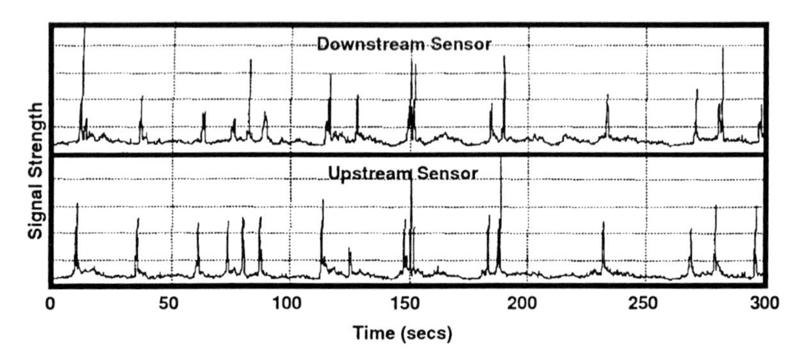

Figure 5 - Wavy Flow Regime

Figure 6 shows a typical trace for a flowline operating in the slug flow regime. In this flowline slugs were detected with a frequency of approximately 30 per hour. Each consists of a sharp rise in signal level as the slug front passes the sensor, a decline to a steady signal level as the main body of the slug passes, and then a further decline to the background level as the rear of the slug passes. The slugs observed in this flowline had a mean velocity and length of approximately 10 metres/sec and 100 metres respectively. The variation in peak signal level between events was smaller for this type of flow than for wavy flow. Also the shape of the trace was similar at the upstream and downstream sensors. This is indicative of stable events that are being driven through the flowline by the gas at a steady velocity.

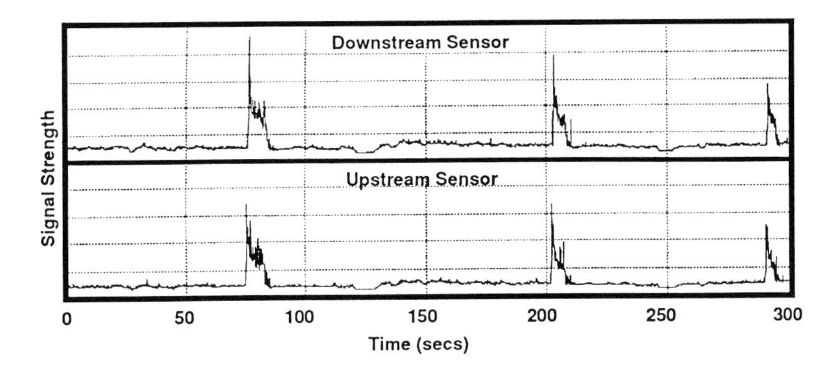

Figure 6 - Slug Flow (Vm = 10 m/s)

Figure 7 shows the trace obtained for a single slug on another large diameter flowline in Alaska. The slug velocity in this case was approximately 7 m/s. This lower velocity leads to a smaller difference between the signal level in the main body of the slug and the background signal level. A small peak is detected as the rear of the slug passes the sensor indicating a turbulent region as liquid is shed from the tail. The length of this slug at the measurement location, half way between the line start and the separator, was approximately 250 metres.

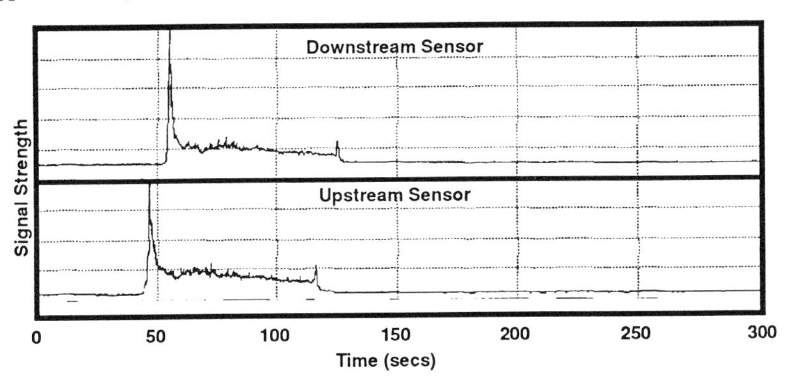

Figure 7 - Slug Flow (Vm = 7 m/s)

The effect of line de-pressurisation on the ultrasonic trace is shown on Figure 8. This trace was obtained at a measurement location close to the end of the flowline. The flowline terminated in a pipe restriction followed by a manifold and a header line into the separator.

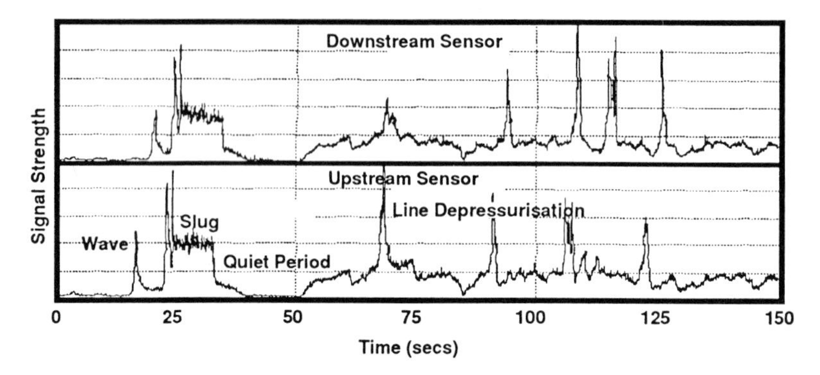

Figure 8 - Flowline Depressurisation Following Slug

The trace initially shows a wave (after 20 seconds) and a slug (25-35 seconds) passing the two sensors. The two events are closer together at the downstream sensor indicating that the slug velocity is significantly higher than that of the wave. Following the passage of the slug there is a period of very low signal level (40-50 seconds), similar in duration to the time taken for the slug to pass the sensor. This occurred as the slug passed through the restriction, and indicates a stalling of the flow. Once the tail of the slug had entered the separator the flowline pressure upstream of the restriction was observed to drop significantly. High gas velocities, associated with this flowline depressurisation, caused a rise in the background noise level (50 seconds) and also led to the formation of high velocity waves (70, 90, 110, 115 and 125 seconds). After a period of time the signal level decayed to the normal background level before the arrival of the next slug.

The maximum increase in signal level due to the depressurisation in the system decreased as the transducer was located further from the line exit. Thus, the trace obtained in Figure 8, approximately 200 metres from the line exit, could easily be analysed whereas interpretation of traces obtained close to the manifold would be more complicated.

The effect of placing the sensor to close to the line exit or on vertical section of pipe is shown of Figure 9. The pipe layout in this case consisted of a short riser, a 5 metre section of horizontal flowline and finally a short downcomer into the manifold. The horizontal section ended in a blind flange. A densitometer was located on the horizontal section of flowline (bottom trace). Ultrasonic emission measurements were obtained on the vertical riser (top trace), just upstream of the blind flange (second trace) and beside the densitometer (third trace). Pressure measurements at the densitometer location were also available.

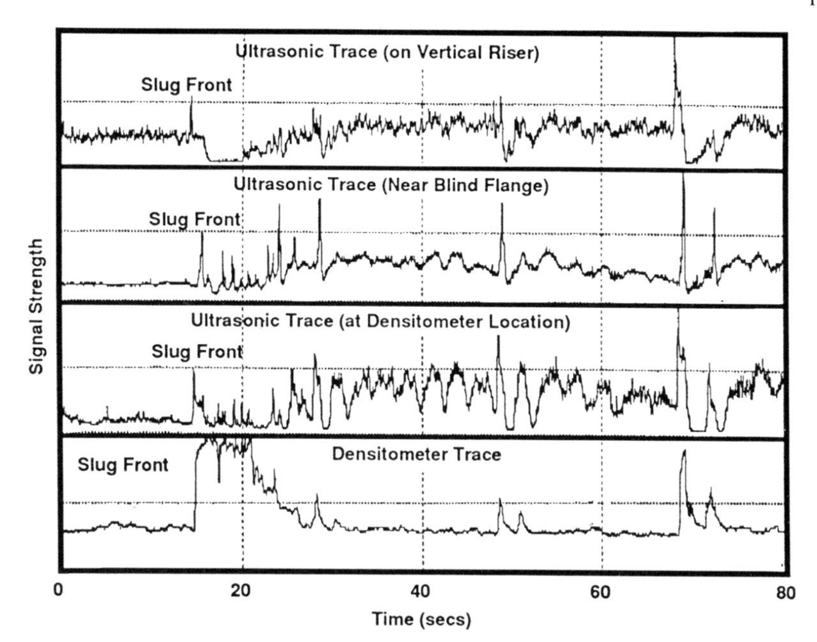

Figure 9 - Densitometer and Ultrasonic Signals Detected Just Upstream of Manifold

The densitometer trace shows the passage of a slug between 17 and 21 seconds. A rapid increase in line pressure was observed as the slug front passed the transducer. Following the passage of the slug the line depressurised over a period of time resulting in the formation of the small waves observed later in the trace.

The ultrasonic emissions for the sensor located next to the densitometer clearly identify the slug front. However, the signal level declines after a short period of time as the slug front reaches the blind flange and the slug stalls. Some further fluctuations in signal level occur as the remainder of the slug passes the sensor at a relatively low velocity and drains into the manifold. Following the exit of the slug into the separator, the line depressurisation leads to high gas velocities, high turbulence and signal levels greater than those obtained during the slug passage. At the blind flange only the slug front can be detected on the trace followed by the noise from the depressurisation stage.

The top trace on this plot indicates that the characteristic turbulence energy fluctuations for a slug passing through a vertical pipe are entirely different to passage through a horizontal pipe. Between slugs churn flow is observed. This flow is chaotic and thus turbulent energy levels are high. The front of the slug can still be identified by a sharp rise in signal level. However, the lower gas content in the slug body leads to a steadier flow and a significant drop in signal level. The end of the slug is then marked by an increase in signal level to that associated with the churn flow.

Measurements obtained 100 metres further upstream show much lower signal levels during the depressurisation stage and are similar to the traces shown on Figure 6, allowing the slugs to be clearly identified.

This field trial work has indicated that the ultrasonic signal is most easily interpreted when the equipment is located on a horizontal or near-horizontal section of flowline preferably 20 metres from the nearest fitting. The transducer should also be located more than 100 metres, or one slug length if this is greater, from the end of the flowline to prevent the depressuriation noise swamping that emitted by slugs.

Although interpretation of the signal is more complicated for a transducer located elsewhere in the system, if interpretation of the signal is possible the data will often yield a greater breadth of information on the flow conditions. However, automatic data analysis in such situations would require the development of techniques that were specific to each sensor location.

The field work has also identified a lower mixture velocity limit of between 2 and 4 m/s, depending on the operating conditions, below which the turbulence energy in the system is too low to obtain useful results.

7. RELATIVE MERITS OF ULTRASONIC EMISSIONS AND DENSITOMETRY SYSTEMS

The ultrasonic emissions system cannot replace all the functions of densitometry. It is therefore important to consider the two systems as complementary. Both systems are non-intrusive and can be used to determine multiphase flow regime, slug frequency, velocity and length.

Densitometry has the benefit of being able to provide a reasonably accurate measurement of liquid hold-up. This measurement is an important requirement in the calculation of slug volumes and is required to estimate the force exerted by slugs. Densitometry is also less restricted in terms of the locations at which measurements can be made and easily interpreted, and can be used without removal of insulation. Finally, the signal to noise ratio is not a function of the mixture velocity and thus densitometer can be used for pipelines operating at low flowrate.

The benefits of the ultrasonic emissions system are primarily in its portability and safety. The transducer and pre-amplifier are light and simple to install and a 2-channel signal processing unit driven by a 12 V DC power supply can be manufactured in a box only 300 mm square and 100 mm deep, allowing the instrument system to be fitted into a briefcase. Additional advantages are that the instrument does not require re-calibration for each flowline on which its used, and that the response time of the ultrasonics system is much shorter than the densitometry system. This leads to improvements in the measurement of slug velocity for flowlines with high mixture velocities. Finally the ultrasonic emissions system is considerably cheaper - an 8-channel system can be supplied for about 25% of the cost of eight densitometers.

8. APPLICATIONS

At present this equipment is employed as a field monitoring system by BP to obtain data on multiphase systems for the development of design methods. A system has also been supplied to BP Alaska for flowline monitoring in association with ongoing corrosion trials (4). This work is aimed at determining the effects of multiphase flow regime on corrosion rates under field conditions.

Another application in which a slug detection instrument, such as the ultrasonic emissions system, could be used is to control the process plant before and during the arrival of a slug could be envisaged.

9. CONCLUSIONS

9.1 A method for monitoring multiphase flow regime and slug characteristics in oil and gas production flowlines has been developed by BP Exploration. The system consists of a suitably located passive ultrasonic emissions detection device combined with data logging and analysis software.

9.2 The ultrasonic emissions system is intrinsically safe, non-intrusive, portable, compact, simple to install and requires no calibration.

9.3 Extensive field testing of the equipment has been conducted over a wide range of flowline diameters and operating conditions. This has enabled the optimum hardware settings and data analysis techniques to be obtained.

9.4 The limitations of the present system have been identified. These consist of ensuring that the sensors are located on a horizontal or near-horizontal section of flowline, away from any fittings and at least 100 metres upstream of the flowline exit. The equipment can only be effectively used on flowlines where the mixture velocity is in excess of 2-4 m/s.

9.5 The technique has been compared to that of gamma ray densitometry - a method used by BP for monitoring multiphase flowlines since 1985. This comparison has identified advantages of both systems indicating that the techniques are complementary.

10. REFERENCES

1. Hill, T.J., Wood, D.G., A New Approach to the Prediction of Slug Frequency, SPE 20629, presented at SPE 1990 Annual Technical Conference and Exhibition, New Orleans, Sep 1990.

2. Hill, T.J., Multiphase Flow Field Trials on BP's Magnus Platform, JERT 109, 142-147, Sep 1987.

3. Hill, T.J., Riser-Base Gas Injection Into The SE Forties Lines, BHRA 4th Int Conf on Multiphase Flow, Nice, France, June 1989.

4. Green, A.S., Johnson, B.V., Choi, H., Flow Related Corrosion in Large-Diameter Multiphase Flowlines, SPE 20685, paper presented at 65th Annual Technical Conference of the Society of Petroleum Engineers, September 1990

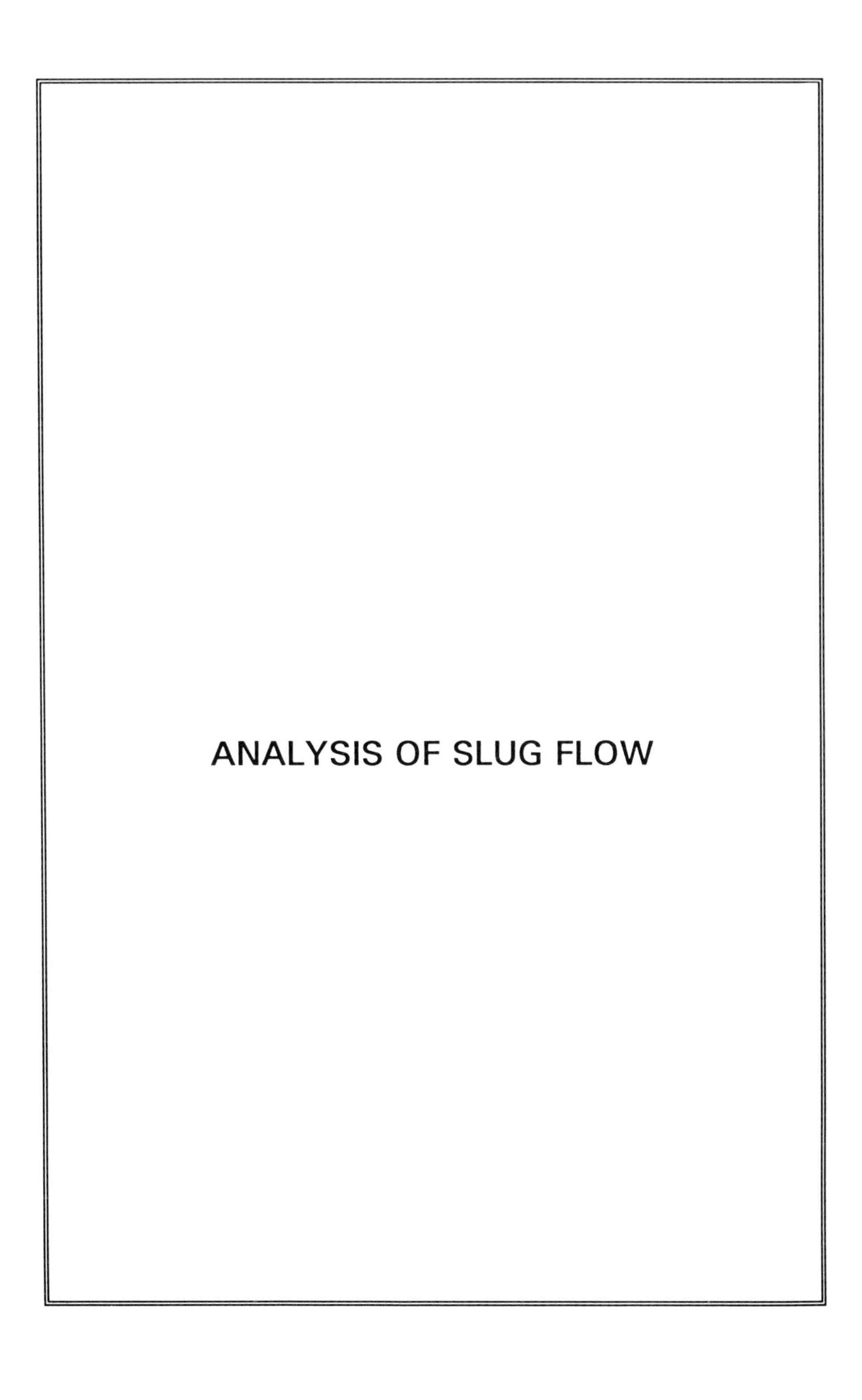

ANALYSIS OF SLUG FLOW

DETERMINATION OF SLUG LENGTH DISTRIBUTIONS BY USE OF THE OLGA SLUG TRACKING MODEL

Egil Hustvedt
Statoil Reasearch Centre
Postuttak
N-7004 Trondheim, NORWAY

ABSTRACT

Slug flow is a very common flow pattern in multiphase pipelines for crude oil. The slug flow regime causes large variations in oil and gas flow rates entering the downstream process. The frequency and size of these slugs must be predicted to ensure that downstream process facilities are correctly designed.

In most Eulerian types of one dimensional, dynamic, flow models (i.e OLGA, PLAC), slug flow has so far been treated by an average approach. This method gives correct pressure drops and liquid holdup, but no information on slug lengths and frequencies. In a new version of OLGA a Lagrangian type of slug tracking model is superimposed on the Eulerian scheme. The propagation velocities of slug fronts and tails and their position are computed. From this it is possible to determine the frequency and size of the slugs.

Statoil and Total have, within the scope of the POSEIDON programme, performed multiphase flow measurements on a 6" pipeline at the Sidi El Itayem field in Tunisia. The present paper compares the measured data with analysis performed with the OLGA slug tracking model.

INTRODUCTION

The main part of the remaining Norwegian petroleum resources are in relatively small fields, located within reasonable distances from existing offshore infrastructures. Available processing capacity on these platforms will increase throughout the 90'ies, and satelite field development based on multiphase technology has a significant potential for cost savings.

Slug flow is a very common flow pattern in multiphase pipelines for crude oil , and operational design tools or simulators must be capable of describing slug flow phenomena. Slug flow is characterized by liquid plugs propagating through the pipeline. To simulate these phenomena accurately, it is required that computer models are capable of tracing and maintaining slug fronts and tails in a non-diffusive way.

Standard one-dimensional, dynamic, two fluid models fail in their treatment of slug flow because of,

- Numerical diffusion of sharp variations in holdup.
- Slug flow treated by an average approach, gives no information on individual slug positions, individual slug lengths and local liquid holdup distribution as functions of time.

The current paper focuses on determination of slug characteristics, i.e. slug lengths and slug frequencies, by use of the OLGA slug tracking model recently developed by IFE and SINTEF (Bendiksen *et.al (1990)* and Straume *et.al (1992)*). The slug tracking model is a hybrid Lagrangian - Eulerian scheme where a Lagrangian type front tracking scheme is superimposed on the Eulerian scheme in the standard OLGA model. Each slug tail and front are described with Lagrangian coordinates, giving the position of the tail and front as functions of time. The results of the computations are compared with multiphase flow measurements from the Sidi El Itayem field, (Todal and Dhulesia, *1992)*, performed within the POSEIDON program.

POSEIDON is a cooperation between Statoil and Total which includes the Flow Measurement project. The objective of this project has been to provide high quality measurements from multiphase pipelines in operation. The results are used in verification of our multiphase flow models and to obtain a better technological base for development and operation of oil/gas fields by use of multiphase flow technology. The project started in 1987. Measurements at the Sidi El Itayem field were performed in June 1991.

THE SIDI EL ITAYEM FIELD - A BRIEF DESCRIPTION

The Sidi El Itayem (SIT) Field is situated on shore near the city Sfax in Tunisia. The field is operated by Compagnie Franco-Tunisienne de Pétroles (C.F.T.P).

There are two parallel pipelines (6" and 4") leading unprocessed well-stream to the production centre. A unit for testing of multiphase pumps is installed at the inlet to the pipelines. This system includes a test separator and a multiphase pump with a complete instrumentation and data logging system. A process flow diagram is shown in Figure 1. The oil-, water- and gas phases are mixed downstream the test separator at the inlet to the pipeline.

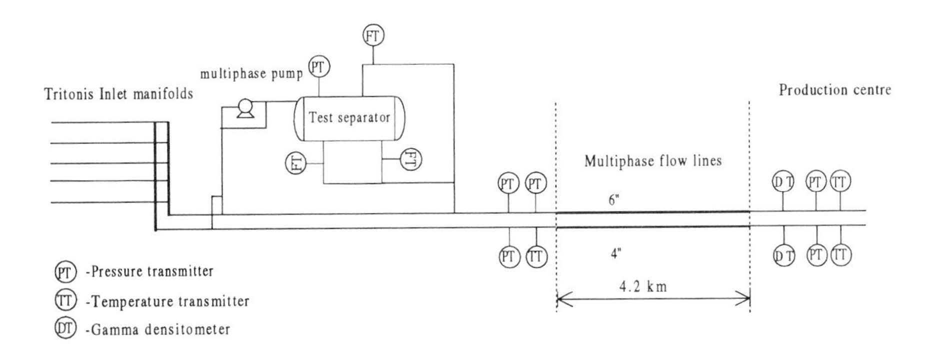

Figure 1 *Sidi El Itayem - Layout of the flow lines with instrumentation*

The two pipelines, together with the lines from other structures enter a manifold at the production centre. The pressure in the outlet manifold is maintained fairly constant at 5 bar by the downstream process.

The POSEIDON Flow measurements were carried out in June 1991. The measurements were based on pressure and temperature transmitters, orifice plates for the flow rates and a gamma densitometer for local liquid holdup measurements. Instrument locations are shown in Figure 1.

The fluid of Sidi El Itayem is a three phase mixture with a very high watercut (~ 75 %)

Main pipeline parameters during the measurement campaigns are listed in Table 1.

Table 1 Sidi El Itayem 6" pipeline - Operational parameters during the measurement campaigns

Oil flow rate	1-3	Sm^3/h
Water flow rate	2.5-12	m^3/h
Gas flow rate	130-2500	Sm^3/h
GLR	17-193	(-)
Inlet pressure	7-13	bar
Outlet pressure	5	bar
Inlet temperature	30-55	°C
Outlet temperature	25-35	°C

THE SLUG TRACKING MODEL

The standard OLGA model (Bendiksen *et.al 1991*) treats slug flow by an average approach, where slug flow is assumed to consist of an infinitely long train of fully developed slug units. This gives a correct average pressure drop and holdup, but no information on slug characteristics as slug length, individual slug velocity or frequency.

The OLGA slug tracking model (Bendiksen *et.al 1990*, Straume *et.al 1992)* is a hybrid Lagrangian - Eulerian scheme where a Lagrangian type front tracking scheme is superimposed on the Eulerian scheme in the standard OLGA model. Each individual slug tail and front is described with Lagrangian coordinates, giving the position of the tail and front as a functions of time. The model has been verified against large scale startup and dynamic terrain slug experiments with good results (Straume *et.al 1992*) .

When the slug tracking model introduces a slug with a front and a tail in a pipeline section, the liquid slug is treated as dispersed bubble flow and the slug bubble as separated flow (stratified or annular). The definition of the slug tail and front are as shown in Figure 2. These are defined as the position along the pipeline where the flow changes between dispersed and separated flow.

The present slug model handles three different slug initiation models. A dynamic terrain slug model, a startup slug model and a hydrodynamic slug model. In the hydrodynamic slug model, slugs may be initiated when the standard OLGA scheme predicts transition to slug flow.

After initiation, terrain slugs, startup slugs and hydrodynamic slugs are treated equally.

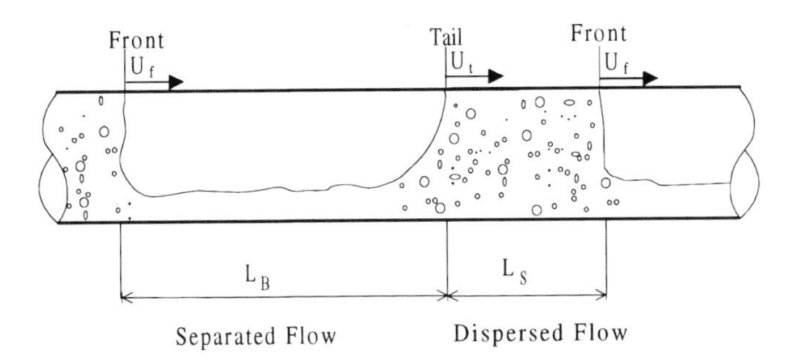

Figure 2 *Definition of slug tail and slug front*

Slug tails and slug fronts are treated separately. They are allowed to propagate in different directions and may independently change propagation direction during the simulation.

There are two procedures for calculating the tail or front velocity. The model distinguishes between a Taylor bubble and a level or braking front type. Transition between the two types are dynamic and controlled by the pipe inclination , propagation direction and liquid velocity in the slug.

In principle,when considering hydrodynamic slug flow, the slug tail is characterized by a slug bubble which propagates into the liquid slug. The velocity is then given by the slug bubble velocity correlation of Bendiksen (1984):

$$U_B = C_0 U_{ls} + U_0$$

Here C_0 is a constant depending on the inclination of the pipe, U_{ls} is the liquid velocity in the slug and U_0 is the inclination dependent bubble nose velocity in stagnant liquid. The slug front is characterized by a breaking front and the slug front velocity, U_f, is given by a volumetric flow balance at the slug front moving to the right in Figure 2, see Straume *et.al* (1992) for details.

Once the tail and front velocities are calculated, their positions are found from the following connection between the Lagrangian and Eulerian coordinates:

$$\frac{dZ}{dt} = U$$

The slug tracking scheme include models for removing and merging of slugs. A slug is removed from the pipeline when the slug has decayed, and merging of two slugs takes place when the slug bubble has decayed. In adition, the present slug tracking model is not able to handle more than one front and one tail in one pipeline section. Due to this merging of slugs will also take place when we have more than one tail and one front in a section volume.

MODELLING OF THE PROBLEM

This chapter will explain how the physical problem is modelled through the input to the program. The multiphase flow simulations were performed with OLGA 91, the 1991 version of the dynamic simulation program for two-phase flow developed by IFE and SINTEF.

To reduce the computer time, minor terrain effects have been neglected, while the total amount of upward and downward elevation is kept constant. A comparison of the original and the simulated profile is shown in Figure 3. The pipeline was divided into 118 sections for the steady state calculations and 222 sections when the slug tracking option was used.

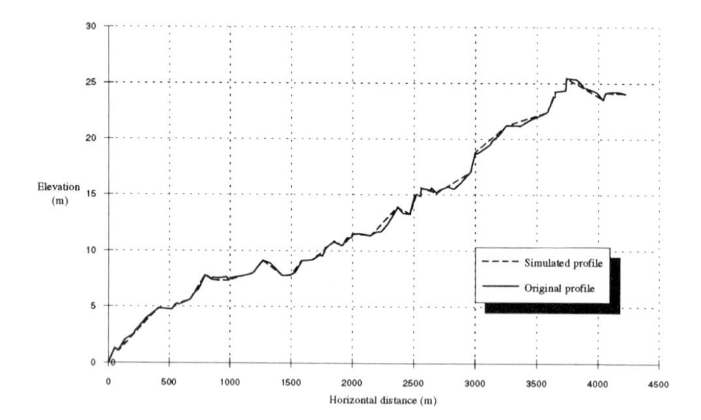

Figure 3 *Sidi El Itayem - Pipeline Topography*

The production on mass basis and the inlet temperature was specified as inlet boundary conditions. The outlet conditions were specified by the pressure at the outlet manifold. For the steady state calculations, the pipeline was simulated with heat transfer through the walls. The wall temperature was initialized by the code. All boundary conditions were kept constant as a function of time.

We have no readings of the temperature outside the pipeline during the measurements, and due to this the outside temperature was tuned so that the temperature drop in the pipeline agreed with the measurements.

Temperature calculation are currently not available in the slug tracking model so the temperature field obtained from the steady state calculations is used.

The fluid property file was generated with the Statoil Thermodynamic Estimation Program, STEP. (Inhouse computer package delivered by CALSEP A/S). The mixture composition is found by remixing the composition of the separator oil, the separator gas and pure water to user specified Gas Liquid Ratios (GLR). Three different production conditions corresponding to GLR of 53, 110 and 193 have been simulated.

During the measurements the liquid flow rate was kept almost constant.

RESULTS

Both steady state calculations with standard OLGA and dynamic simulations with the slug tracking model have been performed.

Special care had to be taken in determination of the effective viscosity of the oil and water mixture, because the mixture was expected to form emulsions. In the thermodynamic estimation program, the liquid viscosity of a mixture of hydrocarbons and water is calculated by linear interpolation between the viscosity of the oil phase and the water. This means that special effects like emulsions must be introduced manually.

The PETRECO 5" flow simulator (Figure 4) was used to find representative values for the viscosity of the water/oil mixture at different temperatures and flow rates. All tests were performed on a live oil system with composition as in the actual flow line.

The principle for such measurements is to expose the fluid to a realistic mechanical and thermal treatment while measuring the torque on the wheel shaft. The effective viscosity is calculated based on results from these torque measurements.

A comparison of the measured and calculated viscosity is shown in Figure 5. In this test the temperature was varied at a constant velocity of 3.5 m/s using a watercut of 70%. This corresponds to the measurements with GLR = 53. The measurements showed that a 70-75% watercut mixture resulted in an effective viscosity of more than 10 times that of oil at 30°C. For higher GLR the mixture velocity is larger, and the viscosity will be even bigger.

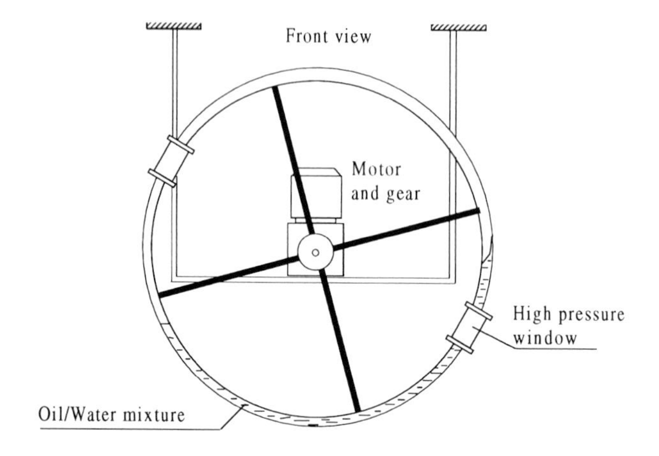

Figure 4 *Sketch of the PETRECO flow simulator.*

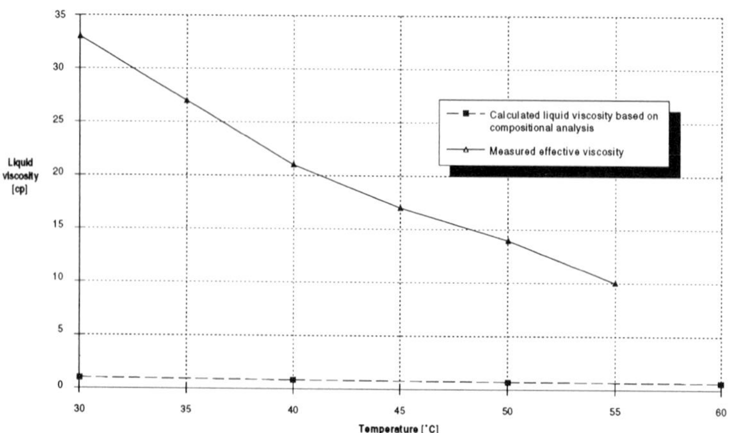

Figure 5 *Effective mixture viscosity vs temperature.* $U_{l,mix}$ *=3.5 m/s, WC=70% and GLR = 53*

Steady State Calculations

The steady state simulations were performed to verify the pressure drop calculations, and to study the influence of the effective viscosity. During the measurements the liquid flow rate was approximately constant but the gas flow rates were changed giving three different gas-liquid ratios.

The results from the pressure drop calculations are shown in Table 2. The results are obtained by using the measured effective viscosity as shown in Figure 5. When the viscosity from a standard thermodynamic estimation program is used the pressure drop is under-estimated by up to 45%.

Table 2 Steady state pressure calculations

GLR [-]	**53**	**110**	**193**
Measured ΔP [bar]	2.95	4.90	7.27
Calculated ΔP [bar]	2.67	4.25	6.67
Error [%]	-9.5	-13.3	-8.3

Standard OLGA predicts slug flow in the whole pipeline, for all the simulated flow rates, except in some pipe sections with downward inclination. Because standard OLGA has a momentary flow regime transition, a slug flow regime will not survive a pipe section where OLGA predicts stratified flow.

Calculations with the OLGA slug tracking model

In the following, calculations with the OLGA slug tracking modell will be compared with data from slug characteristic measurements from the Sidi El Itayem Field. In the measurements, a gamma densitometer was used to obtain slug statistics at the outlet of the 6" pipeline, at the production centre (Figure 1).

Slug duration distribution from the measurement campaign with GLR=110 is shown in Figure 6. As shown in the histogram, the slug length at the outlet of the pipeline seems to follow a unimodal, unsymmetrical, probability distribution, such as a Log Normal or Inverse Gauss distribution. The histogram represents seven and a half hour of measurements.

Figure 7 shows the slug duration histogram after two and a half hours simulated time, with the OLGA slug tracking model, with the same production conditions. As seen, the model almost reproduces the same maximum and mean slug length as measured, but the shape of the histogram and number of slugs are different. The model predicts almost twice as many slugs as detected in the measurements. The present hydrodynamic slug initiation model is preliminary and needs further development. The initiation mechanism for real hydrodynamic slugs is complex and includes a random character.

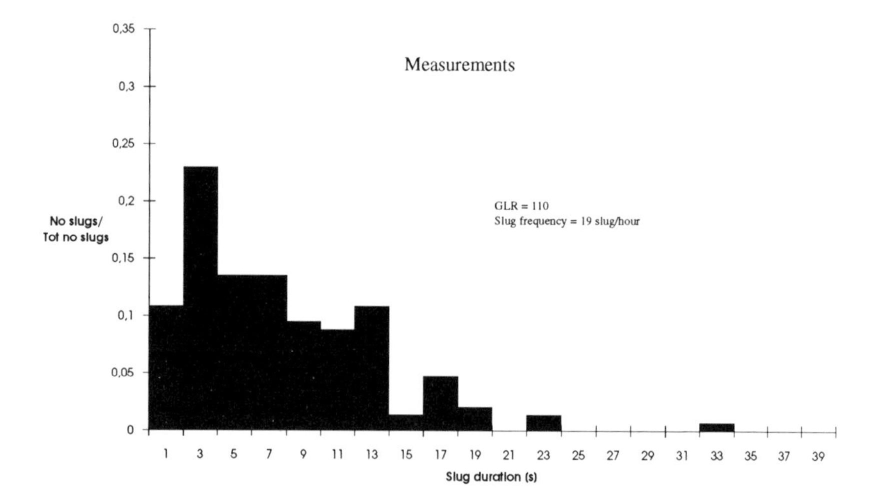

Figure 6 *Measured slug duration distribution.*

The histogram of measured slugs in Figure 6 is based on readings of densities as shown in Figure 8. A special computer program was developed to detect slugs from the measured data files. The program scans the data files and detects slugs having a density above a preset value. The slug duration is recorded as the time the density remains above this value. Slugs with a maximum density below 500 kg/m^3 and/or with a duration shorter than 1 second are rejected. The calculated liquid holdup as a function of time at the outlet of the pipeline is shown in Figure 9. The OLGA slug tracking model gives directly the slug length of different slugs passing a given pipeline position. In the same way as for the measurements, slugs with duration shorter than 1 second are rejected.

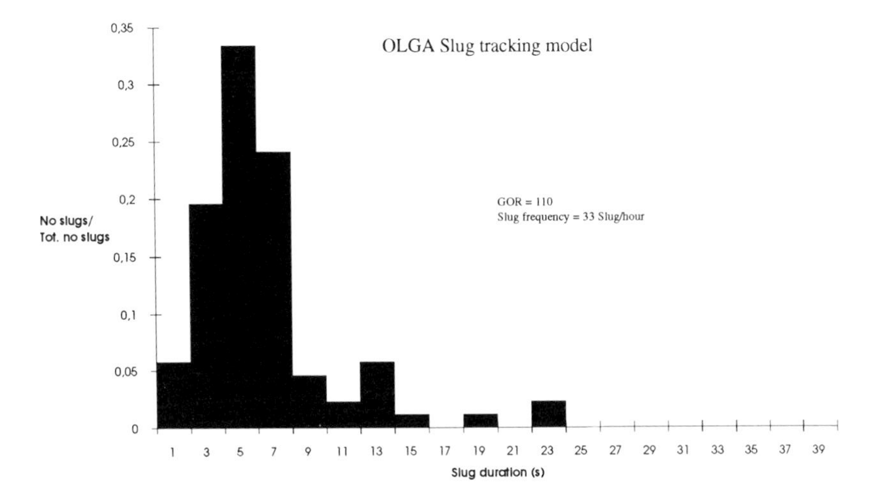

Figure 7 *Slug duration distribution calculated by the OLGA slug tracking model*

The simulated time series in Figure 9 is not intended to be a replica of the measured time series in Figure 8. The purpose of the simulation is to obtain statistical values. Further, the zero time point is chosen independently for the two time series. Therefore, the two time series should not be compared directly slug by slug along the time axis.

Pressure variation at a given position, near the pipeline exit when slugs are passing, are shown in Figure 10 for the measurements and in Figure 11 for the simulations. As seen, the pressure variation over a slug is up to one bar in both the calculations and the measurements. The increased pressure represents the pressure drop over a slug unit, and consequently the pressure variation increases with slug length and velocity.

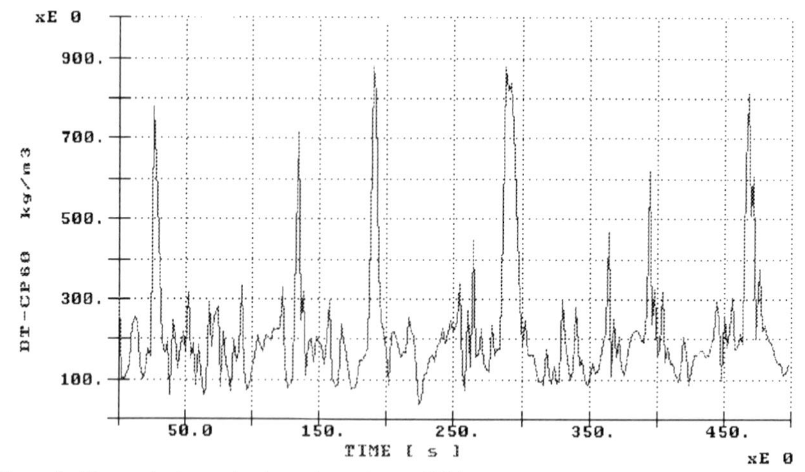

Figure 8 Measured mixture density at the outlet manifold.

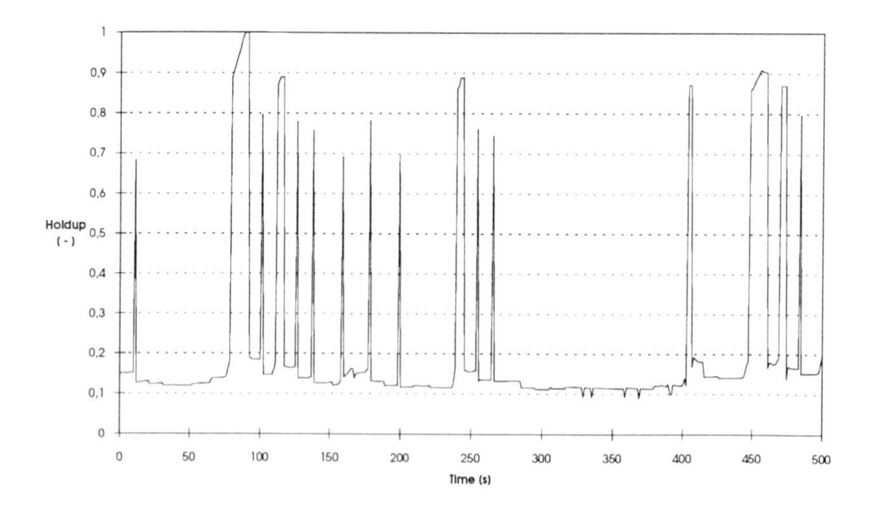

Figure 9 *Calculated Liquid holdup vs time at the outlet of the line*

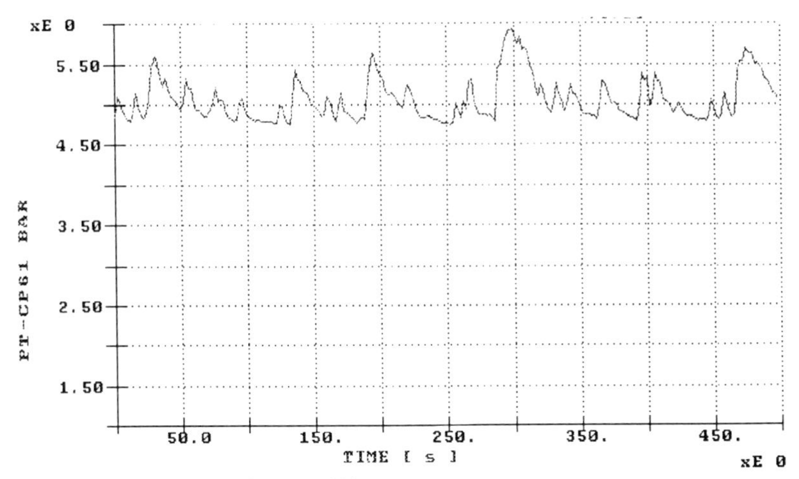

Figure 10 *Outlet Pressure at CP - manifold.*

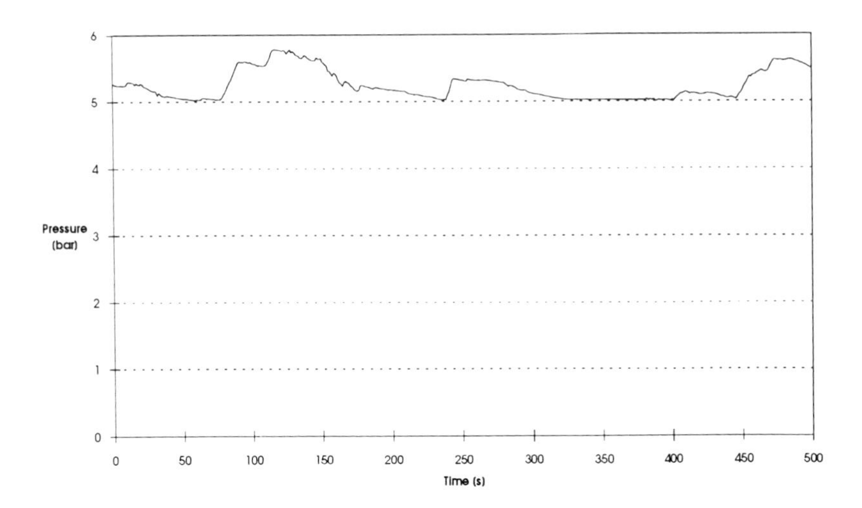

Figure 11 *Calculated pressure close to the outlet of the line.*

The variation of the predicted slug length for a typical slug is shown in Figure 12 as a function of the slug front position. The slug is developed from a very small slug, at about 800 metres, and increases to a maximum length of almost 70 metres before it starts to decrease. Most typically the slug length decreases in upward inclined sections and increase when the slug is passing dips along the pipeline. The slug growth is a function of the holdup and the velocity in the liquid film in front of the slug. Because liquid accumulates in the dips the slug is growing.

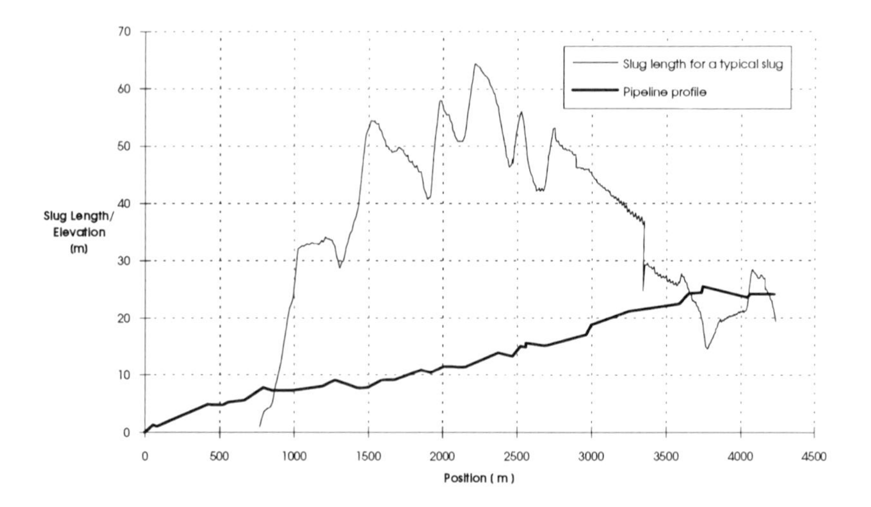

Figure 12 *Slug length for a typical slug at different pipeline positions*

The tail and front velocities of the same slug are shown in Figure 13. As shown in the figure, the tail velocity is more stable than the front velocity which has some peaks probably due to numerical weakness when two slugs are merging.

As shown in Figure 14, the predicted slug front velocity follows a symmetrical distribution. We do not have velocity measurements from Sidi El Itayem, but measurements from other pipelines show that the slug velocity follows a symmetrical distribution.

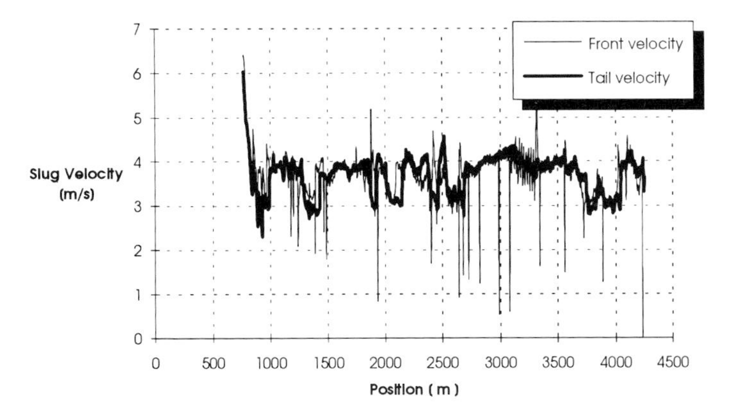

Figure 13 *Front and tail velocity for the same slug at different pipeline positions.*

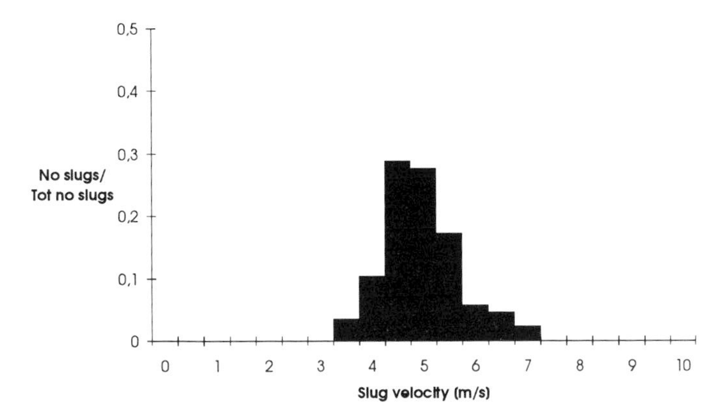

Figure 14 *Calculated slug velocity distribution at the outlet of the pipeline.*

CONCLUSIONS

The OLGA slug tracking model has been compared with multiphase flow measurements obtained at the Sidi El Itayem field in Tunisia.

The main observations can be summarized as follows.

- In pressure drop calculations with OLGA, it is very important to include fluid property effects as emulsions when calculating the effective viscosity.
- The slug tracking simulations show that the OLGA slug tracking scheme is able to initiate and track individual slugs. The model reproduces slug length distributions similar to those obtained in field measurements.
- The OLGA slug tracking model gives higher slug frequency than observed in the measurements.
- The slug velocity distribution predicted by OLGA follows a symmetrical distribution.

ACKNOWLEDGEMENTS

The author would like to thank STATOIL and TOTAL for the permission to publish this paper.

REFERENCES

Bendiksen, K.H. 1984 "An Experimental Investigation of the Motion of Long Bubbles in Inclined Tubes" *Int. J. Mult. Flow,* Vol. 10, No. 4, pp. 467-483.

Bendiksen, K., Malnes, D., Straume, T & Hedne, P. 1990. "A non-diffusive numerical model for transient simulation of oil-gas transportation systems." *European Simulation Multiconference,* Nuremberg, 10-13 June.

Bendiksen, K., Malnes, D., Moe, R. & Nuland, S. 1991. "The dynamic two fluid model OLGA: Theory and applications." *SPE Production Engineering*, May 1991, pp. 171-180

Straume, T., Nordsveen, M. and Bendiksen, K.,1992 "Numerical simulations of slugging in pipelines" *ASME int. symp. on Multiphase flow in wells and pipelines,* Anaheim, 8-13 Nov. 1992

Todal, O. and Dhulesia, H., 1992 "Data acquisition on multiphase pipelines" *Multiphase Transportation III*, Røros, 21-22 Sept.

EVOLUTION OF SLUG FLOW IN LONG PIPE

Jean FABRE, Philippe GRENIER and Emilie GADOIN
Institut de Mécanique des Fluides, UA CNRS 005
Avenue Camille Soula, 31400 Toulouse, France

INTRODUCTION

Gas liquid slug flow is an alternation in space and time of long bubbles and liquid slugs (figure 1). The set of a slug and a long bubble forms a cell which propagates along the pipe. One of the peculiarities of slug flow is its chaotic nature, the bubble and slug length, and at a less extend their velocity, being randomly distributed. It is generally acknowledged that the mechanisms which control the structure of slug flow are not yet satisfactorily understood. Different arguments may given to account for this. On the one hand slug flow is a chaotic process so that no deterministic model is expected to apply. On the other hand the slug flow models devoted to predict the pressure gradient and liquid holdup require limited information upon the flow structure so that little effort has been made for its understanding.

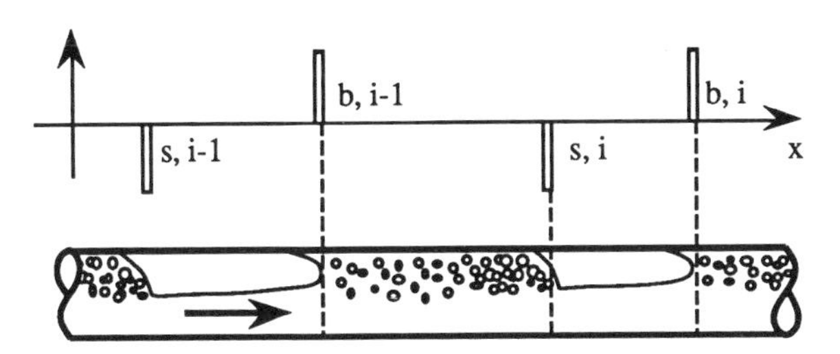

Figure 1. Slug flow: identification of bubble and slug fronts

Among the various characteristics of the flow structure, slug length has probably been the most investigated (see the review by Fabre & Liné, 1992). In horizontal flow, the mean values lies between 20–40 pipe diameters (Dukler & Hubbard 1975; Nicholson et al, 1978; Ferré 1979; Ferschneider, 1982; Nydal et al, 1992). In addition, statistical analyses have been carried out to determine the probability density distribution of slug length: the experimental results of Brill et al (1981) are well fitted with a log-normal distribution whereas the theoretical study of Dhulesia et al (1991) gives credit to an inverse gaussian distribution.

In this paper the following question is addressed. Let us consider, the case of a gas-liquid mixture flowing in a very long pipe, supplied with constant inlet flow rates. This is a classical situation which is encountered in hydrocarbon transportation. When slug flow occurs, the sequence of liquid slugs and long bubbles created near the inlet, are transported by the flow. The flow structure far from the inlet thus results partly from the entrance mechanism and partly from the evolution experienced by the structure during its travel along the pipe. If no evolution happens, the structure keeps the memory of the entrance and one must focus upon the transient flow at this point to predict the statistical quantities. In contrast, if the space evolution is strong, the main issue is to know whether a fully developed solution, independent from entrance, exists. We have carried out air-water experiments in long pipe to know whether slug flow is developing and how fast.

EXPERIMENTAL FACILITY, VOID MEASUREMENT AND PROCESSING

The structure of slug flow has been experimentally investigated in a long horizontal pipe of 5 cm inside diameter and 90 m length. Air and water were injected into the pipe through a mixer made of a Y junction. Water was supplied from a centrifugal pump and air from a rotative compressor. The flow rates of each phase were measured with two sets of orifice plates.

A special attention was paid to the space evolutions of the velocity and length of both long bubbles and liquid slugs. It was thus necessary to determine their occurrence. This was made from the instant output signal of conductive probes placed in each of the 4 test sections, located at 25 m, 45 m, 65 m and 89 m from the mixer: these sections will be further referred to as section 2, 3, 4, 5. The probes are similar to these of Andreussi et al (1988). Each probe is made up of 2 stainless steel rings flush mounted at the wall, 5 cm away from each other. The variation of the conductivity of the liquid with temperature is compensated by using a reference probe fully immersed in the liquid upstream the mixer. The probe conductance is converted to a DC voltage through a differential home made amplifier.

The probe signals were recorded on the hard disk of a HP1000 computer, during a period $T=N/f_e$, where f_e is the sampling frequency and N, the maximum number of samples allowed by the size of memory. This frequency was taken nearly equal to V_e/d, where V_e is the estimated velocity of the cells and d, the distance between the probe rings: this choice leads to a good compromise between the space resolution of the probes (f_e as great as possible) and the choice of the period (T as great as possible). During this period which was nearly equal to 30-60 mn, the passage of about 1000 cells was recorded.

In order to identify the bubble and slug front, a numerical procedure was applied to determine for each run the void threshold above which a long bubble is present: this was made from the probability distribution of void which is clearly bimodal for the cases which were investigated. Thus a preliminary experiment was done for each run, in order to determine this distribution: moreover, it was also used to determine the mean cell velocity V_e from the cross-correlations of the probe signals.

For each run, the occurrences of bubble and slug fronts in each test section can be recognized from the time evolution of void. They form 4 sets of events corresponding to the 4 different probe locations, the goal being to identify the front each event comes from. The method was to find whether for each front i arising at time t_i at the section m, there exists a front j at time t_j at the section n, such as:

$$t_i(x_m) + \frac{(x_n - x_m)}{(V+\Delta V)} < t_j(x_n) < t_i(x_m) + \frac{(x_n - x_m)}{(V-\Delta V)} \quad (1)$$

where ΔV is the range of expected velocity which must be twice to three times the standard deviation of the cell velocities. It is thus possible to determine, from their transit time between 2 successive test sections, the individual velocities V_{bi} of bubble-i and V_{sj} of slug-j. Having the time of passage, the bubble and slug lengths, L_{bi}, L_{sj} are deduced. One obtains a collection of individual lengths and velocities which may be analyzed from statistical point of view, the goal being to determine whether slug flow is developing over a long distance, and if it does, which mechanism controls this development.

In this paper the experimental results corresponding to 3 different runs are presented (see table 1). The superficial gas velocity $U_{GS,at}$ is calculated at atmospheric outlet condition, in contrast to U_{GS} which represents the local superficial velocity accounting for density correction. U_m is the mixture velocity.

	Run 2	Run 4	Run 5
$U_{GS,at}$ (m/s)	1.25	0.37	0.94
U_{LS} (m/s)	0.97	0.50	0.50
U_m (m/s)	2.22	0.87	1.44

Table 1. Run conditions

EXPERIMENTAL RESULTS ON BUBBLE AND SLUG MOTION

Each front moves at velocity V_i: the average $\overline{V}$ and standard deviation σ_v of bubble and slug velocity were determined (see table 2) from:

$$\overline{V} = \frac{1}{N} \sum_{i=1;N} V_i \qquad \sigma_v^2 = \frac{1}{N-1} \sum_{i=1;N} (V_i - \overline{V})^2$$

		Bubble			Slug		
		2—3	3—4	4—5	2—3	3—4	4—5
Run 2	$\overline{V}$	2.390	2.495	2.579	2.395	2.494	2.571
	σ_v	0.0178	0.0121	0.0108	0.0220	0.0167	0.0136
Run 4	$\overline{V}$	1.126	1.144	1.146	1.130	1.143	1.147
	σ_v	0.0077	0.0068	0.0066	0.0110	0.0087	0.0072
Run 5	$\overline{V}$	1.665	1.705	1.717	1.667	1.705	1.716
	σ_v	0.0126	0.0131	0.0107	0.0174	0.0144	0.0109

Tableau 2. Bubble and slug velocities: averages and standard deviations in m/s

It may be shown that the mean velocities of both bubbles and slugs are remarkably equal in each test sections. They increase slightly with the distance from inlet. This effect comes from the pressure drop which tends to decrease the gas density and to increase the superficial gas velocity. It can be predicted provided that the superficial gas velocity is corrected for the variation of gas density. A comparison between calculated and measured values is shown in figure 2 for illustration: the calculations have been performed with the Tacite code (Pauchon et al, 1993).

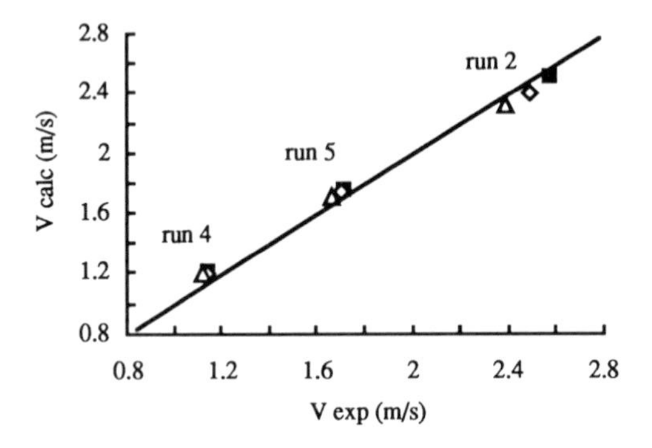

Figure 2. Comparison between experimental and calculated bubble velocity.
△ 2—3; ◇ 3—4; ■ 4—5

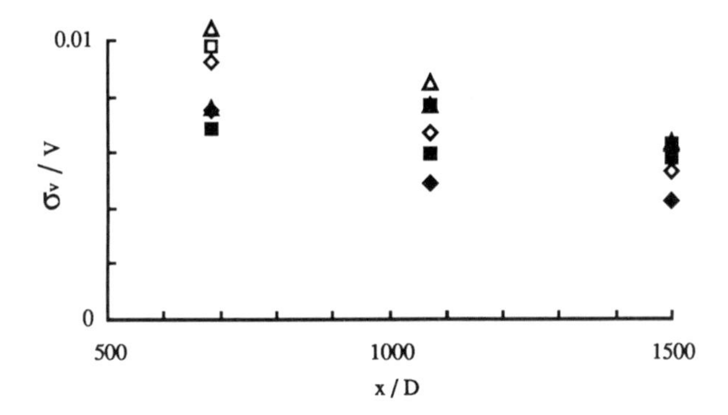

Figure 3. Space evolution of standard deviation of velocities.
Open symbols refer to slugs; solid symbols refer to bubbles; ■□: run 2; ◆◇: run 4; ▲△: run 5

The ratio of the standard deviation to the average σ_V/V decreases slowly with x: it is given in figure 3 versus x/D, where D is the pipe diameter. Although, this ratio is greater for slugs than for bubbles near the inlet, the difference tends to decrease when the distance increases. At the last test section, σ_V/V takes nearly the same value for bubbles and slugs: the equality could be the indication that the flow is fully developed.

That σ_V is greater for slugs than for bubbles is explained from the probability distribution of velocity plotted in figure 4 for one of the runs, at different distances from inlet. It appears clearly that there exist a few slugs whose velocity is greater than the average.

Figure 4 shows that these slugs move at a velocity up to 5% greater than the average (≈ 5 σ_V): they are in fact the signature of coalescence between bubbles as shown below.

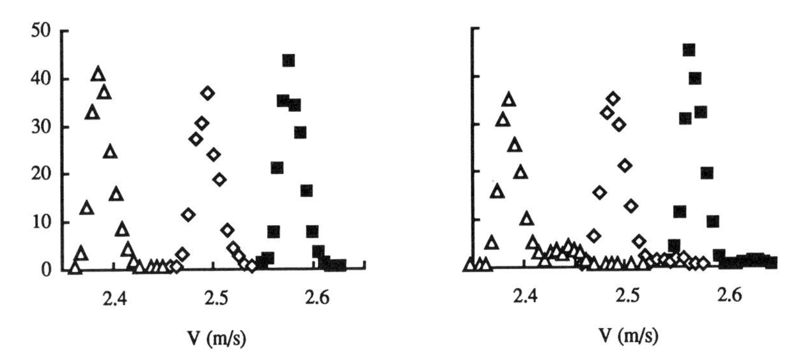

Figure 4. Probability density distribution of velocities for run 2.
Left: bubble velocity; right: slug velocity. △: 2–3; ◇: 3–4; ■: 4–5

The randomness of bubble and slug velocity is not yet understood. Let us define ΔV_i the difference of velocity of the individual i with respect to the average:

$$\Delta V_i = V_i - \overline{V} \tag{2}$$

It is a priori expected that ΔV_i depends upon x. A first possibility could be that the bubble and slug velocities are fixed at the location where the slug are formed (entrance effect), each of them keeping the same difference with respect to the average during their travel (ΔV_i is constant): in such a case the fastest fronts would overtake the slowest and one would expect a decrease of σ_V. A second possibility is that the velocity of each front fluctuates during its travel, in a like random manner. This question may be answered by determining the integral length scale over which the front motion is correlated. This length scale L_V can be defined by:

$$L_V = \int_x^{\infty} R_{VV}(x, x+\xi)\, d\xi \tag{3}$$

$$R_{VV}(x_p, x_q) = \frac{1}{\sigma_{vp}\sigma_{vq}} \frac{1}{N-1} \sum_{i=1;N} \Delta V_{ip}\, \Delta V_{iq} \tag{4}$$

where R_{VV} is a cross-correlation coefficient of ΔV_i, V_{ip} the velocity of the front i at distance x_p, and V_{iq}, the velocity of the same front at distance x_q. The results of R_{VV} are given in table 3 for 2 different locations which is not enough for L_V to be determined. However the values of R_{VV} can be interpreted as follows. If ΔV_i remains constant with x, one expects that R_{VV}=1. If ΔV_i changes over a distance typically smaller than $x_p - x_q$, thus R_{VV} must be 0. The results of table 3 show that there is a tendency for bubbles, and at a less extend for slugs, to keep constant their velocity difference with respect to the average.

	Bubble $R_{VV}(x_3,x_4)$	$R_{VV}(x_4,x_5)$	Slug $R_{VV}(x_3,x_4)$	$R_{VV}(x_4,x_5)$
Run 2	0.41	0.17	0.39	0.27
Run 4	0.60	0.60	0.38	0.61
Run 5	0.62	0.52	0.38	0.49

Table 3. Values of the correlation coefficient $R_{VV}(x_i, x_j)$.

Experimental Results on Bubble and Slug Lengths

As reported by several authors, the slug lengths are in the range 10—40 diameters. They are reported in figure 5 together with the results of Théron, obtained in the same facility. We are concerned here by the space evolution. The results are summarized in table 4. We observe that the mean bubble length increases in contrast to the mean slug length which remains nearly constant. The increase in bubble length results mainly from gas expansion. The standard deviation shows that both bubble and slug lengths are very dispersed about their average. For the bubbles it increases with the distance from the inlet, whereas it decreases for the slugs. The ratio σ_L/L is plotted with x/D in figure 6. It remains constant for the bubbles whereas it decreases for the slugs. This suggests that even at a distance of 1500 D the flow is not yet fully developed, excepted perhaps for the run 4 corresponding to a low mixture velocity.

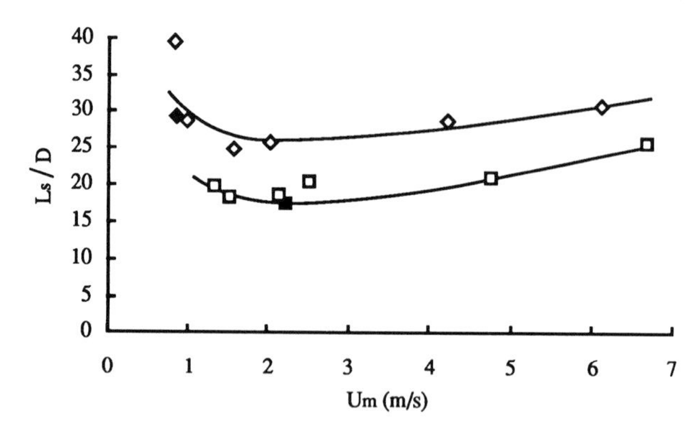

Figure 5. Slug lengths. ◆◇: U_{lS}=0.5 m/s; ■□: U_{lS}=1 m/s
Open symbols: Théron results (1989). Solid symbols: present results

The probability density distributions of bubble and slug lengths are plotted in figure 7. The space evolution is not clearly visible for slug lengths. For bubble the probability distribution seems to be shifted to the right with x. This corresponds to the effect of gas expansion which increases the length of each bubble as the inverse of pressure.

		Bubble			Slug		
		2—3	3—4	4—5	2—3	3—4	4—5
Run 2	L	1.481	1.632	1.816	0.856	0.906	0.876
	σ_L	0.437	0.483	0.525	0.282	0.251	0.229
Run 4	L	1.391	1.478	1.500	1.457	1.466	1.490
	σ_L	0.463	0.477	0.492	0.310	0.269	0.219
Run 5	L	3.128	3.260	3.326	1.083	1.111	1.104
	σ_L	0.992	1.031	1.065	0.304	0.261	0.229

Table 4. Bubble and slug lengths: averages and standard deviations in m.

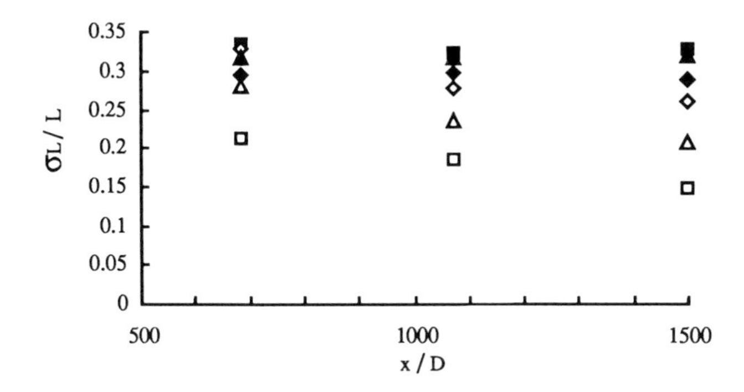

Figure 6. Space evolution of standard deviation of lengths.
Open symbols refer to slugs; solid symbols refer to bubbles; ■□: run 2; ◆◇: run 4; ▲△: run 5

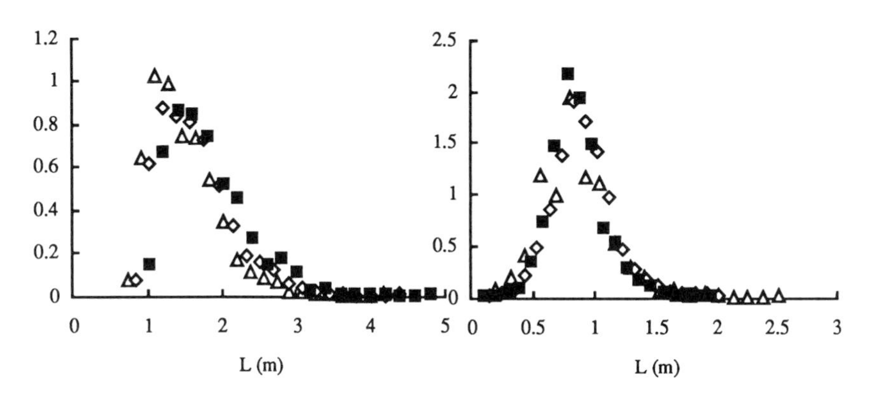

Figure 7. Probability density distribution of lengths for run 2.
Left: bubble length; right: slug length. △: section 3; ◇: section 4; ■: section 5

BUBBLE COALESCENCE

In following the fronts from one test section to another, it can be seen that some of them vanish: bubble coalescence occurs by slug disappearing. The rate of coalescence r defined by the ratio between the number of vanishing fronts over the total front number, is plotted in figure 8 versus the distance. On the one hand the higher the liquid velocity, the greater the rate of coalesce: r can take values as high as 10%. On the other hand the number of vanishing fronts decreases when the distance from the inlet increases: r was found to vary like x^{-2}. If we assume that the effects of both viscosity and surface tension are not dominant, r is a function of x, U_{LS}, D and the gravity g only. Thus we seek for a dimensionless relationship of the form $r=f(U_{LS}^2/gD, D/x)$. Taking into account that r is proportional to x^{-2} this relation was found to reduce to:

$$r = 5.10^4 \frac{D\,U_{LS}^2}{g\,x^2} \tag{5}$$

from the experimental results. Although these results are yet too scarce to confirm the validity of eq. (5), this suggests that the rate of coalescence increases like the square of the superficial liquid velocity. However there is no proof concerning the influence of the pipe diameter.

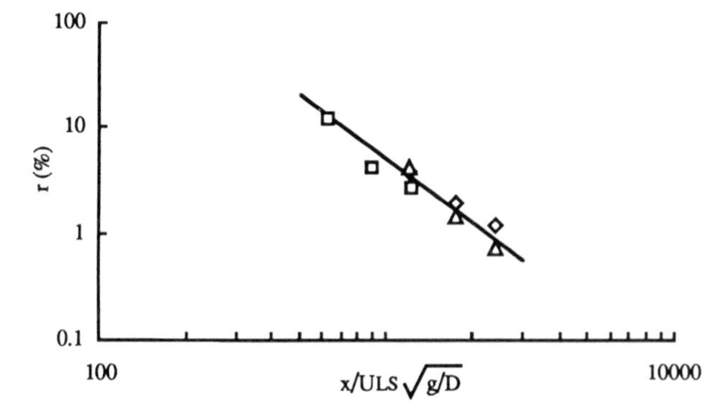

Figure 8. Rate of coalescence versus the distance;
□: run 2; ◇: run 4; △: run 5; —: eq (5)

When coalescence occurs, two bubbles of length L_{b1} and L_{b2} merge for generating a single bubble of length L_b. Figure 9 shows that the length of the resulting bubble is greater than the sum of the lengths of the two initial ones: this suggests that when two bubbles are merging they also capture a part of small gas bubbles carried out by the vanishing liquid slug. When a slug of given initial length L_{s1} disappears, the liquid feeds its next neighbour of initial length L_{s2}. After coalescence has occurred this slug has a length L_s. Figure 9 shows that L_s is smaller than $L_{s1}+L_{s2}$. This confirms that some gas contained initially in the slugs has escaped into the long bubbles. Thus bubble coalescence tends to create larger structures as it is a priori expected.

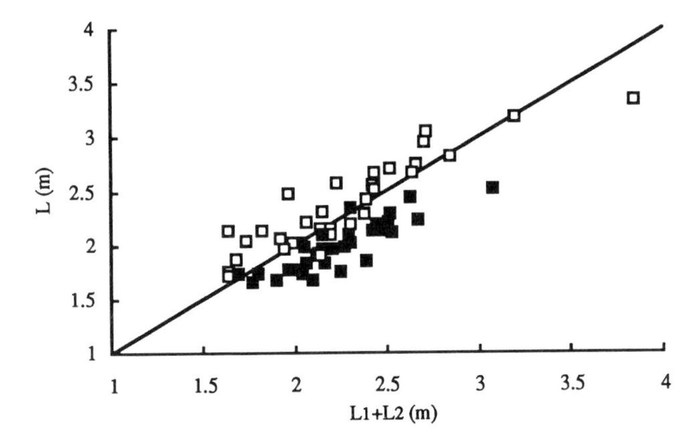

Figure 9. Length of the resulting slug ■ or bubble □
versus sum of lengths of the two combining slugs or bubbles (run 4)

It is of course anticipated that only smallest slugs may disappear between two consecutive test sections. The probability for a slug of given length L_{s1} to disappear between two test sections distant of about 20 m has been plotted in figure 10. This leads to the conclusion that slugs whose length to average ratio is smaller than 0.4 do not survive after this distance contrary to those for which it is greater than 0.7. This is an important result which has to be confirmed for different flow conditions with more contrasted inlet flow rates.

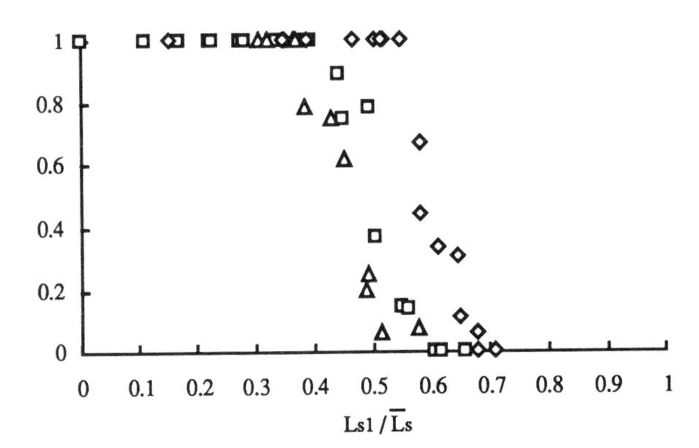

Figure 10. Probability for slug of given length to disappear; □: run 2; ◇: run 4; △: run 5.

Statistical Correlations between Lengths and Velocities

Let us define the correlation coefficient R_{XY} between two physical quantities X and Y of the same event i, measured in the same test section x_p (see figure 1 for the definition of event i). R_{XY} is an evaluation of the statistical dependance between X and Y whose value lies between 0 and 1. The higher R_{XY}, the stronger the interaction between the two quantities.

$$R_{XY}(x_p, x_p) = \frac{1}{\sigma_{Xp}\sigma_{Yp}} \frac{1}{N-1} \sum_{i=1;N} (X_{ip} - \overline{X_p})(Y_{ip} - \overline{Y_p}) \qquad (6)$$

The values of the correlation coefficient are given for different quantities and for run 2 in table 5. The other runs exhibit however the same trends. No correlation is observed between the bubble length and its velocity: this conclusion has been already pointed out by Nicklin et al (1962) for vertical flow. An interesting and unexpected result is the fact that the lengths of the bubble and the following slug are correlated. For the other cases the correlation coefficients are poor to moderate. In particular the velocities of both bubble and next slug are not independent as shows the value of $R_{V_bV_s}$.

	$X=V_b$	$X=V_s$	$X=L_b$	$X=L_s$
$Y=V_b$	1			
$Y=V_s$	0.52	1		
$Y=L_b$	0.04	0.36	1	
$Y=L_s$	0.32	0.56	0.83	1

Table 5. Correlation coefficient between length, velocity of bubble, slug. Run 2.

This coefficient has been plotted in figure 11 versus the rate of coalescence, for the three runs and the different test sections: it can be shown that the smaller the rate of coalescence the stronger the correlation coefficient between the velocity of a bubble and this of the next slug. A similar conclusion holds for the other coefficients, so that when the coalescence process takes place the physical quantities are poorly correlated. This is expected to occur in short pipe at high liquid rate. In contrast, far from the inlet, the coalescence decreases, the flow structure becomes fully developed and some quantities become dependant from each others.

In order to illustrate the correlation between the velocities of the bubble and the next slug, the 3D histogram has been plotted in figure 12. The density is proportional to the number of bubble fronts having the velocity V_b whereas the velocity of the next slug is V_s. There exist some points inside the dashed rectangle which have been identified to the fronts between which coalescence has occurred. For these point the bubble velocity is close to the average whereas the velocity of the next slug is greater than the average: after coalescence the next slug experiences an increase in velocity. These points leave a signature clearly visible in the probability density distribution of figure 4. Apart from these points, a strong correlation is obtained between the bubble velocity V_b and the velocity of the next slug V_s.

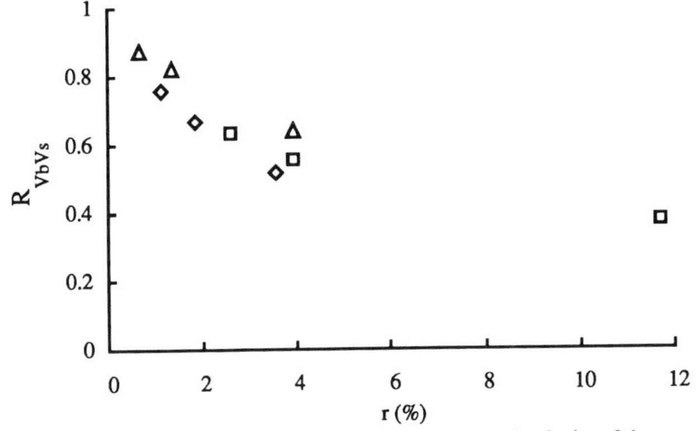

Figure 11. Correlation coefficient between bubble velocity and velocity of the next slug versus rate of coalescence; □: run 2; ◇: run 4; △: run 5.

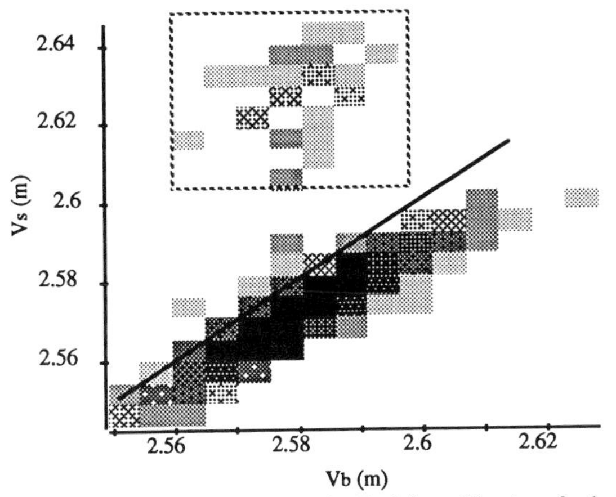

Figure 12. 3D histogram of bubble of velocity V_b followed by slug of velocity V_s for run 2, test section 5 (the density is proportional to the number of events).

CONCLUSION

A statistical investigation of the structure of slug flow has been performed in a long pipe, 90 m long. The space evolution of the various quantities (length and velocity of long bubbles and liquid slugs) has been studied, by determining their average, standard deviation and density probability distribution at different pipe locations. It is shown that the evolution of these quantities with the distance from the pipe inlet is weak, and that it is mainly linked to gas expansion.

The flow development mainly arises from the coalescence process which takes place near the pipe inlet and decreases when x increases: the rate of coalescence is seen to depend upon both the distance and the liquid flow rate through eq (5). Besides one observes that the interaction between the long bubbles and the liquid slugs is strengthened when the rate of coalescence weakens.

ACKNOWLEDGEMENT

This study has been developed with the support of the EvE program. It has been sponsored by TOTAL Exploration Production.

REFERENCES

Andreussi, P., Di Donfrancesco, A., Messia, M. 1988. An impedance method for the measurement of liquid hold-up in two phase flow *Int. J. Multiphase Flow* 14:777–87

Brill, J. P., Schmidt, Z., Coberly, W. A., Herring, J. D., Moore, D. W. 1981. Analysis of two phase tests in large diameter flow lines in Prudhoe bay field. *Soc. Pet. Eng. J.* 271:363–78

Dhulesia, H., Bernicot, M., Deheuvels, P. 1991. Statistical analysis and modelling of slug lengths. *5th Int. Conf. on Multiphase Production, 19-21 June 1991, Cannes France.* pp 80-112. Cranfield BHRG

Dukler, A. E., Hubbard, M. G. 1975. A model for gas liquid slug flow in horizontal and near horizontal tubes. *Ind. Eng. Chem. Fundam.* 14:337–47

Fabre, J., Liné, A. 1992. Modelling of two-phase slug flow. *Annu. Rev. Fluid Mech.* 24:21–46

Ferré, D. 1979. Ecoulements diphasiques à poches en conduite horizontale. *Rev. Inst. Fr. Pét.* 34:113–42

Ferschneider, G. 1982. Ecoulements gaz-liquide à poches et à bouchons en conduite. *Rev. Inst. Fr. Pét.* 38:153–82

Nicholson, M. K., Aziz, K., Gregory G. A. 1978. Intermittent Two Phase Flow in Horizontal Pipes: Predictive Models. *Can. J. Chem. Eng.* 56:653–63

Nicklin, D.J., Wilkes, J.O., Davidson, J.F. 1962. Two phase flow in vertical tubes. *Trans. Inst. Chem. Engs.* 40:61–68

Nydal, O. J., Pintus, S., Andreussi, P. 1992 Statistical characterization of slug flow in horizontal pipes *Int. J. Multiphase Flow* 18:439–53

Pauchon, C., Dhulesia, H., Lopez D., Fabre J. 1993. Tacite: a comprehensive mechanistic model for two-phase flow, *6th International Conference on Multi Phase Production*, Cannes, France:16-18 Juin 1993

Théron, B. 1989. *Ecoulements instationnaires intermittents de gaz et de liquide en conduite horizontale*. Thèse Inst. Natl. Polytech., Toulouse

Tronconi, E. 1990. Prediction of slug frequency in horizontal two-phase slug flow. *AIChE J*. 36:701–9

MEASUREMENT AND ANALYSIS OF SLUG CHARACTERISTICS

IN

MULTIPHASE PIPELINES

H. DHULESIA (TOTAL, FRANCE)

E. HUSTVEDT, O. TODAL (STATOIL, NORWAY)

ABSTRACT :

TOTAL and STATOIL have, within the scope of the POSEIDON co-operation, performed the measurement of operating conditions of multiphase pipelines operated by subsidiaries of TOTAL in Indonesia, Tunisia and Argentina. The objectives of these measurements were to collect good quality data on steady state and transient operations as well as the extensive slug characteristics. A methodology is proposed to obtain good quality data from operating multiphase pipelines which involves the use of high standard pressure / temperature transmitters, special instruments such as gamma-ray densitometers, WOR meter and dedicated data acquisition systems.

Extensive data on slug characteristics have been collected. The analysis of this data reveals the following points:

- length and velocity of a slug can be estimated by monitoring the pressure variation at the outlet of the pipeline.
- liquid hold-up of a slug is a function of superficial gas velocity and the slug length. The existing correlations do not take the effect of slug length into account.
- slug length distribution can be uni or bi-modal depending on the type of slug formation mechanisms.
- uni-modal distribution of slug lengths can be best fitted by an inverse Gaussian law.
- slug velocity at the outlet of the pipeline varies significantly and a large variation in the slug velocities was observed for small slugs.

INTRODUCTION

The multiphase production is emerging as a standard production method for the development of marginal fields. The multiphase production technology consists of transporting untreated well-head effluent through a single pipeline either to an existing processing platform or to onshore processing facilities. Usually such multiphase pipelines operate in the slug regime due to long transport distances, uneven pipeline profiles and the upstream and downstream upsets. In such situations, the prediction of slug characteristics is vital for the design as well as the operation of the pipeline and the downstream process facilities.

Most of the models used in the design and operation of pipelines are based on experimental loop data. As far as the hydraulic modelling of separated flow regimes (i.e. stratified or annular) is concerned, it can be extrapolated, with some precautions, from the experimental loops to an operating pipeline. However, an extrapolation of slug flow regime is not obvious because of the effect of pipeline length, pipeline diameter and pipeline profile on the slug characteristics.

Under the POSEIDON II co-operation between TOTAL and STATOIL (Todal and Dhulesia, 1992), the operating conditions of multiphase pipelines operated by TOTAL were measured under steady state and transient conditions. The objectives of the Flow Measurement Project have been not only to validate available simulation models, but also to improve the prediction capability of the models.

The POSEIDON II project started in 1987. The measurements were carried out on three multiphase pipelines as indicated below :

- BEKAPAI pipeline (Indonesia) from June to October 1990
- SIDI EL ITAYEM pipelines (Tunisia) during June 1991
- HIDRA pipelines (Argentina) during March, May, June and October 1992

Approximately 400 million data points were recorded during the measurement campaigns on the above three pipelines.

The main aim of this paper is to present some results obtained from slug characteristics measurements. First, a brief description of the above three pipelines is given, followed by a description of methodology applied to acquire high quality data. Next, some important findings from the analysis of this data are given which may be applied to improve the operation of certain pipelines.

BRIEF DESCRIPTION OF PIPELINES

Bekapai (Indonesia) :

The Bekapai pipeline is a 42 km 12" sealine. As shown in Fig. 1, the LP wells containing oil, water and gas are sent to a separator. The gas and oil from the LP wells are compressed and pumped respectively before being mixed and fed into the pipeline, while the HP gas wells are directly fed into the pipeline. The free water is removed from the separator and injected to the water injection wells after a proper treatment.

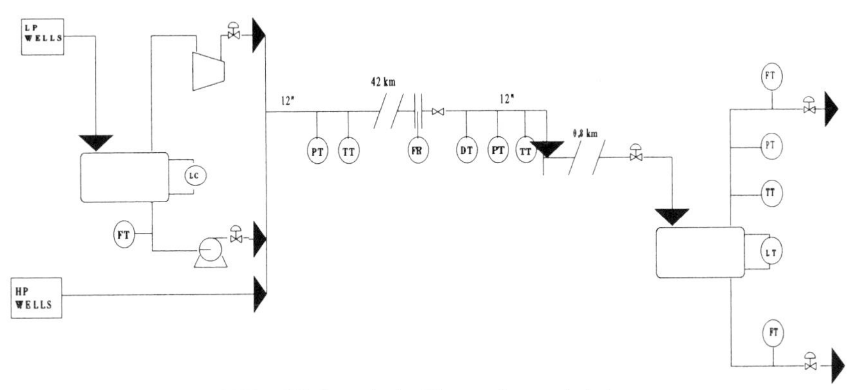

FIG. 1 - Simplified PFD of Bekapai pipeline and associated process

The low pressure wells contribute 340 000 Sm^3/d of gas to a total of 620 000 Sm^3/d. Consequently, the gas flow rate is reduced to 280 000 Sm^3/d when the compressor is out of service.

Table 1 gives the main characteristics of the Bekapai pipeline. The pipeline profile is given in Fig. 2 which shows a 36 m export riser at the inlet, then a nearly horizontal section of 20 km followed by another section of about 20 km with a constant positive slope. The last 2 km of the onshore pipeline has an elevation of about 50 m.

Sidi El Itayem (Tunisia) :

Two on-shore flowlines of 4" and 6" diameter transport oil, gas and water over a distance of about 4.2 Km. The production from different wells is fed to an adequately instrumented three phase test separator situated at the pipeline inlet. The elevation profile of these two parallel flowlines is shown in Fig. 3 which indicates an elevation of 25 m over 4.2 Km. The main characteristics of these flowlines are given in Table 2.

Diameter	inch	12
Length	Km	42
Oil Flow rate	Sm^3/d	1600
Inlet temperature	° C	38
Outlet pressure	bar a	10
Outlet Temperature	° C	28
Low gas flow rate operating conditions		
Gas flow rate	Sm^3/d	280 000
Inlet pressure	bar a	20
High gas flow rate operating conditions		
Gas flow rate	Sm^3/d	620 000
Inlet pressure	bar a	24

Table 1 - Main characteristics of Bekapai pipeline

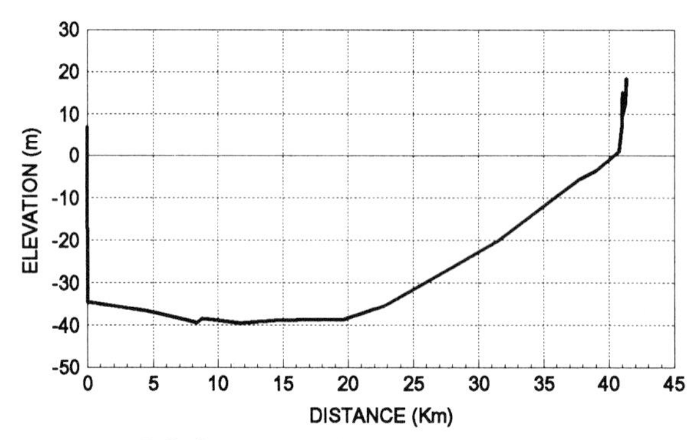

FIG. 2 - Bekapai pipeline topography

HIDRA PIPELINE (Argentina) :

The Hidra field consists of two well-head platforms : Hidra North and Hidra Centre. The production of Hidra North is transported to Hidra Centre via a 14" pipeline of 3.6 Km. The total production of the two platforms is transported to the on-shore process facilities by a 20" pipeline of 14.2 km. The detailed description of the field and the pipelines is given by Bernicot and Dhulesia (1991). The main characteristics of these two pipelines are given in Table 3.

Diameter	inch	4 and 6
Length	Km	4.2
Oil flow rate	Sm^3/h	1 - 3
Gas flow rate	Sm^3/h	130 - 2500
Water flow rate	m^3/h	2.5 - 12
Gas Liquid Ratio	-	17 - 180
Inlet pressure	bar a	7 - 13
Outlet pressure	bar a	5
Inlet temperature	° C	30 - 55
Outlet Temperature	° C	25 - 35

Table 2 - Main characteristics of Sidi El Itayem flowlines

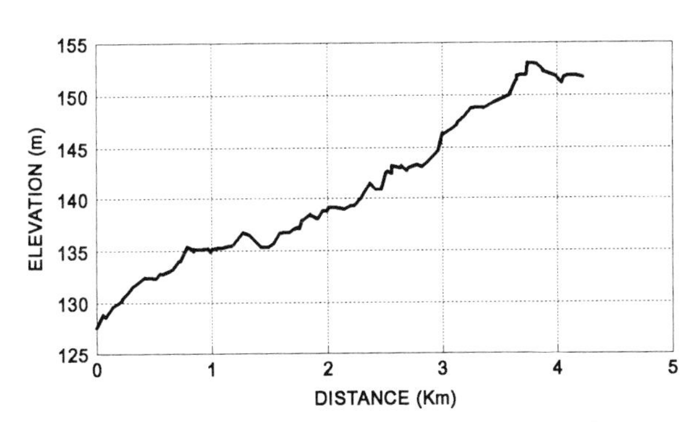

FIG. 3 - Sidi El Itayem pipeline topography

PIPELINE		HIDRA NORTH	HIDRA CENTRE
Diameter	inch	14	20
Length	Km	3.6	14.2
Oil flow rate	Sm^3/d	1 800 - 2 400	3 000 - 3 900
Gas flow rate	Sm^3/d	210 000 - 330 000	350 000 - 550 000
Water flow rate	m^3/d	360 - 480	600 - 800
Gas Liquid Ratio	-	90 - 140	90 - 140
Inlet pressure	bar a	12 - 14	10 - 12
Inlet temperature	° C	60 - 70	60 - 70
Outlet pressure	bar a	10 -12	5 - 5.3
Outlet Temperature	° C	20 - 30	10 - 20

Table 3 - Main characteristics of Hidra pipelines

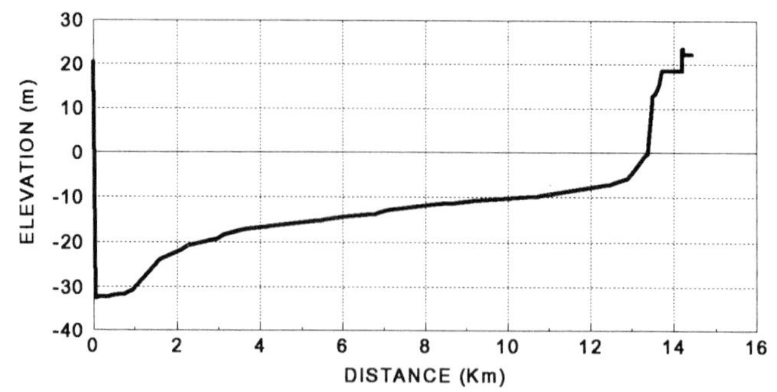

FIG. 4 - Hidra pipeline topography

It must be noted that the pipeline of 14" connecting the Hidra North platform to the Hidra Centre platform has a riser at the inlet and outlet. The outlet riser causes severe slugging instabilities in the 20" pipeline. This problem is described in detail by Bernicot and Dhulesia (1992).

METHODOLOGY TO ACQUIRE GOOD QUALITY DATA

Examinations and analysis of operating data from several pipelines collected with the standard field instruments led to a conclusion that such data are neither complete nor sufficiently accurate to use for the validation of models.

Based on this conclusion, it was decided to establish a methodology to collect high quality and complete data from the operating pipelines. This methodology consists of :

- *IDENTIFICATION OF MEASUREMENT PARAMETERS*

  Minimum number of parameters of the pipeline in question were identified which would be necessary for the validation of any pipeline simulation model. Our objectives were to measure not only the operating conditions of pipelines but also the operating conditions of the downstream process and equipment which are directly related to the operation of the pipeline.

- *INSTALLATION OF NEW SENSORS, ORIFICE FLOW METERS AND TRANSMITTERS*

  For all three pipelines, the list of the above minimum parameters indicated that there were no measurements of some vital parameters such as pipeline departure pressure and temperature. Consequently it was necessary to install new sensors and the corresponding transmitters.

- *REPLACEMENT OF EXISTING TRANSMITTERS AND ORIFICE FLOW METERS*

It was found that some of the existing transmitters did not have satisfactory accuracy. Such transmitters were replaced by HONEYWELL SMART transmitters which were of the best industrial standard with 0.1 % accuracy of calibrated span. Four sets of orifice plates for oil, gas and water were prepared for the measurements in Tunisia. The plates were changed corresponding to the changes of flowrates during the measurement campaigns. For the Bekapai pipeline installations, the gas orifice plate was replaced by a new one.

- *CALIBRATION OF TRANSMITTERS AND ORIFICE FLOW METERS*

The existing and new transmitters required an elaborated calibration procedure especially for the orifice plates according to the ISO 5167 standard. The range of certain existing transmitters was reduced to obtain a better accuracy.

- *USE OF SPECIAL INSTRUMENTS*

Gamma-ray densitometers were used on the pipelines to measure the local liquid hold-up which was necessary for the estimation of slug characteristics. Only one gamma-ray densitometer was used in Indonesia and Tunisia, while two were used in Argentina. These gamma-ray densitometers are clamp-on instruments equipped with a source of Cs137 (300 mCi) at one side of the pipe and a gamma-ray detector at the other side.

By-pass spool pieces were installed on the Bekapai and SIT pipelines in order to carry out a two point calibration (empty pipe and pipe filled with water) of the gamma-ray densitometers.

Measurement of flowrate in the Hidra 20" pipeline turned out to be complicated because of a 6" condensate/water pipeline entering the 20" line upstream the first stage separator. Consequently, it was necessary to install a WOR meter (AGAR OW 101) and a venturi meter in order to provide an estimate of the condensate and water flowrates in the 6" pipeline.

- *DEDICATED DATA ACQUISITION SYSTEM*

A dedicated data acquisition system was used for all three pipelines. Several computers were used to measure the data at a logging frequency from 1 to 20 Hz.

All 4 - 20 mA transmitter signals were converted to 1 - 5 V signals compatible with PC interface. Signals from the existing transmitters were wired in series with the control room instruments and connected to A/D cards through 16 channels interface boxes with low-pass filters with a cut-off at 40 Hz. Some of the interface boxes also included power supply and galvanic isolation.

As the slug characteristics were estimated from the gamma-ray densitometer readings, it was necessary to measure the fluid density at a frequency of 20 Hz. Other parameters such as pressure, temperature and flowrate were generally logged at a frequency of 1 Hz.

- *SPECIFIC PVT SAMPLING AND ANALYSIS*

Physical fluid properties have vital influence on multiphase pipeline hydraulics and are generally derived from the fluid composition data. A usual PVT analysis practice is to take oil and gas samples from the test separator for each well. As the fluid passing through pipelines comes from several wells, the fluid composition based on one particular well cannot be representative. Consequently, oil and gas samples were taken from the first stage separator situated at the outlet of the pipeline. Moreover, some specific measurements were carried out for the liquid phase viscosity which is always difficult to predict from the composition. As the SIT and Hidra pipelines contain a significant amount of water, a water-oil emulsion was suspected. The viscosity of emulsions under dynamic conditions was consequently measured in the PETRECO flow simulator wheel in Norway.

SLUG CHARACTERISTICS

Fig. 5 shows typical measurements of the fluid density (measured with gamma-ray densitometer) and the pressure during the passage of a slug at the outlet of the Bekapai pipeline. It is interesting to notice that the slug front corresponds to a minimum pressure (A) while the slug tail corresponds to a maximum pressure (B). The abrupt pressure increase of roughly 1.1 bar at the slug front is mainly due to dynamic pressure effects caused by several bends located a few metres downstream the pressure transmitter.

The slug length is obtained by multiplying ΔT, time difference between the passage of front and tail of the slug, and the slug velocity. Calculations of ΔT obtained from gamma meter readings and from the pressure measurements are compared in Fig. 6 which shows an encouraging agreement. This observation implies that the slug length and velocity can be calculated from the pressure measurements provided that there is no pressure pulses propagating from the outlet. This technique may not be applicable for a laboratory loop because the pressure drop over small slugs cannot be distinguished from other pressure variations related to the physical instabilities in flow such as liquid waves, inlet / outlet effects etc.

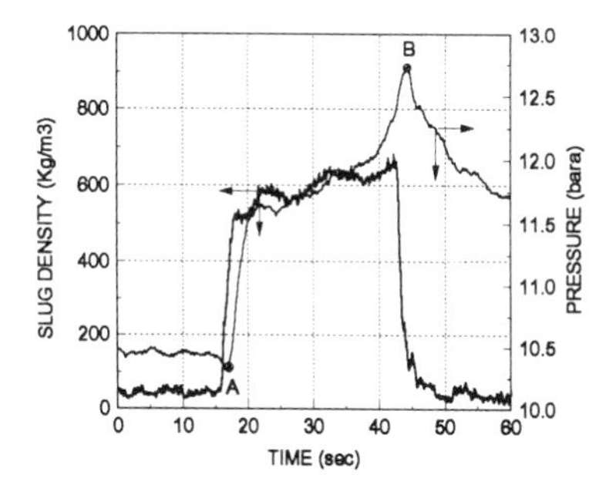

FIG. 5 - **Typical density and pressure profile in a slug**

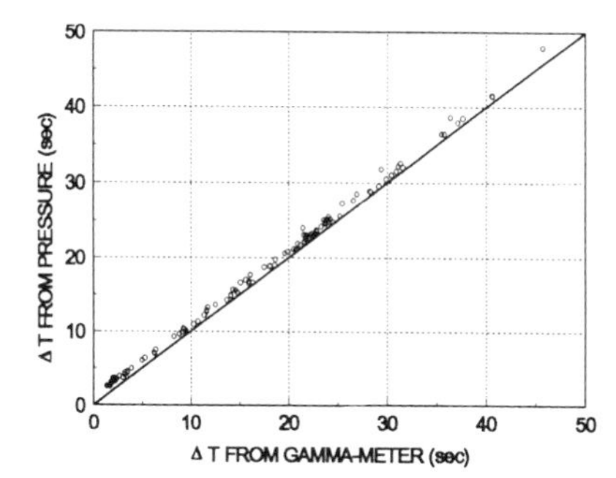

FIG. 6 - **Comparison of pressure and gamma-densitometer techniques for estimation of slug length**

Another important finding is illustrated in Fig 7 where ΔP across the slug is plotted against the slug length. This figure indicates that pressure drop across the slug is proportional to the slug length which means that it is very important to correctly predict the slug length in a mechanistic model. The pressure drop of approximately 0.9 bar for the slug length approaching zero is mainly due to dynamic effects as explained above.

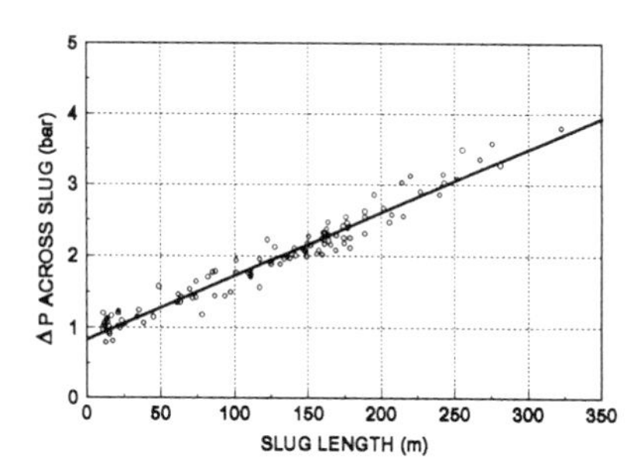

FIG. 7 - Pressure drop across the slug versus slug length

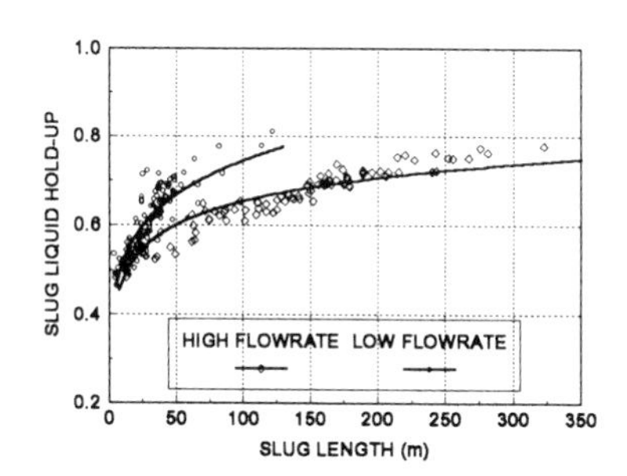

FIG. 8 - Liquid hold-up of slug versus slug length

This finding has a potential application for the detection of slugs by monitoring the pressure. This technique can be used within a slug control system in order to reduce the size of the slug catcher.

The average density of a slug is obtained by integrating the measured density profile of the slug as shown in Fig. 5. When the average density of a slug is divided by the liquid density, the liquid hold-up of the slug is obtained (provided there is an homogeneous mixture in the pipeline). Fig. 8 illustrates the slug liquid hold-up versus the slug length for the Bekapai pipeline for two different superficial gas velocities (with a constant liquid superficial velocity). This figure clearly indicates that the slug liquid hold-up is function of slug length as well as of gas superficial velocity. The slug liquid hold-up varies between 0.5 and 0.8 depending on the slug length. In a test loop, however, the slug lengths are quite constant and small. None of the available correlations from the literature predicts the slug liquid hold-up as a function of slug length. Some of the available correlations (Fabre et al, Gregory et al, Sylvester) are compared with measured data in Table 4 which clearly indicates that these correlations developed from test loop data cannot be satisfactorily applied to a pipeline.

SLUG LIQUID HOLD-UP CORRELATION	LOW FLOWRATE	HIGH FLOWRATE
Fabre et al	**0.27**	**0.54**
Gregory et al	**0.58**	**0.67**
Sylvester	**0.64**	**0.65**

Table 4 - Prediction of slug liquid hold-up by different available correlations

Collection of slug length distribution data from operating pipelines was one important objective of the project. In order to perform a proper statistical analysis, slug length data were collected for a long duration. The slug length distributions measured for the Bekapai pipeline are shown in Fig. 9. As it can be seen from this figure, the slug length distribution is uni-modal for the low flowrate and bi-modal for the high flowrate. An appropriate explanation to the bi-modal distribution occurring at the high flowrate has not yet found.

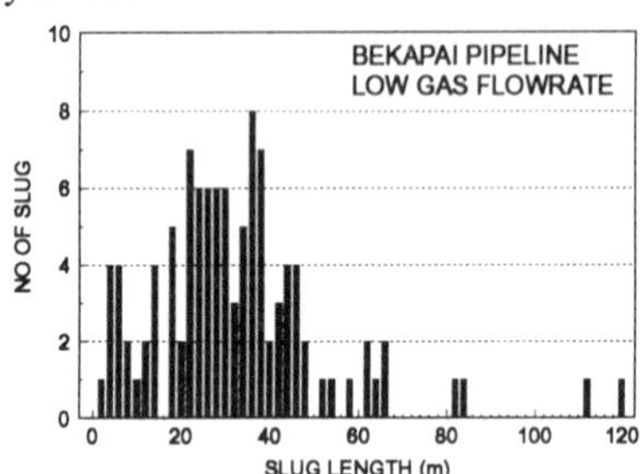

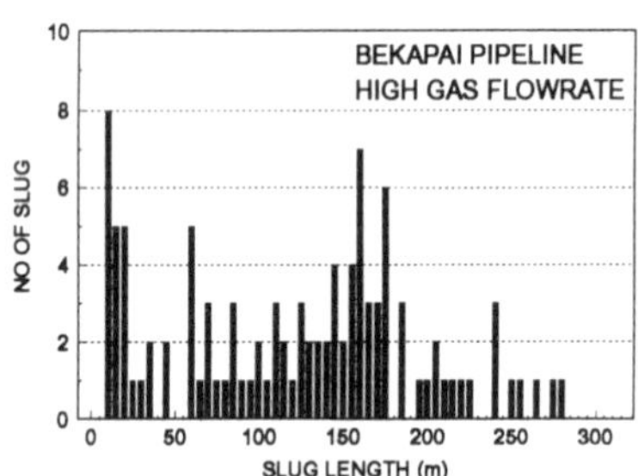

FIG. 9 Slug length distribution for the Bekapai pipeline

In Fig. 10, the slug length distributions for the Hidra 20" pipeline are shown for two different measurement campaigns. The distribution for campaign 1 is uni-modal while that for campaign 26 is bi-modal. A distribution of long slugs in both campaigns may be due to the severe slugging occurring in the 14" riser at the Hidra North platform. It is believed that the first peak of the distribution is due to slug formation in the last section of the pipeline.

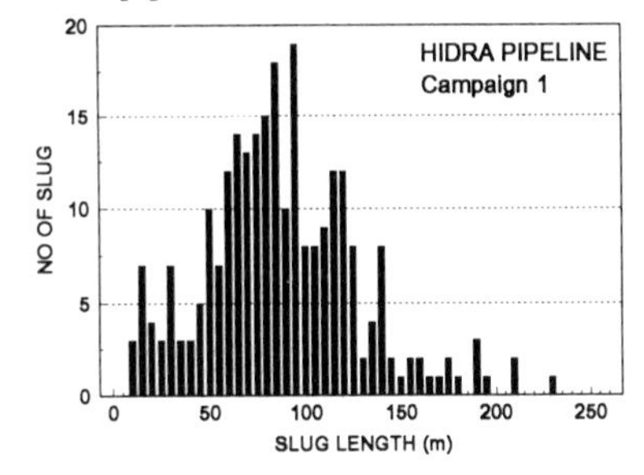

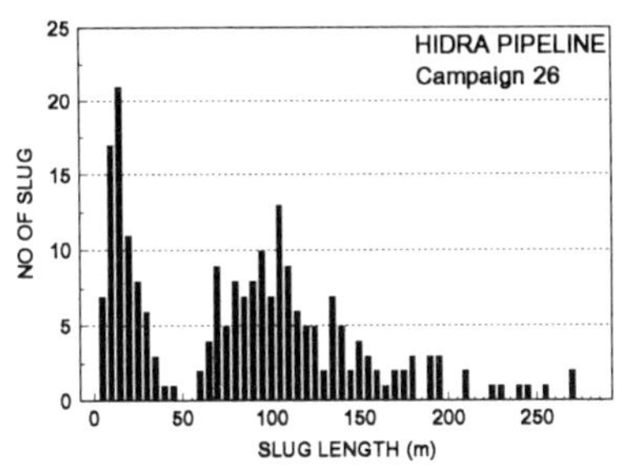

FIG. 10 Slug length distribution for the Hidra 20" pipeline

The uni-modal distributions of slug lengths can be well fitted by Inverse Gaussian distributions (Dhulesia et al, 1991). Fig. 11 shows a comparison of an inverse Gaussian and a log-normal distribution (Brill et al, 1981) laws for the 20" Hidra pipeline which indicates that the best fit is obtained by an inverse Gaussian distribution.

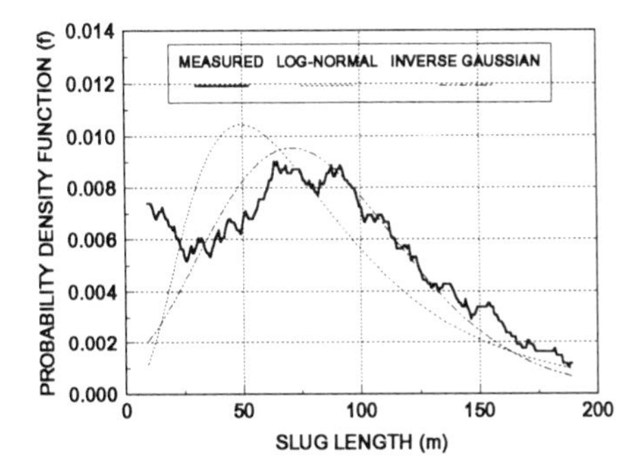

FIG. 11 - Comparison of Inverse Gaussian and Log-normal distribution laws for the 20" Hidra pipeline

Extensive data on the distribution of slug velocities at the outlet of the 20" Hidra pipeline using two gamma-ray densitometers have also been collected. Fig. 12 shows a typical slug velocity distribution. It is interesting to notice a significant variation in slug velocity i.e. from 3 to 9 m/s. An attempt was made to investigate whether the slug velocity is a function of the slug length. Fig. 13 clearly indicates that there is no apparent correlation between them, but the small slugs travel at varying velocity and long slugs travel at constant velocity. This finding is also an important parameter in a two-phase flow model because most of the work published on the test loop data demonstrates that the slug velocity is relatively constant (Nydal et al, 1992). Some of the points in Figs. 12 and 13 should be interpreted as "waves" and not as "slugs" as their propagation velocity is less than the mixture velocity.

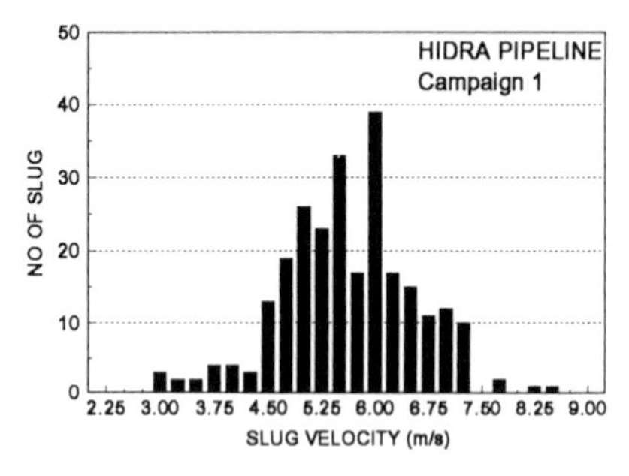

FIG. 12 - Typical slug velocity distribution

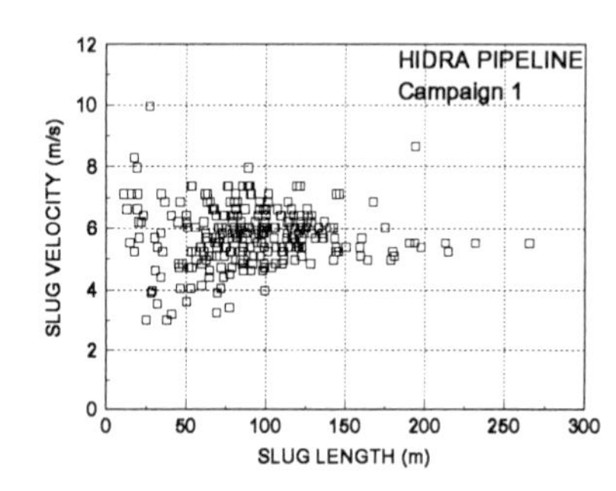

FIG. 13 - Slug velocity versus slug length

CONCLUSIONS

- Acquisition of good quality data from operating pipelines requires careful preparation which involves the use of high standard pressure / temperature transmitters, special instruments such as gamma-ray densitometers and dedicated personal computer based data acquisition system.

- PVT sampling from the first stage separator, i.e. at the outlet of pipeline, is necessary to measure the composition of effluent passing through the pipeline. As the liquid phase viscosity is a vital parameter, specific tests are necessary to measure the viscosity of the water/oil/gas mixture.

- Data acquisition of long duration is necessary to carry out some meaningful statistical analysis of slug characteristics.

- The slug length and the slug velocity can be estimated by use of pressure measurements at the pipeline outlet.

- The liquid hold-up of a slug is a function of the superficial gas velocity as well as the slug length. The effect of slug length on the liquid hold-up is significant and none of the existing correlations takes this effect into account.

- Slug length distributions can be uni or bi-modal. A uni-modal distribution is obtained when there is only one mechanism of slug formation, but a bi-modal distribution is obtained when the hydrodynamic slugging is superimposed by either the severe slugging or the terrain induced slugging.

- A uni-modal slug length distribution can be well fitted by the Inverse Gaussian distribution law.

- The slug velocity at the outlet of a pipeline varies significantly contrary to observations made on a test loop. The variation in the slug velocity increases with a decrease in slug length.

ACKNOWLEDGEMENT

The authors would like to thank TOTAL INDONESIE, TOTAL AUSTRAL and la Compagnie Franco-Tunisienne des Pétroles for their useful assistance throughout the different phases of POSEIDON Flow Measurement project.

REFERENCES

Bernicot, M. and Dhulesia, H., "How to simulate the dynamics of the HIDRA 20" Sea-line system" presented at *Multiphase Operation Offshore* organised by IBC, London (11-12 Dec. 1991)

Brill, J.P., Schmidt, Z., Coberly, W. A., Herring, J. D. and Morre, D. W., " Analysis of two-phase tests in large diameter flow lines in Prudhoe Bay field" , *Soc. Pet. Eng. J.*, 271, 363-378 (June 1881)

Dhulesia, H., Bernicot, M. and Deheuvels, P., "Statistical analysis and modelling of slug lengths" In : *Multiphase Production* (Edited by A. P. Burns) Elsevier, London, 80-112 (1991)

Fabre, J., Ferschneider, G. and Masbernat, L.,"Intermittent gas liquid flow modelisation in horizontal and slightly inclined pipes", International Conference on Physical Modelling of Multiphase Flow (BHRA), Coventry, England, F2 (April 1983)

Gregory, G. A., Nicholson, M. K. and Aziz, K., "Correlation of the liquid volume fraction in the slug for horizontal gas-liquid slug flow", *Int. J. Multiphase Flow* 4, 33-39 (1978)

Nydal, O. J., Willes, J. O. and Andreussi, P. "Statistical characterization of slug flow in horizontal pipes", *Int. J. Multiphase Flow* 18, 439-453 (1992)

Sylvester, N. D., "A mechanistic model for two phase slug flow in pipes", *ASME J. Energy Res. Tech.*, 109, 206-213 (Dec. 1987)

Todal, O. and Dhulesia, H., "Data acquisition on multiphase pipelines" presented at *Multiphase Transportation III* organised by Norwegian petroleum Society, RØROS, (21-22 Sept.1992)

THE EFFECT OF FLUID PROPERTIES AND GEOMETRY ON VOID DISTRIBUTION IN SLUG FLOW

A.Paglianti, G.Trotta, TEA, Italy
P.Andreussi, University of Udine, Italy
O.J.Nydal, IFE, Norway

ABSTRACT

New data relative to the measurement of the void fraction in liquid slugs in near horizontal pipes are presented and compared with existing correlations. The data have been obtained for air-water mixtures at varying the operating pressure in the range 1-26 Bar and for air-light oil mixtures at atmospheric pressure. The effects of pipe diameter and liquid superficial velocity has also been assessed both for air-water and air-light oil systems at atmospheric pressure. Present data and other published data have been compared with existing correlations, showing that the correlation proposed by Andreussi and Bendiksen (1989) gives an acceptable overall fit to available data. However, also in this case, some of the trends showed by the data are not well reproduced.

INTRODUCTION

Available slug flow models are based on the assumption that slug flow can be described as a sequence of slug units traveling at a constant translational velocity (Dukler and Hubbard, 1975). The unit is formed by a liquid slug of given length and uniform void followed by a long gas bubble flowing above a stratified liquid layer. The aerated slug region is treated as fully developed bubbly flow and in near-horizontal pipes it is usually assumed that the slip between the gas and the liquid in the slug is negligible.

In slug flow models, the average void fraction in the slug body is obtained by means of empirical closure relations, based on fairly limited data sets. Moreover, the data obtained in different laboratories have been taken adopting different experimental methods (typically, either impedance probes or γ-densitometers) and are only relative to a given geometry and fluid pair. Recently, Nydal (1991) has shown that impedance probes and γ-densitometers give very similar results, when these instruments are properly calibrated. However, in a thorough statistical analysis of slug flow, Nydal et al. (1991) have shown that data analysis to obtain the average flow parameters in slug flow should be very careful because at least three different flow structures may exist: regular, fully developed slugs; short, very aerated slugs traveling at the same velocity of regular slugs; and waves. For these reasons, it may be questionable to compare data obtained in different laboratories, with different instrumentation and data analysis procedures in order to assess the effects of geometry, physical properties and operating conditions.

In the present work, a close cooperation between TEA and IFE allowed to obtain a consistent set of data relative to the average void fraction in liquid slugs, adopting the same instrumentation and experimental procedures. The data are relative to air-water mixtures at varying operating pressure in the range 1-26 Bar, and to air-light oil mixtures. The effects of pipe diameter and liquid superficial velocity have also been assessed both for air-water and air-light oil systems at atmospheric pressure. Present data and other published data have been compared with existing correlations.

LITERATURE REVIEW

First data relative to the slug void fraction have been obtained by Hubbard (1965). However, as noticed by Gregory et al. (1978), the scatter of these data is so large that it is impossible to use them to develop a predictive model. In their work Gregory et al. (1978) used two different test sections, with inner diameters of 25.8 and 51.2 mm, and air-light oil mixtures as test fluids. These authors proposed the following empirical correlation of their measurements

$$\alpha_S = 1 - \frac{1}{1 + \left(\frac{U_m}{8.66}\right)^{1.39}} \tag{1}$$

where α_S is the void fraction in the slugs and U_m is the mixture velocity. Malnes (1982) proposed another empirical correlation based on the same data of Gregory et al. (1978),

$$\alpha_s = \frac{U_m}{C_c + U_m} \tag{2}$$

where the coefficient C_c is a function of physical properties given by

$$C_c = 83 \cdot \left(\frac{g \cdot \sigma}{\rho_l}\right)^{1/4} \tag{3}$$

In this equation σ is the surface tension and ρ_l the liquid density.

Fershneider (1983) reported a set of data obtained in a high pressure loop operating at 15 Bar using natural gas and condensate as working fluids. The pipe diameter in these experiments is 146 mm. Fershneider (1983) proposed the following correlation of his measurements

$$\alpha_s = 1 - \frac{1}{\left(1 + \frac{Fr^2 \cdot Bo^{0.2}}{625}\right)^2} \tag{4}$$

where the Froude and Bond numbers are defined as

$$Fr = \frac{U_m}{\sqrt{\left(1 - \frac{\rho_g}{\rho_l}\right) \cdot g \cdot D}} \tag{5}$$

$$Bo = \frac{(\rho_l - \rho_g) \cdot g \cdot D^2}{\sigma} \tag{6}$$

In these equations, D is the pipe diameter and ρ_g the gas density.

Barnea and Brauner (1985) suggested that the void fraction in liquid slugs is the maximum which can be sustained in dispersed bubble flow, before the transition to slug flow occurs. This assumption gives a simple method to determine the void fraction in the slugs, when the transition between slug and bubble flow is known or can be predicted. In fact, for each couple of gas and liquid superficial velocities $(U_{gs})_{bs}$ and $(U_{ls})_{bs}$ along the transition, according to these authors it can be assumed

$$\alpha_s = \frac{(U_{gs})_{bs}}{(U_{gs})_{bs} + (U_{ls})_{bs}} \tag{7}$$

Barnea and Brauner (1985) also developed a correlation to determine the transition between slug and bubble flow and have been able to show that Eq. (7) gives a good fit to the data of Gregory et al. (1978).

Recently, Andreussi and Bendiksen (1989) have developed a theoretical investigation of the phenomenon of gas entrainment in an advancing liquid slug, which has also been supported by specific experiments carried out by Nydal and Andreussi (1991) and Nydal (1991). According to Andreussi and Bendiksen (1989), in fully developed slug flow the gas entrainment at the slug front equals the gas outflow from the slug tail, if it can be assumed that the aeration of the liquid layer in the stratified flow region is negligible. By means of simple physical representations of the various fluxes, the following correlation has been derived by these authors:

$U_m \rangle U_0$:

$$\alpha_s = \frac{U_m - U_0}{(U_m + U_1)^n}$$

$U_m \le U_0$: (8)

$$\alpha_s = 0$$

where

$$U_0 = 2.6 \cdot \left(1 - 2 \cdot \left(\frac{d}{D}\right)^2\right) \cdot \sqrt{g \cdot D \cdot \frac{(\rho_l - \rho_g)}{\rho_l}} \tag{9}$$

with d = 2.5 cm, and

$$U_1 = 2400 \cdot \left(1 - \frac{\sin\theta}{3}\right) \cdot Bo^{-3/4} \cdot \sqrt{g \cdot D \cdot \frac{(\rho_l - \rho_g)}{\rho_l}} \tag{10}$$

In these equations, θ is the pipe inclination with respect to the horizontal and the exponent n is introduced to take into account the effect of gas density

$$n = 1 - 3 \cdot \frac{\rho_g}{\rho_l} \tag{11}$$

The correlation developed by Andreussi and Bendiksen (1989) has been tested against a wide range of data relative to air-water mixtures flowing at atmospheric conditions and, as well, against the data of Gregory et al. (1978) and Fershneider (1983).

Besides the data considered by Andreussi and Bendiksen (1989) and the data presented in this paper, in recent years other data became available and appear to be fully reliable, namely the data obtained by Kouba (1985) and the measurement taken at SINTEF Two-Phase Flow Laboratory and presented by Fuchs and Brandt (1989) and Linga (1992).

EXPERIMENTAL SET-UP

Present hold-up measurements have been performed in two loops designed and assembled at the University of Pisa in collaboration with TEA . The first loop, described by Andreussi and Bendiksen (1989) operates at atmospheric pressure and includes an inclinable bench 17 meters long which supports the test section. The pipeline consists of trasparent Perpex tubes with inner diameters of 50 or 90 mm. The test fluids are air and light oil and the void fraction has been measured by a capacitance probe made of two ring electrodes mounted flush to the pipe wall. A detailed description of the behaviour of this probe is reported by Andreussi et al. (1988).

The second loop (see Fig. 1) includes a test section 70 metres long with inner diameter of 53 mm. This section is made of two straight pipes 36 m long, which are connected by a semicircumference whose diameter is 4 m. The pipeline is made of AISI 304, 6 m long tubes connected by carefully machined flanges, with no discontinuity between consecutive tubes. In the lower part of each tube, in the center, pressure taps have been mounted. The test fluids are water and air and the operative pressure can be varied from atmospheric up to 30 bar. The liquid is circulated by a multistage pump and is measured by two magnetic flowmeters before entering in the test section. For the air circulation a booster, able to give a differential pressure of 5 bar, is used and the gas flowrate is metered by two vortex meters. Since pressure and temperature conditions change, air pressure and temperature are measured just upstream of the vortex and the signals are used to compensate the signal of the vortex. At the end of the test section the two phase mixture is conveyed to an inertial separator, from which the air is recirculated to the booster and the water to the pump. Also in these experiments, the instantaneous liquid hold-up has been determined with a capacitance probe similar to that adopted in the low pressure loop.
The IFE experimental rig consists of acrylic pipes, of 31 mm inner diameter supported by a horizontal aluminium bench 13 meters long. This loop operates at atmospheric conditions with oil and air as test fluids.

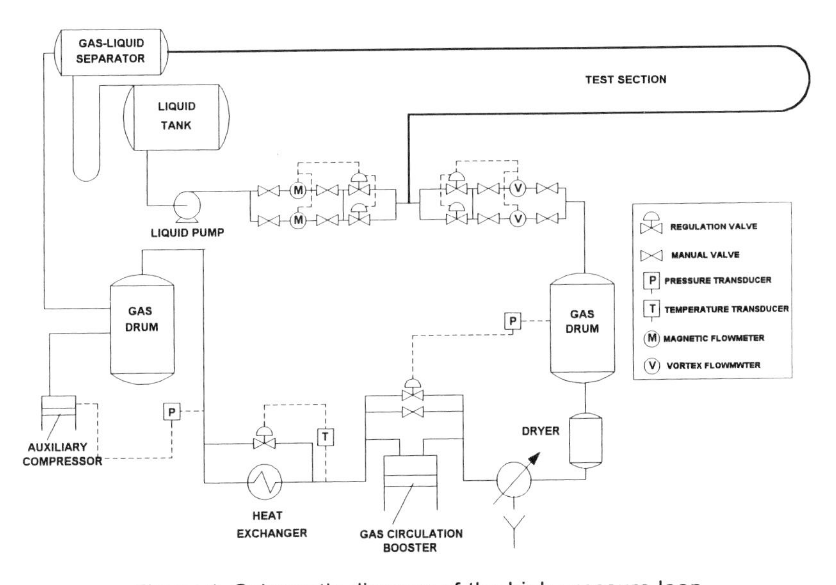

Figure 1. Schematic diagram of the high pressure loop

RESULTS AND DISCUSSION

The new data available allow to assess the effects of important parameters on the mean void fraction in liquid slugs, namely the gas density, the liquid surface tension, the liquid velocity, the pipe diameter. Quite interestingly, the first correlation proposed in the literature (Gregory et al., 1978) assumes that α_S is independent of these parameters. The most recent one (Andreussi and Bendiksen; 1989) only neglects the effects of liquid velocity.

The results obtained with air-water mixtures at pressures of 1-6-16-26 Bar are compared in Fig. 2 with the correlation of Andreussi and Bendiksen (1989). For the sake of clarity, only data relative to P = 16 Bar are compared with all available correlations in Fig. 3. As can be seen from Fig. 2, the experimental measurements clearly indicate an appreciable effect of pressure on α_S. This experimental observation confirms the results reported by Linga (1991) and relative to the flow of nitrogen and light oil mixtures. These data are compared in Fig. 4 with the correlation proposed by Andreussi and Bendiksen (1989).

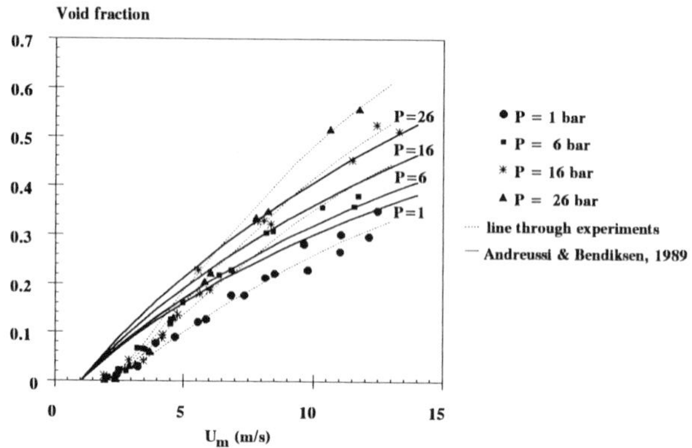

Figure 2. The slug void fraction plotted vs the mixture velocity. Effect of pressure

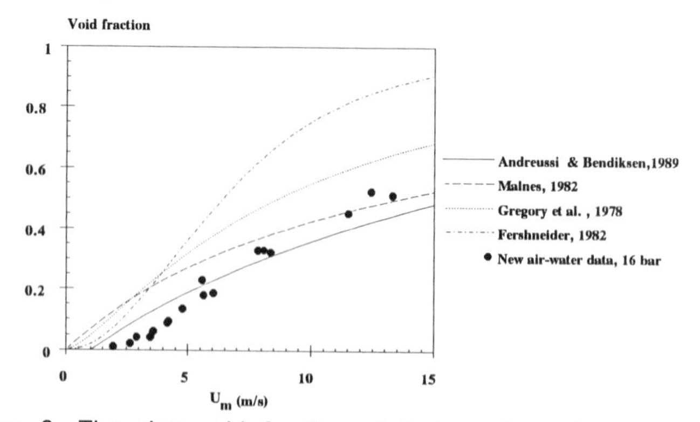

Figure 3. The slug void fraction plotted vs the mixture velocity. Comparison with existing correlations. (air-water at 16 bar)

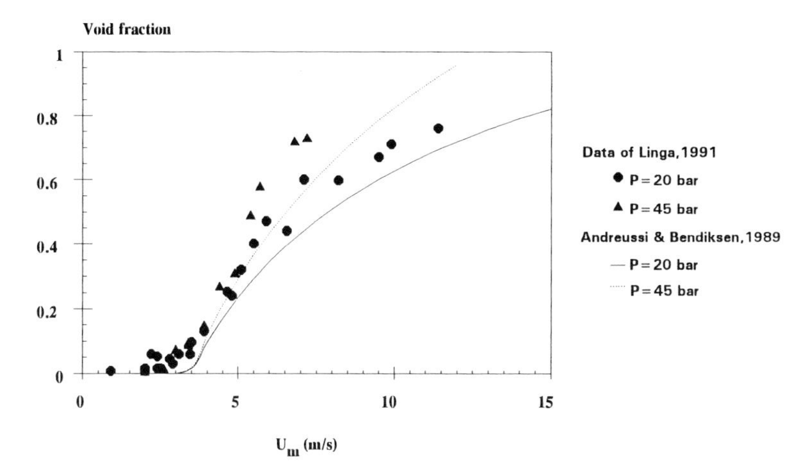

Figure 4. The slug void fraction plotted vs the mixture velocity. Data of Linga (air-oil i.d. 189 mm),

The comparison of data presented in Fig. 3 with available correlations indicates that all of them are quite unsatisfactory, with the possible exception of the one due to Andreussi and Bendiksen (1989) which, in any case, understimates the effects of pressure.

Present measurements relative to air-light oil mixtures are compared in Fig. 5 with air-water data taken in the same loop and in Fig. 6 with the predictions of available models. As shown in Fig. 5, the effects of surface tension appear remarkable. Fig. 6 indicates that, in this case, the correlations of Gregory et al. (1978), Malnes (1982) and Andreussi and Bendiksen (1989) are able to predict quite well the experimental measurements relative to air-oil mixtures. However, the first two of these correlations have been developed in order to fit data relative to similar fluid pairs and are not adeguate when used with air-water mixtures.

The data relative to air-water flow for three different pipe diameters, reported in Fig. 7, do not indicate any appreciable diameter effect. In this case the results are quite surprising, as they partly contradict previous data taken in the same laboratory, on which the correlation of Andreussi and Bendiksen is based. Moreover, a comparison, for instance, between data relative to light oil-air or nitrogen mixtures flowing in small diameter (Fig. 5) or large diameter pipes (Fig. 4) indicates a strong diameter effect on α_S, in particular on the value of the mixture velocity U_0 at which α_S departs from zero. The data represented in these figures are relative to different pressures, besides to different diameters, but it should be noticed that present measurements relative to air-water mixtures do not indicate any effect of pressure on the velocity U_0.

The data represented in Fig. 7 are compared with the correlation of Andreussi and Bendiksen (1989). As can be seen, the influence of D on α_S in the correlation is strong, but present measurements indicate that this effect should be removed, at least for small diameter/large surface tension systems.

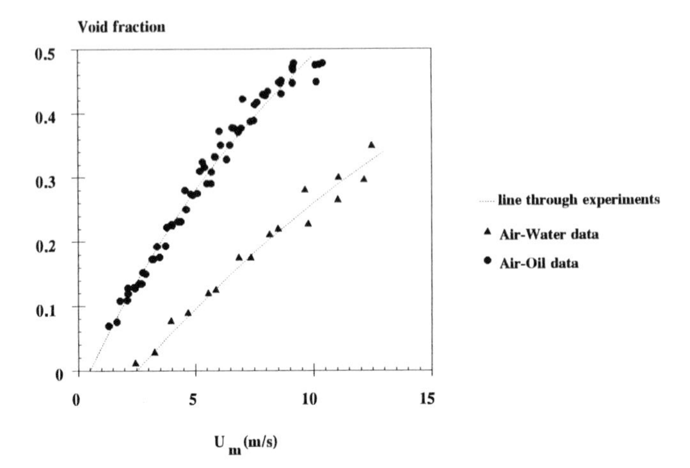

Figure 5. The slug void fraction plotted vs the mixture velocity. Comparison of present data relative to air-water and air-oil flow.

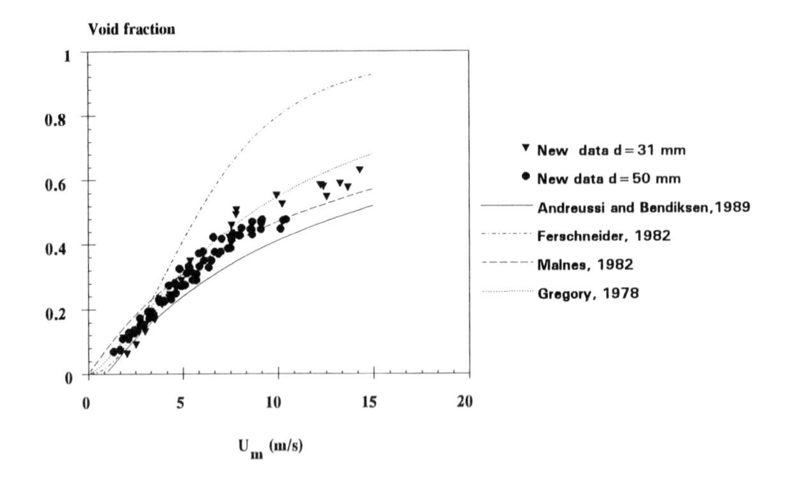

Figure 6 The slug void fraction plotted vs the mixture velocity. Comparison of air-light oil data with existing correlations

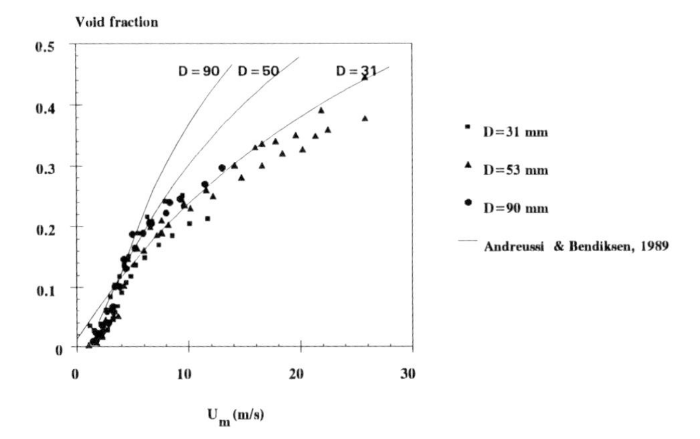

Figure 7 The slug void fraction plotted vs the mixture velocity. Data of Nydal (1991)

In Fig. 8, data relative to four different values of the liquid superficial velocity are represented. It can be noticed that at larger values of U_{ls} the effects of this parameter are appreciable and have not been noticed before. As a conseguence, none of the available correlations predicts any effect of the liquid velocity. It is interesting to notice that a dependence of α_s on U_{ls} and not only on U_m is in contradiction with the hypothesis of Barnea and Brauner (1985) that void in the slugs is equal to the value at the transition between slug and bubble flow, for the given mixture velocity.

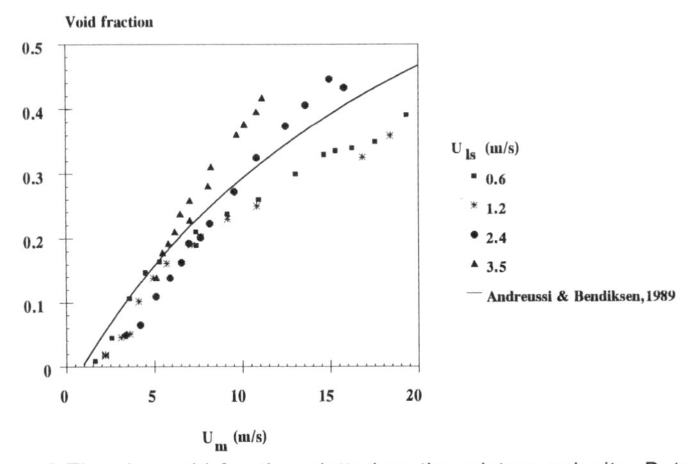

Figure 8 The slug void fraction plotted vs the mixture velocity. Data of Nydal (1991) (air-water i.d. 5 cm)

CONCLUSION

The data presented allow to assess the effects of physical properties (gas density, liquid surface tension), geometry (pipe diameter) and operative conditions (liquid flowrate) on the slug void fraction. The main experimental observations can be summarized as follows:

- At a given mixture velocity, the slug void fraction increases appreciably with the gas density.
- For small pipe diameters (D<100 mm), the slug void fraction is always higher for oil than for water. This effect should be attributed to surface tension.
- For large pipe diameters (D>100 mm), the mixture velocity, U_o, at which α_s first departs from zero is larger for oil than for water, then α_s increases rapidly to values above those relative to air-water mixtures.
- For a liquid like water and small pipe diameters, the velocity U_o seems to be indipendent of pipe diameter, while for low surface tension liquids, U_o increases with pipe diameter. It should be noticed that data used for this comparison are relative to different values of the gas density.
- At low liquid superficial velocities, α_s is apparently only a function of U_m, for a given fluid pair and geometry. However, at increasing values of U_m, α_s becomes an increasing function of U_{ls}.

None of the existing correlations is able to follow the trends of available data, with the partial exception of the correlation proposed by Andreussi and Bendiksen. (1989) which is also the most recent and the one with the largest data base. Present data suggest that a number of significant changes should be performed also to this correlation in order to obtain a better prediction of the slug void fraction.

REFERENCES

Andreussi, P.,Didonfrancesco,A., Messia, M. 1988 An impedence method for the measurement of liquid hold-up in two-phase flow. *Int. J. Multiphase Flow* **14**, 777-785.

Andreussi, P. Bendiksen K. 1989 An investigation of void fraction in liquid slugs for horizontal and inclined gas-liquid pipe flow. *Int. J. Multiphase Flow* **6**, 937-946

Barnea, D. Brauner, N 1985 Holdup of the liquid slug in two phase intermittent flow. *Int. J. Multiphase Flow* **1**, 43-49

Dukler,A.E. Hubbard, M.G. 1975 A model for gas-liquid slug flow in horizontal tubes. *Ind. Engng. Chem. Fundam.* **14,** 337-347.

Ferschneider, G. 1983 Ecoulement diphasiques gas-liquide à poches et à bouchons en conduites. *Revue Inst. Fr. Petrole* **38**, 153

Fuchs, P. Brandt, I. 1989 Liquid holdup in slugs. Some experimental results from the SINTEF Two-Phase Flow Laboratory. *4th. Int. Conf. on Multiphase Flow (BHRA), Nice, France* Paper F3

Gregory, G.A., Nicholson, M.,K. & Aziz,K. 1978 Correlation of the liquid volume fraction in the slug for horizontal gas-liquid slug flow. *Int. J. Multiphase Flow* **4**, 33-39

Hubbard, M. G. 1965 An analysis of horizontal gas-liquid slug flow. Ph.D. thesis, Univ. of Houston, Texas

Kouba, G. E. 1986 Horizontal slu flow modeling and metering. Ph.D. Thesis, Univ. of Tulsa, Oklahoma

Linga, H. 1991 Flow pattern evolution; some experimental results obtained at the Sintef multiphase flow laboratory. *4th. Int. Conf. on Multiphase Flow (BHRA), Nice, France*

Malnes, D. 1982 Slug flow in vertical, horizontal and inclined pipes. Report IFE/KR/E-83/002 V. Inst. for Energy technology, Kjeller, Norway.

Nydal, O. J. 1991 An experimental investigation of slug flow, Ph.D. thesis, Univ. of Oslo, Norway

AN INVESTIGATION OF SEVERE SLUGGING CHARACTERISTICS IN FLEXIBLE RISERS

V.TIN, M.M.SARSHAR

CALTEC, BHR GROUP
Cranfield, Bedford MK43 0AJ
England

ABSTRACT

An experimental and analysis programme has been carried out to look at the severe slugging and pressure cycling characteristics of free hanging catenary and 's' shaped risers. The experimental programme involved tests on a 60-metre long, 2-inch bore pipeline connected to a 33-metre high flexible riser through which an air/water mixture was passed.

A preliminary model on severe slugging based on the physical mechanism has been developed to predict pressure fluctuation and liquid sweep out characteristics. The method required to calculate the above values is based on time-dependent pressure balance, liquid and gas mass balance equations. This approach depends strongly on the boundary conditions defined for each stage of the cycle, pipeline liquid hold-up and liquid fall-back predictions.

The predicted values for a free hanging catenary riser were in good agreement with the experimental data. However, a model on the variation of severe slugging-1 type cycle mechanism in an 's' shaped riser is not considered at this early stage as data analysis reveals that further study on the trapped gas behaviour in the downward limb is required for a more accurate prediction of severe slugging characteristics.

NOMENCLATURE

Symbol	Description	Unit
a	- Pipe radius	[m]
A	- Pipe cross section area	$[m^2]$
Fr	- Froude number	[-]
g	- Gravitational acceleration	$[m/s^2]$
H_L	- Liquid Holdup	[-]
l	- Length	[m]
m	- Mass	[kg]
m	- Mass flow rate	[kg/s]
P	- Pressure	[Bar]
t	- Time	[s]
U	- Bubble or gas front velocity	[m/s]
X	- Length of inclined pipe section where liquid accumulated	[m]
Y	- Length of riser section where liquid accumulated	[m]
Z	- Height of liquid column in riser	[m]
ρ	- density	$[kg/m^3]$
β	- Pipe inclination angle from vertical	[°]
θ	- Pipe inclination angle from horizontal	[°]

Subscript

Subscript	Description
1	- Lower riser
2	- Upper riser
F	- Liquid fallback
m	- Mixture
l	- Liquid
p	- Pipeline
r	- Riser
s	- Separator
t	- Translational
bp	- Bubble penetration stage
gb	- Gas blowdown stage
sbu	- Slug build-up stage
sp	- Slug production stage
SL	- Superficial liquid
SG	- Superficial gas
(t_0)	- Time: Beginning of slug build-up stage
(t_1)	- Time: Beginning of slug production stage
(t_2)	- Time: Beginning of bubble penetration stage
(t_3)	- Time: Beginning of gas blowdown stage

1. INTRODUCTION

Unstable behaviour of two-phase flow in a pipeline/riser system is frequently encountered in oil production systems. The most extreme form of unstable flow is the severe slugging which can occur in particular pipeline/riser configurations. In the case of severe slugging, liquid slugs build up at the base of the riser and block off the flow of gas until the upstream gas pressure is high enough to drive the liquid out of the riser. Such conditions may lead to problems with the downstream processing plant due both to excessively high gas velocities and large quantities of liquid which have to be processed rapidly. Accurate prediction of slug characteristics caused by the riser configuration is therefore essential for the safe and efficient design of the downstream processing system.

Although a number of models for vertical risers have been developed in the past, uncertainties still exist over severe slugging characteristics in other types of riser, in particular the 's' shaped configuration. The severe slugging cycles observed from an 's' shaped riser are different from the severe slugging cycles observed in a free hanging catenary riser and the cycles described by previous researchers on vertical risers. An 's' shaped riser may be considered as two free hanging catenary risers connected together, with each section having independent effects on the cycle.

The work on the occurrance of severe slugging for different flexible riser configurations has been published in Ref 1. In this paper, the severe slugging characteristics of free hanging catenary and 's' shaped risers are compared and the effects of riser geometry on severe slugging characteristics is reviewed. Also, a preliminary model for predicting the liquid sweep out during severe slugging in free hanging catenary and 's' shaped riser is developed based on the physical phenomena.

2. DESCRIPTION OF THE TEST FACILITY

A schematic diagram of the main test facility is shown in Figure 1. The test is conducted on a 2-inch diameter pipeline inclined downwards by 2 degrees which runs in three stages for a total distance of 60 metres from the mixing section to the riser base. Water is fed into the mixing section by a centrifugal pump. The flow is controlled by adjusting one of the two valves located in the parallel lines of different diameters. The air is supplied at constant pressure from a compressor and receiver via three different diameter parallel lines, each line is fitted with a flow control valve. Before entering the mixing section, the air is passed through an accumulator vessel which is used to simulate free gas storage in a pipeline upstream of the riser. The air/water mixture flows along the pipeline and then up a riser which is 33-metres high.

The top of the riser is connected to the return pipework via a flexible hose. The return pipework consists of a 6-inch diameter down pipe, the top of which is open to atmosphere so as to act as a simple air/water separator.

Further separation takes place in the water tanks at the bottom of the downpipe, the water tank completing the water circuit and acting as a supply tank for the pump.

The pipeline/riser system is equipped with six pressure transducers, eight slug progression probes and an electro-magnetic flowmeter which is installed at the top of the riser. This flowmeter measures the velocity of liquid slugs. The bottom and the top of the riser pipe are fitted with gamma ray densitometers which are used to measure the liquid hold up.

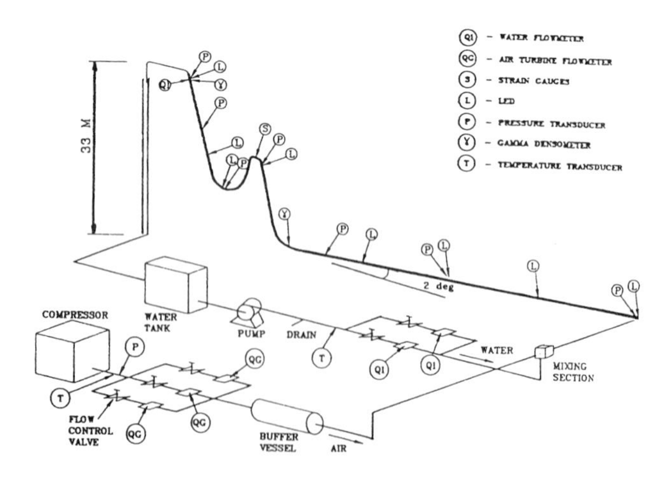

Figure 1 Schematic diagram of the test facility

3. DESCRIPTION OF SEVERE SLUGGING PHENOMENON

3.1 Severe Slugging-1 Cycle in Free Hanging Catenary Riser

The severe slugging and transition cycles observed from the free hanging catenary riser are very similar to the cycles described by previous research (Ref 2 and Ref 3), therefore, only the severe slugging-1 cycle is described in this paper.

The important points for defining severe slugging-1 cycle are:

- Liquid slug length is greater than one riser height.
- No gas penetrates the riser during the slug build-up stage.
- Maximum pipeline pressure is equal to the hydrostatic head of the riser.
- The increase in the pipeline pressure per unit time is less than the increase in the hydrostatic pressure due to the slug build up in the riser.

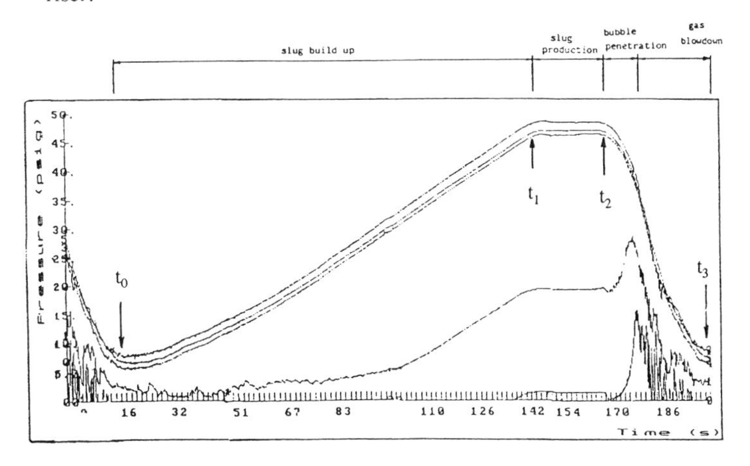

Figure 2 Free hanging catenary riser - severe slugging 1 pressure history plot

The above figure shows the pressure cycle data and visually, the cycle may be described as follows:

Slug Build-up (t_0-t_1): Stratified flow is observed in the downward inclined pipeline and the liquid fall-back in the lower limb blocks the riser foot. The liquid begins to accumulate in both the riser and the pipeline. Further incoming liquid builds up hydrostatic head in the riser until the riser is full of liquid and the maximum hydrostatic head is obtained.

Slug Production (t_1-t_2): Further gas input builds up the gas pressure in the pipeline which begins to push the severe slug front forwards from the bottom of the riser, thus causing liquid to be produced from the top of the riser.

Bubble Penetration (t_2-t_3): Once the gas front is at the riser foot, bubbles penetrate the riser from the gas front and ascend the riser.

Gas blowdown (t_3-t_4): These bubbles increase in frequency and velocity as the hydrostatic head decreases and the gas expands. Once the gas has escaped, a minimum gas pressure is obtained and the remaining liquid in the riser falls back to aid in blocking the riser foot and initiating the cycle again.

3.2 Severe Slugging-1 Cycle in Lazy's' Riser

The severe slugging-1 cycles observed from an 's' shaped riser are different from those observed in a free hanging catenary riser and the cycles described by previous researchers on vertical risers. The severe slugging-1 cycle can be sub-divided into the different categories described below:

3.2.1 Severe slugging-1p (pipeline gas penetration)

The severe slugging-1p cycle in an 's' shaped riser is considered to be similar to the severe slugging-1 cycles described by previous researchers on vertical risers and the experimental results obtained from the free hanging catenary riser described earlier. The only difference here is that the gas trapped at the downward limb during the slug build up stage is compressed as the hydrostatic head builds up in the upper limb. Hence a long liquid slug in the riser is separated by the trapped gas is observed.

3.2.2 Severe slugging-1t (trapped gas penetration)

The severe slugging-1t cycle is characterised by the trapped gas in the downward limb penetrating the upper limb before the pipeline gas penetrates the bottom of the riser during the bubble penetration stage. The cycle occurs at relatively high mixture flow rates when the trapped gas penetrates the upper limb due to the forces created by the surface tension between the gas and liquid at high liquid velocities.

Figure 3 shows the pressure cycle. When the liquid reaches the top of the riser, the slug tail in the pipeline begins to move forward causing the front of the slug to be produced from the riser. During the slug production stage, the trapped gas in the downward limb moves forward and penetrates the upper limb. This marks the end of the slug production stage. The cycle can be identified from the output of the pressure transducer located at the upper limb (shown in Figure 3). The pressure transducer recordes two pressure increments during the bubble penetration stage. The first pressure increment is caused by the trapped gas penetration and the second increment is caused by the pipeline gas penetration.

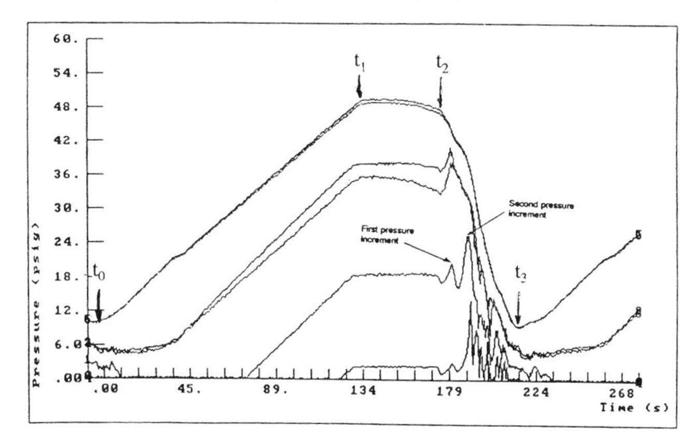

Figure 3 Severe slugging 1t pressure history plot - lazy's' riser

3.2.3 Severe slugging-1 with intermediate cycle on lower limb

The experiments indicate that operating within the severe slugging region a lazy's' riser can be considered as two free hanging catenary risers connected together, with each riser having independent effects on the cycle. Figure 4 shows a severe slugging cycle with an intermediate cycle on the lower limb. The lower limb cycle can be identified by the stepped pressure increment recorded from the pressure transducer located on the pipeline.

Consider two free hanging catenary risers connected together, a severe slugging-1 cycle occurs in the lower riser and the long liquid slug moves into the upper riser. The liquid accumulates in the upper riser, causing an increase in the outlet pressure of the lower riser, hence further severe slugging on the lower riser is suppressed by the increase in the outlet pressure.

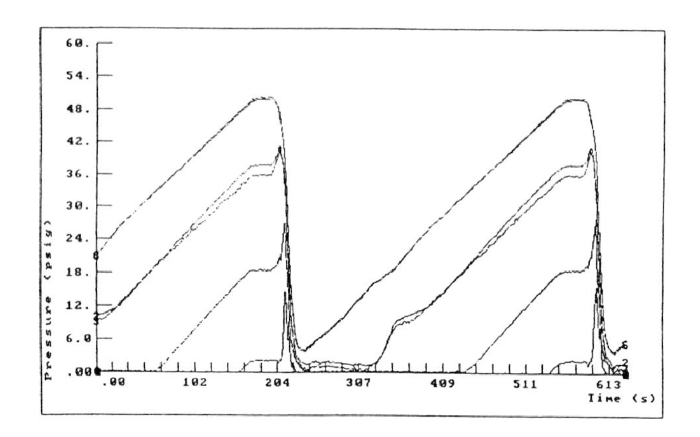

Figure 4 Severe slugging cycle with intermediate cycle

3.2.4 Severe slugging-1b

Figure 5 shows a slight variation on the severe slugging-1p cycle. There is no slug production stage in this cycle and no liquid builds up in the pipeline during the slug build-up stage. The characteristics of this cycle are the short slug length (just over one riser height) and short cycle time.

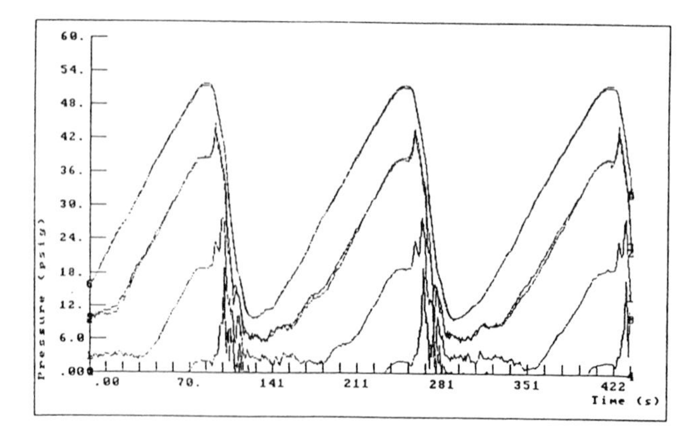

Figure 5 Severe slugging 1b cycle

4. THEORETICAL ANALYSIS

The pipeline/riser system considered is shown in Figure 6. The system consists of downward inclined pipeline with length l_{p1} connected to the lower limb of the 's' shaped riser with length l_{r1}, and the downward limb and upper limb of the riser with length l_{p2} and l_{r2} respectively.

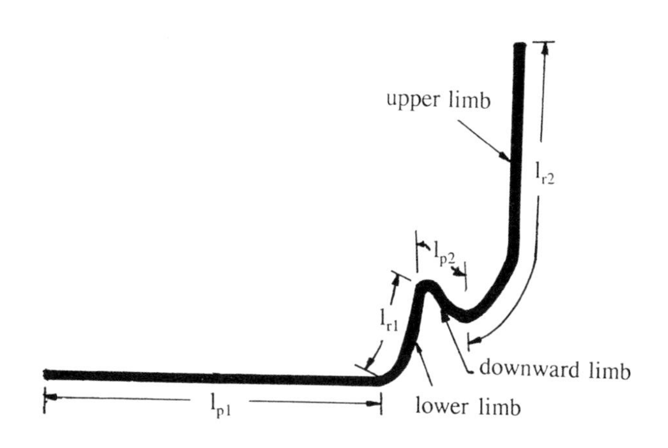

Figure 6 Pipeline - 's' shaped riser system

The preliminary model presented is for predicting the slug length and the liquid sweep-out during severe slugging-1 cycle in free hanging catenary and lazy's' risers. The analysis required to calculate the above values is based on time-dependent pressure balance, liquid and gas mass balance equations. The input gas and liquid flowrates are assumed to be constant throughout the cycle.

The prediction of liquid sweep out include the modelling of the slug build-up, slug production and bubble penetration stages of the cycle where boundary conditions are defined based on physical phenomenon, or the values are taken from the final boundary conditions calculated for the previous stage of the cycle.

4.1 Basic governing equations

Assuming the liquid distribution in a lazy's' riser/pipeline system is as shown in Figure 7, and the gas weight is relatively smalland is neglected, the pressure balance equation is:

$$P_P = P_s + \rho_L g\ [\ Z_1 + Z_2 - X_1 Sin\theta_1 - X_2 Sin\theta_2\ ] + \Delta P_{friction} \qquad [1]$$

where P_p is the pipeline pressure, P_s is the separator pressure. The liquid distribution in the system is given in terms of the values X and Z, X_1 and X_2 are the lengths of downward inclined pipe section where liquid accumulates, Z_1 and Z_2 are the heights of the riser pipe section where liquid accumulates.

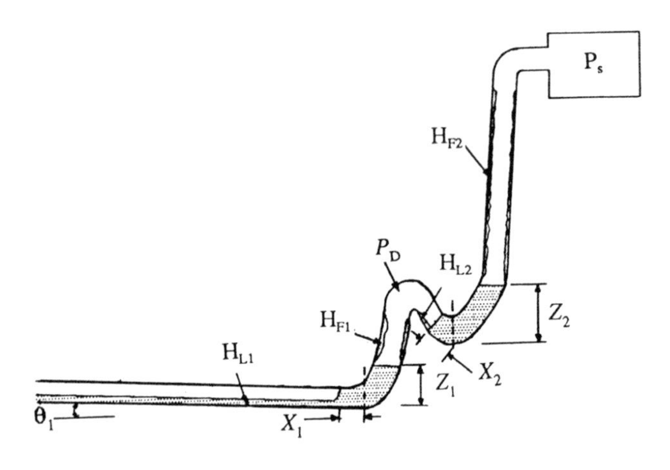

Figure 7 Liquid distribution in a lazy's' riser/pipeline system

The total liquid mass in the pipeline/riser system should satisfy the following liquid mass balance equation.

$$m_L = \rho_L A\ [\ H_{L1}(l_{p1}-X_1) + X_1 + H_{L2}(l_{p2}-X_2) + X_2 + H_{F1}(l_{r1}-Y_1) + Y_1 + H_{F2}(l_{r2}-Y_2) + Y_2\ ] \qquad [2]$$

where A is the pipe cross section area, H_{L1}, H_{L2}, H_{F1} and H_{F2} are the average holdup at different pipe sections. Y_1 and Y_2 are the riser lengths where liquid accumulate and represneted with liquid heights of Z_1 and Z_2 in Figure 7.

Assuming ideal gas behaviour and the temperature distribution across the system is constant, the gas mass balance is simply:

$$m_G = A\left[\frac{P_P}{RT}(l_{p1}-X_1)(1-H_{L1}) + \frac{P_D}{RT}[(l_{p2}-x_2)(1-H_{L2}) + (l_{r1}-Y_1)(1-H_{F1})] + \frac{P_s}{RT}(R_2-Y_2)(1-H_{F2})\right]$$

[3]

The equations [1] to [3] are the basic governing equations for predicting the severe slugging-1 cycle characteristics in an 's' shaped riser. The equations for free hanging catenary or vertical riser can be simply obtained by removing the terms for the downward limb and the upper limb of the 's' shaped riser which are denoted by subscript '2' in the terms.

Based on experimental results and to a good approximation that terms H_{F1} and H_{F2} are equal to zero during slug production stage. Differentiating equations [2], the liquid mass conservation equation is:

$$\dot{m}_L = \rho_L A\left[(\frac{dX_1}{dt})(1-H_{L1}) + \frac{dY_1}{dt}(1-H_{F1}) + \frac{dX_2}{dt}(1-H_{L2}) + \frac{dY_2}{dt}(1-H_{F2})\right] \quad [4]$$

4.2 Slug build-up stage

The liquid distribution in a lazy's' riser/pipeline system at the beginning of this stage (time: t_0) is shown in Figure 8.

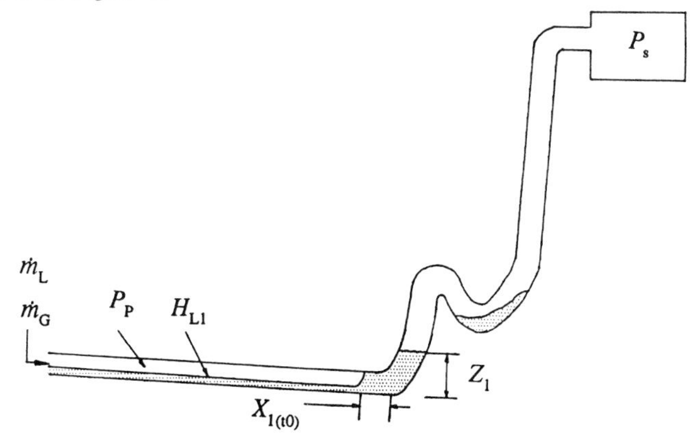

Figure 8 Liquid Distribution in the system at t_0

$Y_{1(t0)}$ is the riser pipe length where liquid accumulate and it is represented with liquid height of Z_1 in Figure 8, this value can be obtained by equation [5].

$$Y_{1(t0)} = a(l_{r1} + l_{r2})\ V_{SL} \qquad [5]$$

where l_{r1} and l_{r2} are the length of the lower limb and the upper limb respectively. Based on three different riser configurations, the value of constant 'a' is 0.156 for 2 degrees declined pipeline, but the range of application of this correlation is restricted by limited experimental data on pipeline inclination angle. However, the value of $Y_{1(t0)}$ is relatively small in comparison to the total length of the riser, hence overall prediction error caused by this value would not be significant.

Based on experimental measurements, only a very small amount of liquid is trapped at the dip of the 's' shaped riser and the liquid fall back is minimum. Hence, $Y_{2(t0)}$, $X_{2(t0)}$, H_{F1} and H_{F2} are approximately equal to zero. The slug build-up time can be obtained by integrating equation [4] from t_0 to t_1:

$$t_{sbu} = A\frac{\rho_L}{\dot{m}_L}\left[(X_{1(t1)} - X_{1(t0)})(1-H_{L1}) + (X_{2(t1)}-X_{2(t0)})(1-H_{L2}) + (l_{r1}-Y_1) + (l_{r2}-Y_2)\right] \qquad [6]$$

4.3 Calculation Procedure for the slug production stage

The lower and upper limbs are full of liquid at the beginning of this stage ($Y_{1(t1)} = l_{r1}$ and $Y_{2(t1)} = l_{r2}$), Figure 9 illustrates the liquid distribution in the system at time: t_1.

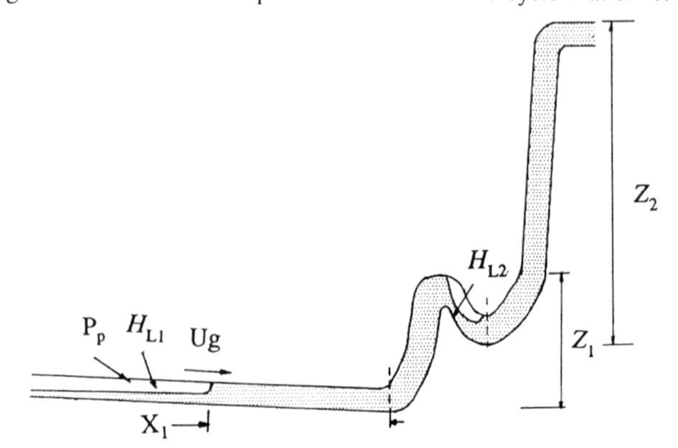

Figure 9 Liquid distribution in the system at t_1

Differentiating equation [3] and inputting the boundary conditions, the bubble front velocity (Ug, $Ug = dx_1/dt$) is obtained as follows:

$$U_g = \frac{\dot{m}_G}{A(1 - H_{L1})} \left[\frac{RT}{P_p}\right] \qquad [7]$$

Since the gas front reaches the riser foot at the end of this stage, $X_{1(t2)} = 0$. Value of $X_{1(t1)}$ can be obtained from slug build up stage calculation results. Hence, the slug production time t_{sp}.

can be obtained by solving [7] and [1]. Also, the liquid sweep out during the slug production stage can be calculated by solving equations [1], [4] and [7] using the calculated value of X_1.

4.4 Bubble Penetration Stage

The liquid distribution in the system at the beginning of this stage (time t_2) is shown in Figure 10. Figure 11 shows the liquid distribution after bubbles penetrate the riser.

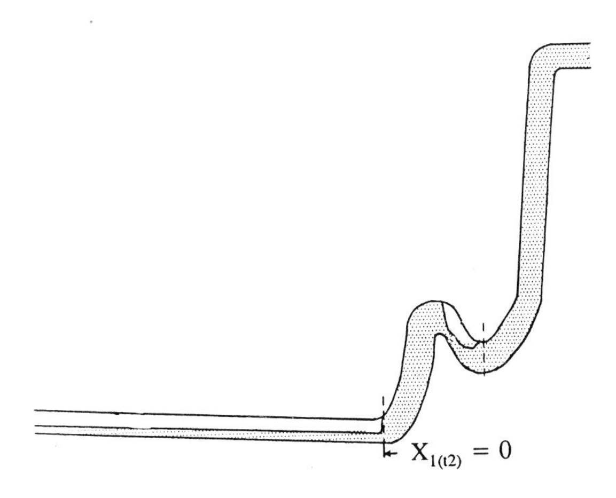

Figure 10 Liquid distribution in the system at time t_2

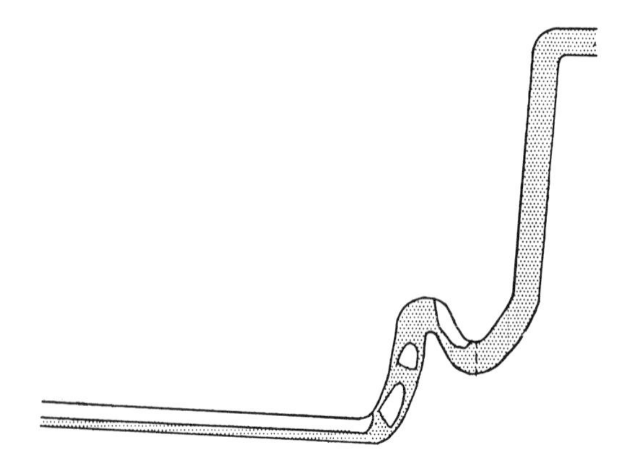

Figure 11 Liquid distribution during bubble penetration stage

4.4.1 Equation for the bubble front velocity

The rate at which the gas bubbles penetrate into the liquid slug is very important in determining the bubble penetration time and liquid slug velocity produced into the separator. The gas that penetrates the bottom of the riser is in the form of either small bubbles or Taylor bubbles. In both cases the translational velocity of the bubbles U_t can be expressed in the following form:

$$U_t = C\, U_s + U_d \qquad [8]$$

where U_s is the mixture velocity and U_d is the drift velocity. For bubble flow, we can assume that $C = 1$ and U_d is given by Couent (Ref 6):

$$Fr' = \frac{U_d}{(2ga\,\frac{\Delta\rho}{\rho})^{0.5}} \qquad [9]$$

where Fr' = Fr as $\Delta\rho = \rho$ for air/water system and 'a' is the radius of the pipe. Hence [9] reduces to:

$$Fr = \frac{U_d}{(2ga)^{0.5}} \qquad [10]$$

As the pipe is inclined at an angle **β** from the vertical. The dependence of Froude number (Fr) on inclination angle (**β**) for three different fluids is shown in Figure 12. The value of U_d can be calculated based on the value of Fr in Figure 12.

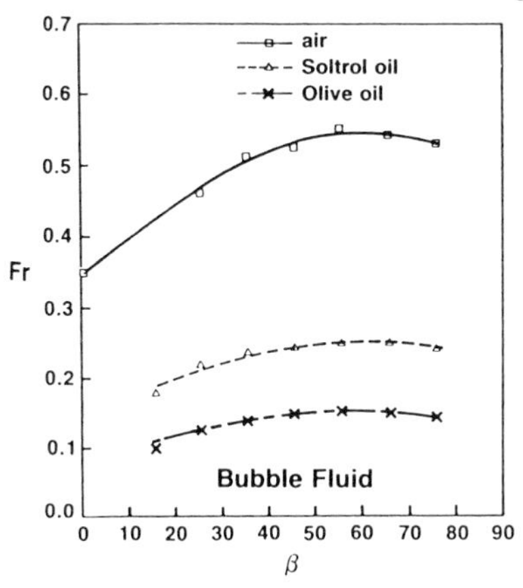

Figure 12 Fr versus β for three different bubble fluids.

4.4.2 Liquid sweep out during bubble penetration stage

The calculation of the bubble penetration stage assumes a bubble penetrating the riser at each time step (Δt). This assumption appears to be unrealistic, but the distribution and the sizes of the bubbles are not used in the calculation, only the average holdup value is used. Hence, the prediction result is not strongly dependent on the division of the time step as it appears in the above assumption.

Refer to Figure 13, the initial values of the variables : U_g and U_s at time t_2 are obtained from the slug production stage calculation results.

Assuming the first bubble penetrates the riser at t_2 and reach location l_b at $t_2 + \Delta t$,

$$l_b = \Delta t \, U_t = \Delta t \, [cU_s + U_d] \qquad [11]$$

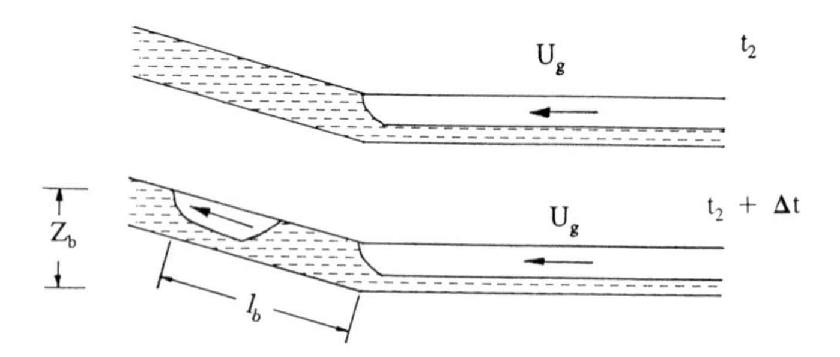

Figure 13 Bubble penetration

Based on pressure balance and gas mass balance equations and assuming that the pipeline liquid holdup remains constant, the volume of the first gas bubble (V_b) should follow the equation:

$$V_{b1} = U_g \, \Delta t \, A \qquad [12]$$

Let the average liquid hold up from the riser foot to riser length l_b be Φ_b,

$$\Phi_b = \frac{U_g}{U_t} \qquad [13]$$

Hence, the average liquid holdup in the lower limb of an 's' shaped riser (Φ) should follows:

$$\Phi = \frac{[(l_{r1} - l_b) + (l_b \, \Phi_b)]}{l_{r1}} \qquad [14]$$

Referring to Figure 13,the first bubble would reach vertical height of Z_b, this value can be obtained based on the value of l_b and the riser geometry. The liquid mass production rate from time t_2 to $t_2 + \Delta t$ can be obtained by solving the liquid and gas mass conservation equations. The mass flow rate m_s should follows:

$$m_s = \rho_l A (V_{sl} + \frac{d\Phi}{dt}) \quad [15]$$

In order to predict the velocity of second gas bubble, it is important to determine the pipeline gas expansion effect due to the decrease in hydrostatic head. Also, the frictional pressure drop should be taken into consideration in this case. The pressure balance equation for free hanging catenary riser becomes:

$$P_s = P_p + \Delta P_{friction} + \rho_L g \Phi l_{r1} \quad [16]$$

Based on the pressure balance, gas and liquid mass balance equations, the liquid sweep-out at each time step can be calculated.

4.4.3 Frictional pressure drop calculation

Based on experimental results, the liquid production during the bubble penetration could be ten times higher than the input liquid flowrate. Due to the high liquid velocity, the frictional pressure drop becomes significant, although the frictional pressure drop can be neglected for the previous stage.

The calculation of the frictional pressure drop along the riser during bubble penetration stage is divided into two parts. The flow pattern from the riser foot up to the location of the first bubble (l_b) is considered as two-phase bubble flow. From point l_b up to the riser exit, it is considered as a single phase liquid flow.

By solving the gas and liquid mass conservation equations, from the riser foot to point l_b, the superficial gas velocity is $U_g / (1-H_L)$ and the superficial liquid flow rate is V_{sl}. From point l_b to the riser exit, single phase liquid pressure drop is used for the calculation. The liquid velocity is simply $m_s / A\rho_L$.

The friction factor used for this calculation is obtained by the Moody diagram and the Beggs and Brill two-phase pressure drop prediction method is used in this work. Ref 9 gives the details of the frictional pressure drop calculation.

5. RESULTS AND DISCUSSION

The experimental and analytical results presented in this section are focused on the liquid sweep out during severe slugging-1 cycle. The severity of the liquid sweep out during a severe slugging cycle is quantified in two ways. First, there is the slug length, which determines the volume of the liquid that the downstream processing facilities would be receiving. Second, there is the maximum slug velocity which introduces hydrodynamic forces in the system, this affects the requirements for the separator performance and the pipework loading.

Experimental results obtained from the 2-inch pipeline/riser test loop are compared with predictions. Figures 14 to 17 present the comparison between the theoretical results and the experimental measurements for slug build-up time, slug production time, bubble penetration time and slug length respectively.

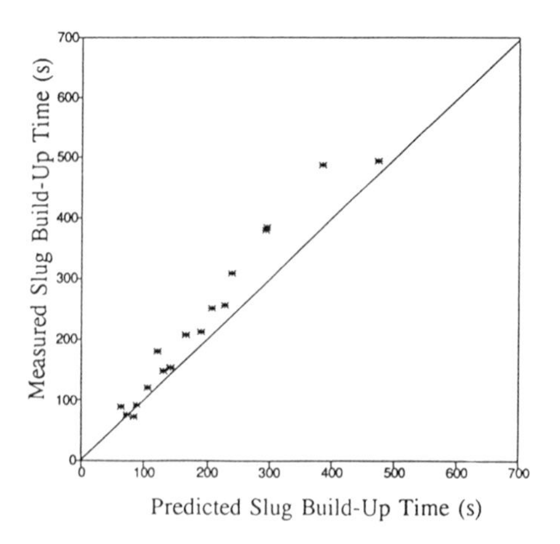

Figure 14 Free hanging catenary riser - Measured Vs Calculated slug build up time

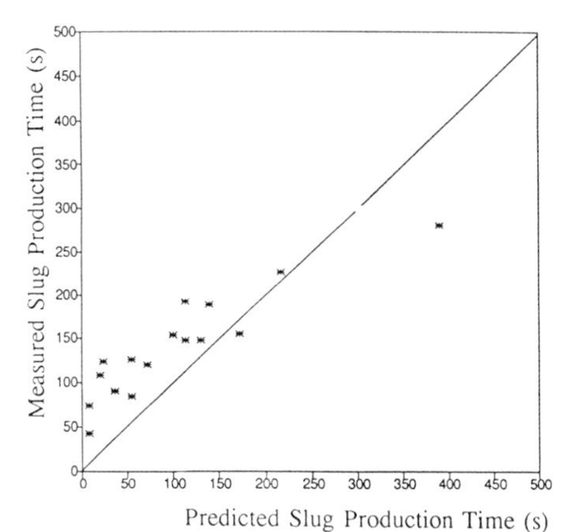

Figure 15 Free hanging catenary riser - Measured Vs Calculated slug production time

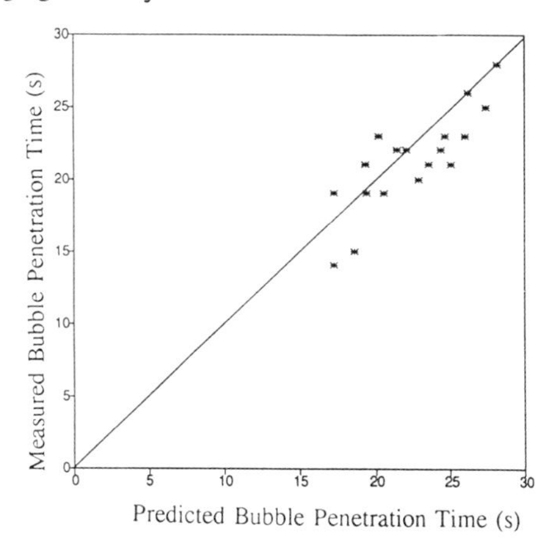

Figure 16 Free hanging catenary riser- Measured Vs Calculated bubble penetration time

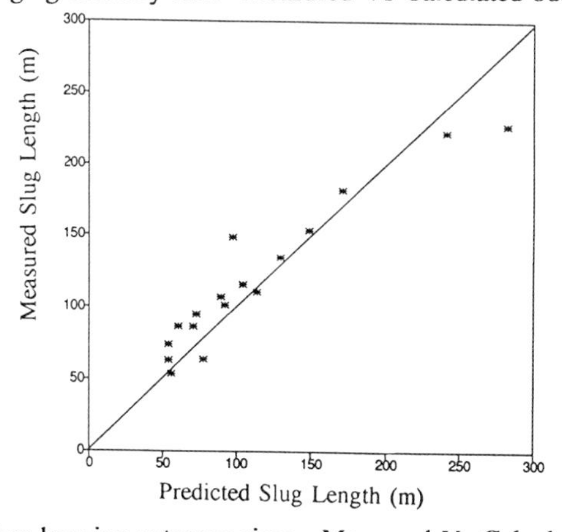

Figure 17 Free hanging catenary riser - Measured Vs Calculated slug length

The model developed has been used successfully to predict the liquid sweep out on severe slugging-1 cycle. Figure 18 shows the comparison between measured and predicted liquid velocity at the riser outlet. The slug velocity is measured during the slug production and bubble penetration stages, and the measurements from the gamma densitometer confirms that only pure liquid flowing through the outlet of the riser.

The predicted liquid sweep out calculation uses a time step of 0.5 second, the time step used for the end of the bubble penetration stage calculation is based on the predicted time taken for the first bubble to reach the top of the riser. The measured time for the end of this stage is based on the readings from a gamma densitometer which is installed at the outlet of the riser.

The slug velocity during the slug production stage is very accurately predicted in the free hanging catenary, since this prediction involves very few assumptions. The measured maximum slug velocity is approximately 4.6 m/s and the predicted value is slightly higher at 5.0 m/s. Although there is a small discrepancy for slug acceleration prediction, the results clearly show that this approach can generally predict the bubble penetration phenomenon reasonably well.

The above-mentioned discrepancy may also be due to the inaccuracies caused by use of the magnetic flowmeter on liquid production measurements, since the flowmeter require approximately 1 second to respond to any changes in flowrate.

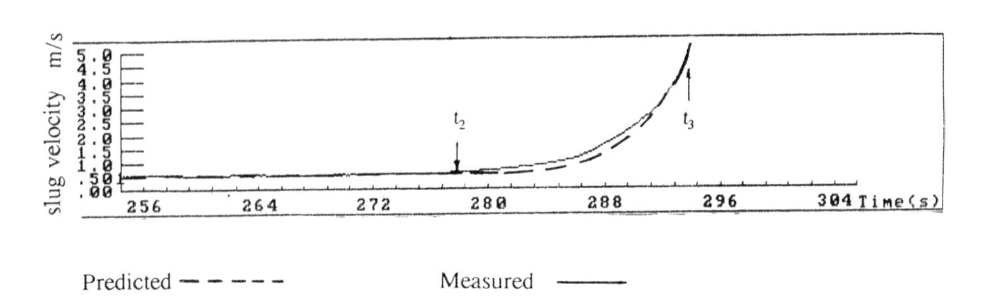

Figure 18 Free hanging catenary riser - Comparison between measured and predicted liquid sweep out.

Figure 19 shows the comparison between measured and calculated slug length for the lazy's' riser. Agreement seems to be reasonable, although the model slightly overpredicts the slug length. The comparison also clearly shows that the occurrence of severe slugging-1t creates shorter slug length than the severe slugging-1 cycle.

It is believed that the main reason for the slight overprediction on the slug length is due to the inaccurate prediction on the length of the trapped gas in the downward limb and also because the 's' shaped riser geometry is simplified in the calculation.

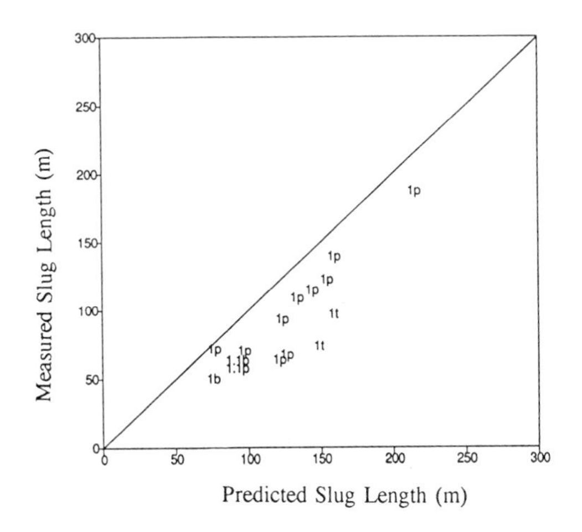

Figure 19 Lazy's' riser - Measured Vs Calculated slug length

Evidently, the severe slugging cycle behaviour in an 's' shaped riser is highly dependent on the trapped gas behaviour. For instance, the gas blow down can be initiated by the trapped gas penetrating the upper limb. Also, the trapped gas behaviour appears to be highly dependent on the gas and liquid physical properties. It is possible that, with a system using different fluids, the trapped gas in the downward limb would significantly affect the cycle behaviour.

One of the assumptions made in the model is the pipeline liquid hold-up prediction. The calculation is based on constant liquid hold-up for different mixture flowrates. However, the experimental results show that the pipeline liquid hold-up fluctuates during the cycle. It is believed that a more accurate prediction for holdup would improve the accuracy of the model.

Moreover, the pipeline liquid holdup values in a free hanging catenary riser are found to be significantly different from the values obtained from previous research on vertical risers (Ref 8). This preliminary model is based on only severe slugging-1p cycle. The severe slugging-1t cycle and the lower limb cycling occurring in an 's' shaped riser described in the previous section are not considered at this early stage. Also, a complete model for predicting severe slugging characteristics in flexible riser system would require accurate predictions of pipeline liquid hold-up, bubble behaviour and liquid fall-back in the riser. Each of these phenomena involves complex processes and needs separate studies to achieve accurate predictions.

5. CONCLUSION

1. Using a number of simplifying assumptions, a mathematical model based on basic physical principles has been developed for the severe slugging 1 cycle. The model can be used to predict pressure fluctuation and liquid sweep-out during severe slugging 1 cycle. The model is tested against experimental data and good agreement is recorded on free hanging catenary riser predictions. The validation results on 's' shaped riser show a relatively higher error.

2. Severe slugging which occurs in an 's' shaped riser is found to be significantly dependent on the behaviour of the trapped gas in the downward limb. The trapped gas behaviour is a complicated phenomena and it needs separate studies to form a more accurate model.

3. Although the range of application of this preliminary model is limited, it clearly indicates that this simple approach can be further developed for accurate field applications. Based on this analytical work, a few other areas have been identified where a better understanding of the physical phenomena is required. Additional work in the following areas is recommended:

    1. Trapped gas behaviour in the downward limb.
    2. Pipeline liquid holdup fluctuation during the cycle.
    3. Holdup in liquid slug.

4. Liquid fall back along the riser.
5. The effect of pressure fluctuation downstream of the riser as maintained by the topside separator. Although the present model assumes a constant separator pressure.

6 ACKNOWLEDGEMENT

The experimental and part of the analytical work was funded by the Health and Safety Executive (HSE) as part of the Multi-Phase Pipelines and Equipment Programme (MPE) run by CALtec Ltd. The author appreciates their permission to publish this paper. However, the views expressed in the paper are those of the author and do not represent the views of the HSE or MPE sponsors.

7 REFERENCES

1. Tin V., - "Severe Slugging in Flexible Risers" - Pro. The 5th international Conference on Multi-Phase Flow. 1991

2. Schmidt Z. - "Experimental Study of Two-Phase Flow in a Pipeline-Riser System" - PhD thesis, U. of Tulsa, 1977.

3. Pots B.F.M., Bomilow I.G., Konijn M.J.W.F., - "Severe Slug Flow in Offshore Flowline/Riser Systems" - SPE 13723, 1985.

4. Fuchs P. - "Pressure Limit for Terrain Slugging" - Proc. The 3rd International Conference on Multiphase Flow, May 1987

5. Couet B., Strumolo G.S., Ziehl W., - "The Effects of Fluid Properties, Inclination and Tube Geometry on Rising Bubbles" - Proc. The 3rd International Conference on Multi-Phase Flow, May 1987.

6. Schmidt Z., Doty D.R., Dutta Roy K., - "Severe Slugging in Offshore Pipeline Riser-Pipe Systems" - SPE Journal, February 1985.

7. Fabre J., Peresson L.L., Corteville J., Odello R., Bourgeois T., - "Severe Slugging in Pipeline/Riser Systems" - SPE 16846, 1987.

8. Taitel Y., - "Stability of Severe Slugging" - Int. J. Multiphase Flow Vol 12, No. 2, 1986.

9. Beggs H.D., Brill J.P. -"A Study of Two-Phase Flow in Inclined Pipes" - Journal of Petroleum Technology, May 1973.

SLUG TRANSLATIONAL VELOCITIES

Malcolm Watson
BP Research & Engineering Centre
Chertsey Road
Sunbury-on-Thames
Middlesex TW16 7LN

ABSTRACT

Slug translational velocities are an important feature of pipeline multi-phase flows. Along with slug lengths and frequencies, translational velocities are the most often recorded properties in experimental and field data. The accurate prediction of slug translational velocities is therefore a crucial test of any mathematical model of multi-phase flow and is an important step towards the development of flow regime maps.

A formula for slug translational velocities has been derived by considering the front of a slug as a moving hydraulic jump. The jump conditions lead to a formula whose coefficients depend on the film and slug hold-ups. Limiting cases for unaerated slugs, low and high mixture velocities are in very good qualitative agreement with well established correlations. The formula also gives results in very good quantitative agreement with experimental data from the BP Exploration Sunbury air/water 6 inch rig.

1 INTRODUCTION

For pipeline two-phase flows operating in the slug regime, one of the important properties is the slug front translational velocity. Any mathematical model of two-phase flows needs the slug translational velocity in order to predict slug lengths and frequencies. Slug translational velocities have been found, from both experimental and field data, to depend linearly on the mixture velocity over a wide range of pipe diameters and fluid properties. However, no universal relationship has been found in that the coefficients differ depending on the data set. Therefore these correlations cannot be used with great confidence outside of the range of the data from which they have been obtained.

There is a need to develop reliable mathematical models that are as free from empirical correlations as possible so that they can be used with confidence to predict the important features of multi-phase flows. Although the flows are very complex and three-dimensional, in order to produce useful engineering tools, it is only feasible to concentrate on simple one-dimensional models. However, it is believed that such models can capture the key physical mechanisms involved and lead to valuable predictions.

Based on the one-dimensional averaged equations of motion for the conservation of mass and momentum for each phase, mathematical models have been developed for transient two- (Watson, 1990) and three-phase flows (Watson, 1992). These models have not included any effects of gas entrainment so that slug hold-ups have been assumed to be one. It has been observed from numerical simulations that slug translational velocities are smaller than those typically observed. It is our belief that the major reason for this discrepancy is the neglect of gas entrainment in the slug.

The modelling approach that we have adopted leads to the front of a slug being treated as a moving hydraulic jump. The jump conditions can be obtained very easily from the governing equations and just relate quantities either side of the jump, and in particular the film and slug hold-ups, leading to a formula for the slug translational velocity. The formula can be validated by comparing it with experimental data for which the film and slug hold-ups have been recorded. This has been done for experimental data from the BP Exploration Sunbury 6 inch rig for air/water flows. Agreement between theory and experiment will help to validate the treatment of the front of the slug as a moving hydraulic jump. Ruder et al (1989) also considered the front of a slug to be a hydraulic jump and examined the jump conditions. However, they did not include the effects of entrainment and only obtained inequalities for the translational velocities rather than explicit expressions.

The formula brings in the effect of gas entrainment by considering slug hold-ups of less than one. The effect of gas entrainment is likely to be more important for high pressure hydrocarbon flows than for low pressure air/water flows which are often used as the basis for correlations that are used in the industry. It is essential therefore that any model should include the effect of gas entrainment if it is to be used with confidence as a predictive tool.

2 PREVIOUS WORK

2.1 Correlations

It has been observed over a wide range of pipe diameters, flowrates and fluid properties that experimental and field data of slug front translational velocities (V_t) are well represented by linear relationships of the form

$$V_t = C_0 V_m + V_0 \qquad (1)$$

where

$$V_m = U_{LS} + U_{GS} \tag{2}$$

is the mixture velocity, U_{LS} and U_{GS} are the liquid and gas superficial velocities respectively and C_0 and V_0 are fitted parameters.

For example Bendiksen (1984) obtained

$$C_0 = 1.05 + 0.15 \sin^2\theta, \quad V_0 = (0.35 \sin\theta + 0.54 \cos\theta)(gD)^{1/2}, \qquad \text{for } Fr < 3.5 \tag{3}$$

$$C_0 = 1.20, \quad V_0 = 0.35 \sin\theta \, (gD)^{1/2}, \qquad \text{for } Fr > 3.5 \tag{4}$$

where $Fr = V_m/(gD)^{1/2}$ is the Froude number.

Brandt and Fuchs (1989) have used large scale, high pressure data from the SINTEF Two-Phase Flow Laboratory in Norway (D=0.189 m) to produce a fit of the form (1) with

$$C_0 = 1.05, \quad V_0 = 0.75 \text{ m/s} \qquad \text{for } V_m < 5 \text{ m/s} \tag{5}$$

$$C_0 = 1.20, \quad V_0 = 0.0 \text{ m/s}. \qquad \text{for } V_m > 5 \text{ m/s} \tag{6}$$

For horizontal flows (θ=0), Bendiksen's results agree very well with those of Brandt and Fuchs.

DG Wood (BP internal report) has produced a fit of the form (1) for air/water data from the BP Exploration 6 inch (0.15 m) rig at Sunbury. For horizontal flows, the best fit is given by

$$V_t = 1.09 \, V_m + 0.79 \text{ (in m/s)}, \tag{7}$$

which is compatible with the results of Bendiksen, and Brandt and Fuchs above.

Russell (1990) considered the usual linear fit (1) for experimental data obtained from the BHRA large pipeline test facility (LPTF) for diameters of 0.2 and 0.4 m. He reported that it was generally necessary to take non-zero values of V_0 and that unaerated slugs could best be represented by taking $C_0 = 1.0$. He went on to suggest that the effect of slug aeration at higher mixture velocities could be represented by a quadratic term.

2.2 Dukler and Hubbard Model

A different approach was adopted by Dukler and Hubbard (1975) who developed a phenomenological model for slug flow. Conservation of mass leads to the equation

$$V_t = V_m + x/(\rho_L A H_s) \tag{8}$$

where x is the rate of pickup at the front of the slug, ρ_L is the liquid density, A is the pipe cross-section and H_s is the slug hold-up. They rewrote (8) in the form

$$V_t = (1+C)V_m \tag{9}$$

where C is the ratio of the rate of pickup to the rate of flow in the slug. For slugs of constant length the rate of pickup at the front of the slug equals the rate of shedding at the rear. They obtained an integral expression for C based on the velocity distribution for turbulent pipe flow. The integral has to be evaluated numerically but for slug Reynolds numbers in the range $3\times10^4 < Re < 4\times10^5$ they showed that C was well represented by the approximation

$$C = 0.021 \ln(Re) + 0.022 \tag{10}$$

so that $0.238 < C < 0.293$ over this range of Reynolds numbers.

3 MODELLING

3.1 Basic Equations

We consider incompressible, two-phase flow of liquid and gas in a circular pipe of diameter D and inclination θ to the horizontal. The one-dimensional averaged equations describing the conservation of mass for the liquid layer and the entrained gas bubbles are

$$\frac{\partial}{\partial t}(\rho_L A_L(1-\phi)) + \frac{\partial}{\partial x}(\rho_L A_L(1-\phi)U_L) = -\Gamma_L, \tag{11}$$

$$\frac{\partial}{\partial t}(\rho_G A_L \phi) + \frac{\partial}{\partial x}(\rho_G A_L \phi U_B) = \Gamma_G, \tag{12}$$

where ρ_L and ρ_G are the liquid and gas densities, U_L and U_B are the liquid and bubble velocities, and Γ_L and Γ_G are algebraic representations of the entrainment of liquid droplets in the gas layer and gas bubbles in the liquid layer, respectively. The cross-sectional area occupied by the liquid layer is A_L of which a fraction ϕ is occupied by the entrained gas bubbles. There is a similar pair of equations for the gas layer with entrained liquid droplets.

If we assume that the bubbles travel at approximately the same speed as the liquid layer, we can add the mass equations (11) and (12) to the corresponding equations for the gas layer to give

$$\frac{\partial}{\partial t}(\rho_L^* A_L + \rho_G^* A_G) + \frac{\partial}{\partial x}(\rho_L^* A_L U_L + \rho_G^* A_G U_G) = 0, \tag{13}$$

expressing conservation of total mass and where $\rho_L^* = \rho_L(1-\phi)+\rho_G\phi$ is the mixture density of the liquid layer with a corresponding expression for the gas layer mixture density ρ_G^*.

Conservation of momentum for the liquid layer with the entrained gas bubbles gives

$$\frac{\partial}{\partial t}(\rho_L^* A_L U_L) + \frac{\partial}{\partial x}(\rho_L^* A_L U_L^2) = -A_L\frac{\partial p_I}{\partial x} - \rho_L^* g \cos\theta A_L\frac{\partial h}{\partial x}$$

$$- \rho_L^* g \sin\theta A_L + F_L, \qquad (14)$$

where F_L is an algebraic representation of the wall and interfacial friction terms. There is a similar equation for the gas layer: see, for example, Bendiksen et al (1991).

Because solutions with shocks are possible, it is necessary to recast the equations into a form suitable for numerical solution. In particular, they must be rewritten in so-called divergence form otherwise numerical methods which rely on shock capturing techniques may give shock strengths and speeds which depend on the numerical method used rather than the physics.

It is possible to put the momentum equation in divergence form by eliminating the interfacial pressure p_I from the liquid and gas layer equations giving a difference of momentum

$$\frac{\partial}{\partial t}(\rho_L^* U_L - \rho_G^* U_G) + \frac{\partial}{\partial x}\left(\frac{1}{2}\rho_L^* U_L^2 - \frac{1}{2}\rho_G^* U_G^2 + (\rho_L^* - \rho_G^*) g \cos\theta\, h\right)$$

$$= -(\rho_L^* - \rho_G^*) g \sin\theta + \frac{F_G}{A_G} - \frac{F_L}{A_L} \qquad (15)$$

These two partial differential equations (pdes) are supplemented by two algebraic equations

$$A_L + A_G = A, \qquad (16)$$

$$A_L U_L + A_G U_G = A(U_{LS} + U_{GS}) \; (= AV_m). \qquad (17)$$

Equation (16) just says that the fluids fill the pipe. Equation (17) says that the total volumetric flowrate is constant and has been obtained from the mass conservation equations by making use of the incompressibility of the phases.

The original system of governing equations has been reduced to two pdes along with two algebraic equations. This is quite consistent because the original system only admits two characteristic directions. Equations (13) and (15) are in the vector form

$$\frac{\partial \mathbf{U}}{\partial t} + \frac{\partial \mathbf{F}}{\partial x} = \mathbf{G}, \qquad (18)$$

which is suitable for application of shock capturing finite difference methods (Watson, 1990). The jump conditions at a discontinuity in $\mathbf{U}$ may be written down from (18) and they are

$$-c[\mathbf{U}] + [\mathbf{F}] = \mathbf{0}, \tag{19}$$

where [.] denotes the difference of the variable either side of the jump and c is the speed of propagation of the jump.

As shown by Whitham (1974), in general, a whole hierarchy of equations in divergence form may be built up from a hyperbolic system. At a discontinuity in the solution the jump conditions (19) will be different for different forms of the equations. For the two-phase flow equations, an alternative divergence form may be found (Watson, 1990). Although the jump conditions for this alternative set of equations differ from those for (13) and (15), in practice the shock speeds and strengths are very similar.

3.2 Slug Translational Velocities

Slug flow occurs for sufficiently large flowrates when an instability on the interface between the liquid and the gas leads to the rapid growth of small disturbances. These disturbances grow and the liquid may bridge the pipe forming a slug. In general some gas will be entrained at the front of the slug giving a slug hold-up of less than one.

The modelling approach that is being adopted here considers the front of the slug to be a shock or moving hydraulic jump. The speed of propagation of the jump is given by (19) above and is the slug translational velocity. The slug/bubble is shown schematically in Figure 1.

The slug front translational velocity is V_t, the liquid velocity in the film immediately ahead of the slug front is V_f and the liquid hold-ups in the slug and the film are H_s and H_f respectively. For a slug of constant length the back of the slug must move at the same speed as the front and since we assume that the liquid bridges the pipe we can take the speed of the gas in the bubble region to be constant and equal to V_t.

Using equation (17) in the film/bubble region we have

$$H_f V_f + (1-H_f)V_t = V_m . \tag{20}$$

The jump conditions at the front of the slug are obtained by applying (19). For the total mass equation we have

$$-V_t [\rho_L^* H_L + \rho_G^* H_G] + [\rho_L^* H_L U_L + \rho_G^* H_G U_G] = 0 \tag{21}$$

and for the momentum equation we have

$$-V_t [\rho_L^* U_L - \rho_G^* U_G]$$

$$+ \left[\frac{1}{2}\rho_L^* U_L^2 - \frac{1}{2}\rho_G^* U_G^2 + (\rho_L^* - \rho_G^*) g \cos\theta \, h\right] = 0 \qquad (22)$$

We ignore the effect of entrainment of liquid droplets into the gas layer so that $\rho_G^* = \rho_G$. In order to simplify the algebra we shall assume that the gas density is very much smaller than the liquid density $\rho_G/\rho_L << 1$. This is usually a good approximation for pipeline two-phase flows. It would be straightforward to retain terms of order ρ_G/ρ_L, but it adds little apart from confusion in the expressions.

In the slug the liquid bridges the pipe but with entrained gas bubbles so that the slug hold-up H_s is, in general, less than one. Hence, in the slug we take

$$\rho_L^* = \rho_L H_s ; \; H_L = 1; \; U_L = V_s \qquad (23);(24);(25)$$

In the film we assume that there is no (or very little) entrained gas, so that

$$\rho_L^* = \rho_L ; \quad H_L = H_f ; \; U_L = V_f \qquad (26);(27);(28)$$

Applying (21) and (22) with (23) to (28) we have

$$V_t (H_f - H_s) = H_f V_f - H_s V_s \qquad (29)$$

$$V_t (V_f - H_s V_s) = \frac{1}{2} V_f^2 + g h_f \cos\theta - \frac{1}{2} H_s V_s^2 - g D H_s \cos\theta \qquad (30)$$

where h_f is the height of the liquid film and D is the pipe diameter.

Elimination of V_f and V_s from (20), (29) and (30) gives a quadratic equation relating the slug translational velocity V_t and the mixture velocity V_m

$$\alpha V_t^2 - 2\beta V_t V_m + \beta V_m^2 = \gamma g D \cos\theta \qquad (31)$$

where

$$\alpha = H_s - H_f^2 - H_s (1-H_s) H_f^2 \qquad (32)$$

$$\beta = H_s - H_f^2 \qquad (33)$$

$$\gamma = 2 H_s H_f^2 (H_s - h_f/D). \qquad (34)$$

Although there is a non-linear relationship between the film height h_f and the film hold-up H_f, for typical film heights in the range 0.4D to 0.6D say, it is a good approximation to take a linear relationship, so that we replace (34) by

$$\gamma = 2H_s H_f^2 (H_s - H_f). \tag{35}$$

The equation (31) has a simple solution and taking the positive root we have

$$V_t = \lambda V_m + \{\lambda(\lambda-1)V_m^2 + \sigma g D \cos\theta\}^{1/2} \tag{36}$$

where

$$\lambda = \beta/\alpha; \quad \sigma = \gamma/\alpha. \tag{37}$$

For slug hold-ups of less than one, we have $\alpha < \beta$ so that $\lambda > 1$ and equality for slug hold-ups of one.

If we allow for drift of the entrained gas bubbles through the slug, we obtain slightly different expressions for α, β and γ but the resulting expression for V_t is again given by (36).

3.3 Limiting Cases

If there is no entrainment of gas bubbles in the slug, so that $H_s = 1$, then the above expressions for α, β and γ simplify and we have for the slug translational velocity

$$V_t = V_m + \{\sigma g D \cos\theta\}^{1/2}. \tag{38}$$

where now

$$\sigma = 2H_f^2 (1-H_f)/(1-H_f^2). \tag{39}$$

The form of (38) is exactly that observed by Russell (1990) for unaerated slugs, in that the gradient with respect to the mixture velocity V_m is one. The second term on the RHS of (38) is not constant and depends on the film hold-up through (39). Entrainment of gas bubbles in the liquid slug leads to $\lambda > 1$. Expanding (36) for small values of $\lambda-1$ gives a term quadratic in V_m, again in agreement with the observations of Russell.

For small flowrates, such that

$$\lambda(\lambda-1)V_m^2/(\sigma g D \cos\theta) \ll 1 \tag{40}$$

we have for the slug translational velocity

$$V_t = \lambda V_m + \{\sigma g D \cos\theta\}^{1/2} \tag{41}$$

where λ and σ are given by (39). This expression is very similar to (38) obtained for unaerated slugs. However, the gradient is now $\lambda > 1$, in accord with correlated fits.

For large flowrates, such that the inequality in (40) is reversed, we can approximate the slug translational velocity by

$$V_t = \mu V_m \tag{42}$$

$$\mu = \lambda + \{\lambda(\lambda-1)\}^{1/2}. \tag{43}$$

We note that the two limiting cases given by (41) and (42) with (43) are in qualitative agreement with the correlations of Bendiksen (1984) and Brandt and Fuchs (1989).

4 RESULTS

We have approximated the expression for the slug translational velocity and shown it to be in qualitative agreement with existing correlations. However, there is no problem in retaining the full expression (36) as it is straightforward to evaluate numerically. The expression for V_t requires knowledge of both the slug (H_s) and film (H_f) hold-ups. Often both of these are not reported in data sets. However, the work by DG Wood at Sunbury on the 6 inch rig for air/water flows does record both of these hold-ups over a considerable number of slug flow measurements. Wood quotes 83 data points in the slug flow region for which both the slug and film hold-ups are recorded. The data cover the ranges

$$0.62 \leq V_m \leq 8.87 \text{ (in m/s)}$$

$$0.19 \leq H_f \leq 0.67$$

$$0.58 \leq H_s \leq 0.96$$

$$1.57 \leq V_t \leq 10.95 \text{ (in m/s)}.$$

Wood produced a straight line correlation given by (7) for these data. A comparison of the measured and correlated results is shown in Figure 2, with the diagonal straight line indicating perfect agreement. There is quite good agreement between the measured and correlated results, but this might be expected since the correlation has been obtained from the measured data. Although straight lines of the form (1) are usually fitted to sets of data, the coefficients differ from case to case, so a correlation obtained from one set of data can only be expected to be of limited predictive capability when applied to other cases, such as different diameter, or different inclination pipelines, or different fluid properties.

The theory of Dukler and Hubbard (1975), described above in 2.2, also provides a prediction for slug translational velocities. A comparison of the measured and predicted results is shown in Figure 3. The predictions are not as good as those obtained from the correlation shown in Figure 2, especially for small mixture velocities, where the velocities tend to be underpredicted.

The model described in this paper leads to the slug translational velocity given by (36). For flows with no gas entrainment, so that the slug hold-up is one, the appropriate expression is given by (38).

Predictions from (38) and (36) are compared with the measured results in Figures 4 and 5 respectively.

Looking at Figure 4 we can see that for no entrainment, (38) gives very good predictions for low mixture velocities, but underpredicts for larger mixture velocities. This might be expected since entrainment is more likely to occur for high flowrates. Figure 5 shows predictions obtained from (36) and includes the effect of gas entrainment. These results are the best of all and demonstrate the importance of a model for gas entrainment. The slug hold-ups that have been used in the formula (36) are those that were observed in the experiments.

In order to provide a useful engineering tool, the slug entrainment needs to be predicted. One of the simplest correlations in form is that of Gregory et al (1978) obtained from small diameter pipes with air and light oil as the operating fluids. Their correlation for slug hold-up is

$$H_s = [1 + (V_m/8.66)^{1.39}]^{-1}. \tag{44}$$

We note that (44) does not incorporate any diameter or inclination effects and is not even based on dimensionless groups, but it is often used. Predicted slug hold-ups from (44) are shown in Figure 6. As can be seen the results are poor, with the slug hold-ups tending to be underpredicted, and shows the danger of using correlations.

Although the Gregory et al slug hold-ups compare very badly with Wood's experiments, it was decided to use these values in the the formula (36) rather than the measured slug hold-ups. The results are shown in Figure 7. For small mixture velocities the results are good, but overpredict the slug translational velocities for larger mixture velocities. For small mixture velocities there is little entrainment, but for large mixture velocities the underprediction of the Gregory et al hold-up leads to overpredictions of the translational velocities.

5 DISCUSSION

The best results for slug translational velocities are those obtained from (36) and shown in Figure 5. Equation (36) requires knowledge of the slug and film hold-ups in order to calculate the coefficients. The results shown in Figure 5 were obtained with the measured slug and film hold-ups. Using the well known slug hold-up correlation of Gregory et al (1978) gives the results shown in Figure 7. Not surprisingly, these results are not as good as those in Figure 5, but they are an improvement on the results with no entrainment. These results show the importance of slug hold-ups in the predictions of slug translational velocities, but a very accurate representation of the slug hold-up is probably not necessary.

Slug hold-ups are normally predicted using correlations such as Gregory et al. The entrainment mechanism at the front of the slug is not well understood and it is doubtful if even three-dimensional CFD calculations could be used reliably without a lot of 'tuning'. However, in the spirit of the one-dimensional modelling approach discussed in this paper, it is believed that semi-empirical models could be developed with the amount of gas entrainment related directly to a suitable dimensionless measure of the strength of the jump at the front of the slug.

However, even without reliable models for slug hold-ups, the formula (36) obtained for slug translational velocities has been validated. For small mixture velocities, there is little entrainment and the approximation (38) works well. In addition, the qualitative form of (38) is in agreement with the observations of Russell (1990). For larger mixture velocities, we again have qualitative agreement with Bendiksen (1984) and Brandt and Fuchs (1989).

Bendiksen (1984) produced empirical relationships including the effects of inclination. However, the dependence on inclination that he obtained seems to have little physical basis. The momentum conservation equations couple the trigonometric terms $\cos\theta$ and $\sin\theta$ with the acceleration due to gravity g, and they are not independent parameters as implied by his results, even though his expressions may fit data reasonably well. In the present model, there is the explicit dependence on inclination through $g \cos\theta$ in (36), but there is also an implicit dependence through the parameters λ and σ which depend on the slug and film hold-ups. These hold-ups will depend on the inclination also: more sophisticated modelling will be needed to determine this.

6 CONCLUSIONS

A formula for slug translational velocities has been obtained by treating the front of a slug as a moving hydraulic jump and taking into account gas entrainment.

The coefficients in the formula are not constant but depend on the slug and film hold-ups.

The limiting cases for unaerated slugs, low and high mixture velocities are all in very good qualitative agreement with existing correlations obtained over a wide range of pipe diameters and fluid properties by independent workers.

Results obtained using this formula are in very good quantitative agreement with data from the Sunbury 6 inch rig for air/water flows.

Results from the formula show the importance of including the effects of gas entrainment in the slug.

The translational velocity is an important feature of slug flow and its succesful prediction is a crucial step in the development of a flow regime map.

7 ACKNOWLEDGEMENTS

I wish to thank Doug Wood and Trevor Hill for the use of the experimental data and The British Petroleum Company plc for permission to publish this paper.

8 REFERENCES

Bendiksen, K, 1984, An experimental investigation of the motion of long bubbles in inclined tubes, Int J Multiphase Flow, **10**, 467.

Bendiksen, K, Malnes, D, Moe, R and Nuland, S, 1991, The dynamic two-fluid model OLGA: theory and application, SPE Prod Eng, May, 171.

Brandt, I and Fuchs, P, 1989, Liquid holdup in slugs: some experimental results from the SINTEF Two-phase Flow Lab, Proc 4th Int Conf on Multi-Phase Flow (Nice, France, 19-21 June, 1989, BHRA)

Dukler, AE and Hubbard, MG, 1975, A model for gas-liquid slug flow in horizontal and near horizontal tubes, Ind Eng Chem Fundam, **14**, 337.

Gregory, GA, Nicholson, MK and Aziz, K, 1978, Correlation of the liquid volume fraction in the slug for horizontal gas-liquid slug flow, Int J Multiphase Flow, **4**, 33.

Ruder, Z, Hanratty, PJ and Hanratty, TJ, 1989, Necessary conditions for the existence of stable slugs, Int J Multiphase Flow, **15**, 209.

Russell, D, 1990, Mechanistic modelling of slug flow characteristics in horizontal and near horizontal pipelines, BHRA Report No MPE046.

Watson, M, 1990, Non-linear waves in pipeline two-phase flows, Proc 3rd Int Conf on Hyperbolic Problems (Uppsala, Sweden, 11-15 June, 1990)

Watson, M, 1992, A model for pipeline three-phase flows, Presented at: 4th Int Conf on Hyperbolic Problems (Taormina, Italy, 3-8 April, 1992)

Whitham, GB, 1974, Linear and nonlinear waves, John Wiley & Sons, New York.

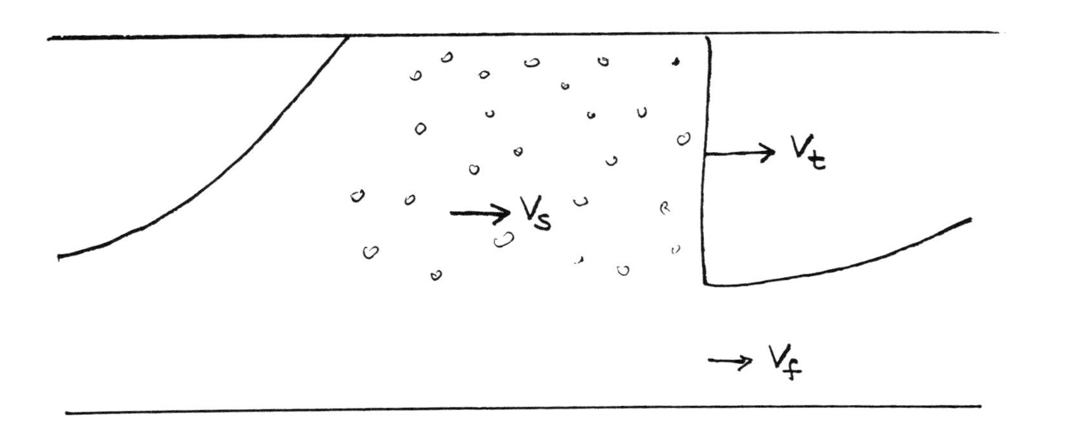

Figure 1 Schematic diagram of slug

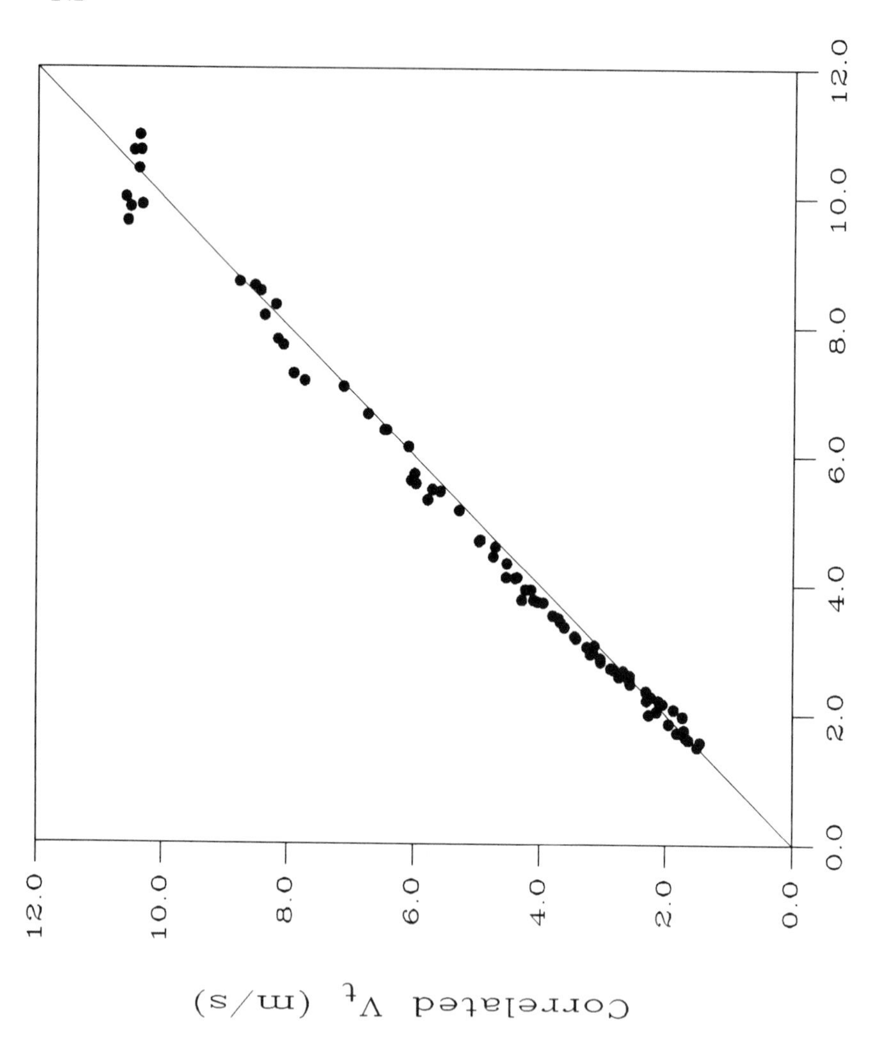

Figure 2 Comparison of BP correlation and measured slug translational velocities

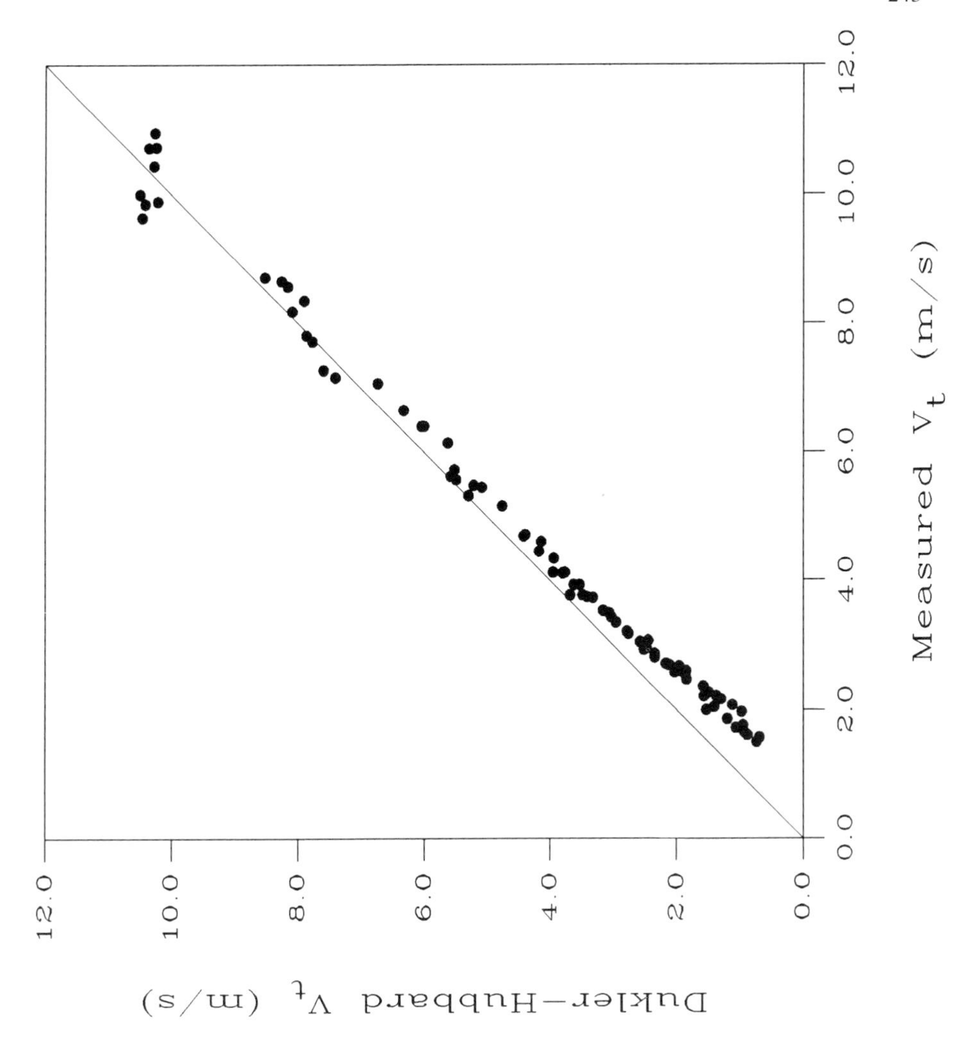

Figure 3 Comparison of Dukler & Hubbard and measured slug translational velocities

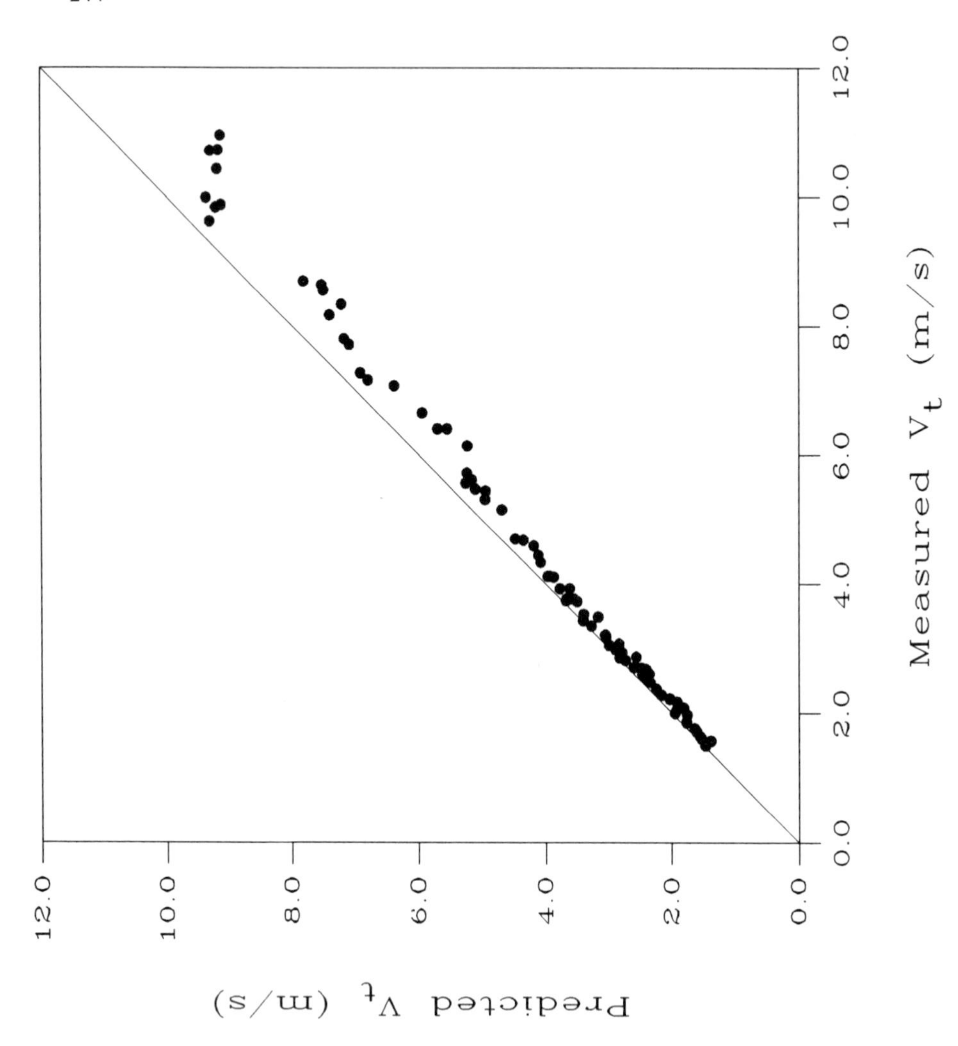

Figure 4 Comparison of formula for no entrainment and measured slug translational velocities

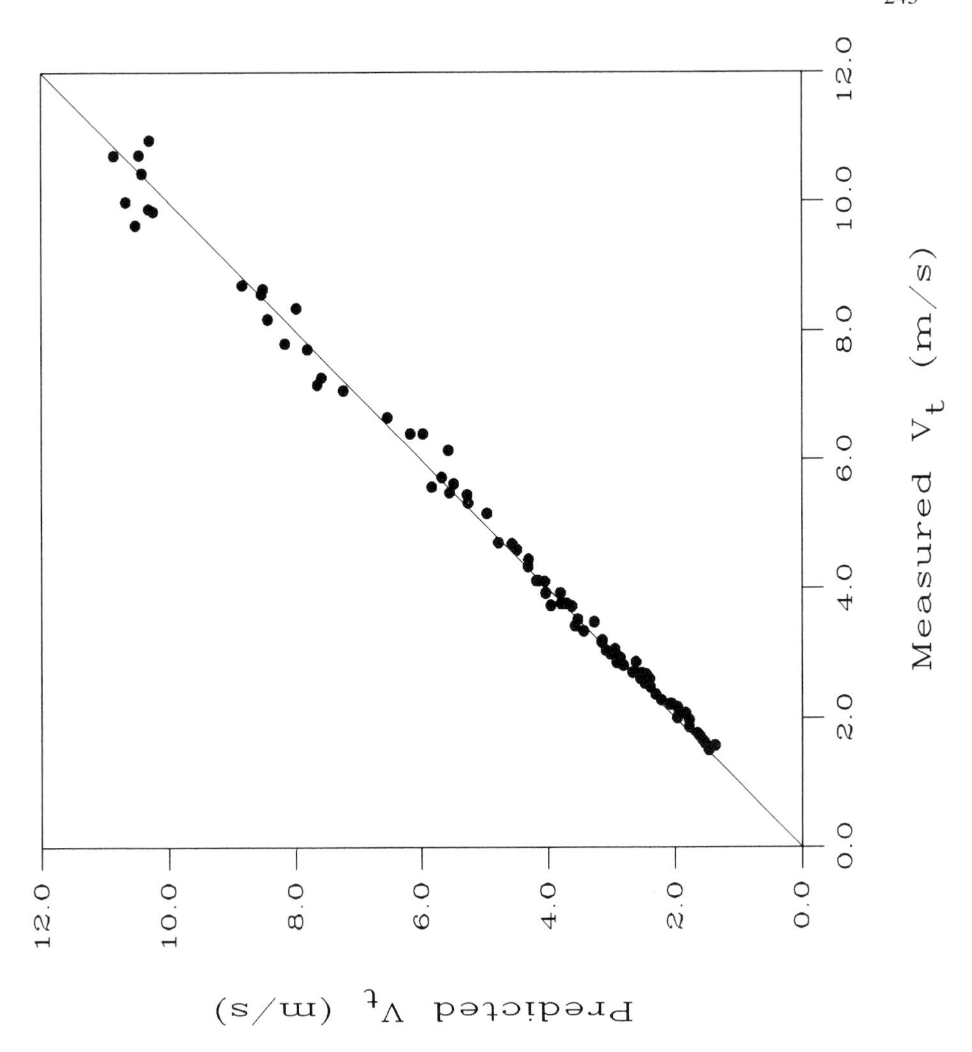

Figure 5 Comparison of formula including gas entrainment and measured slug translational velocities

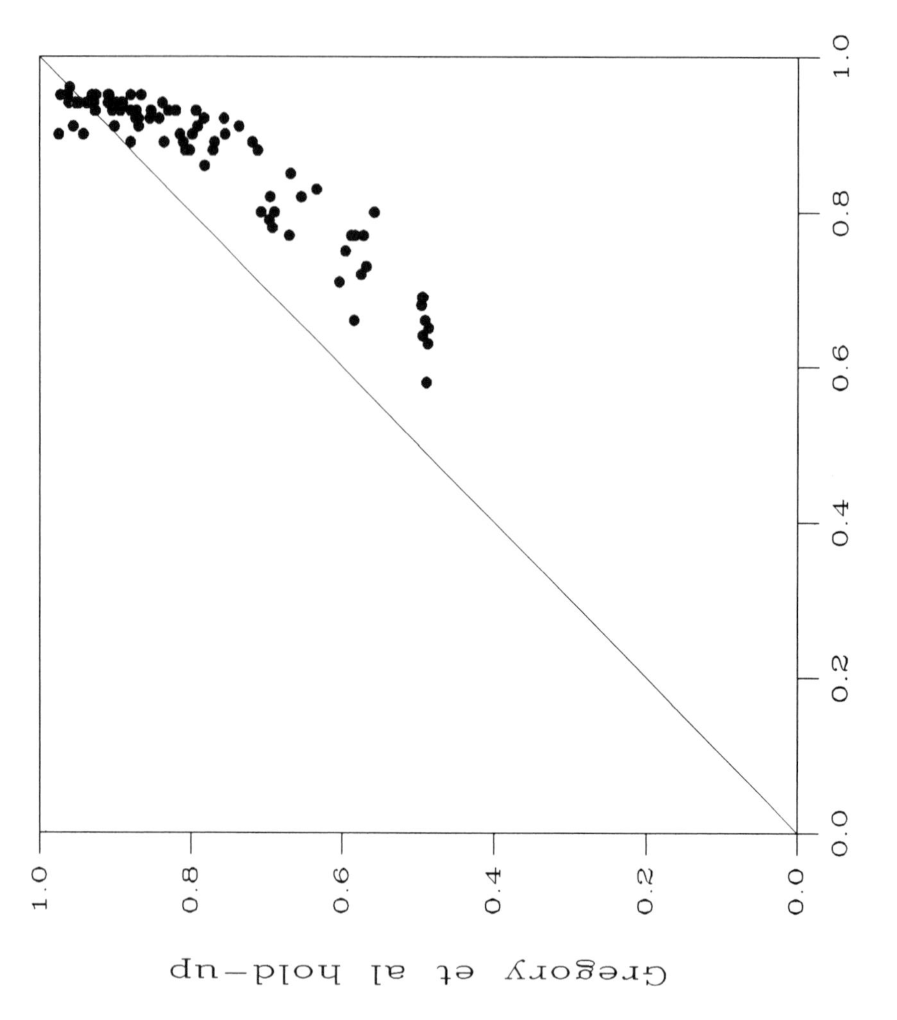

Figure 6 Comparison of Gregory et al hold-up correlation and measured slug hold-ups

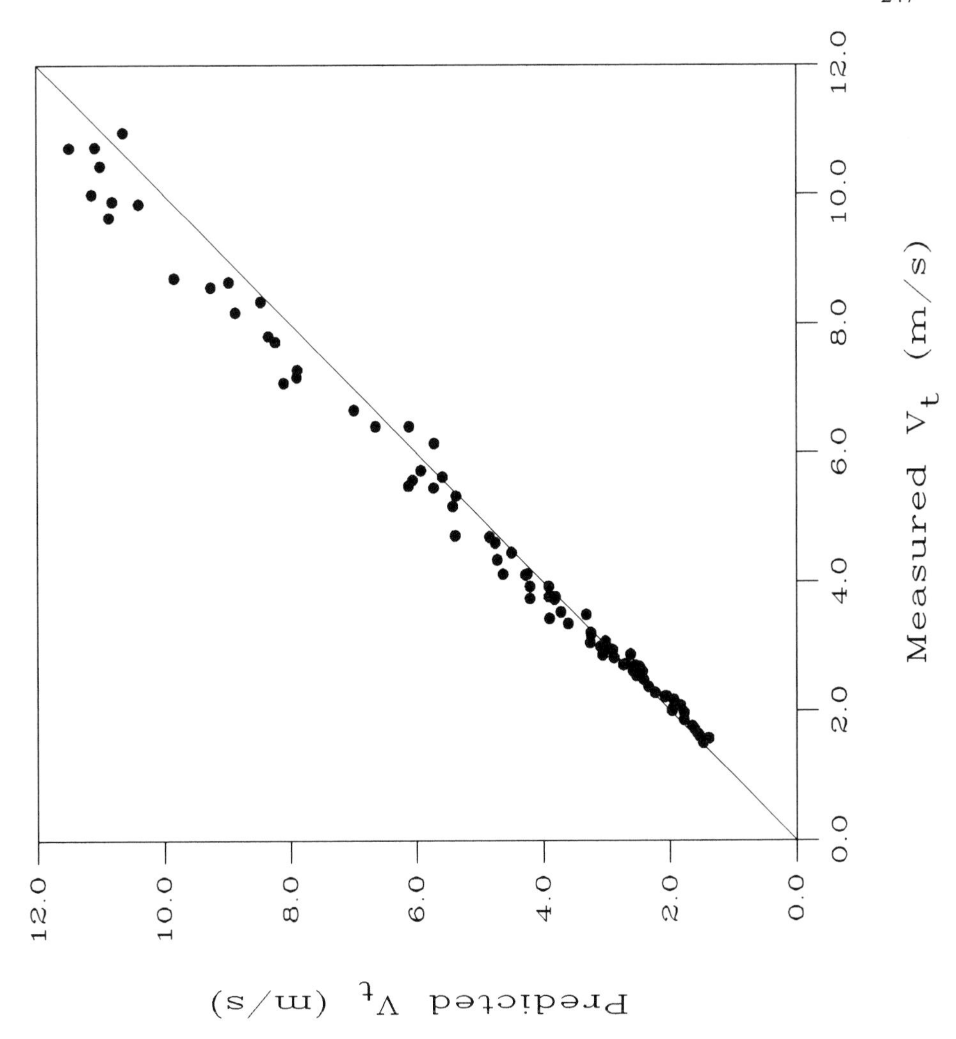

Figure 7 Comparison of formula including Gregory et al hold-up correlation and measured slug translational velocities

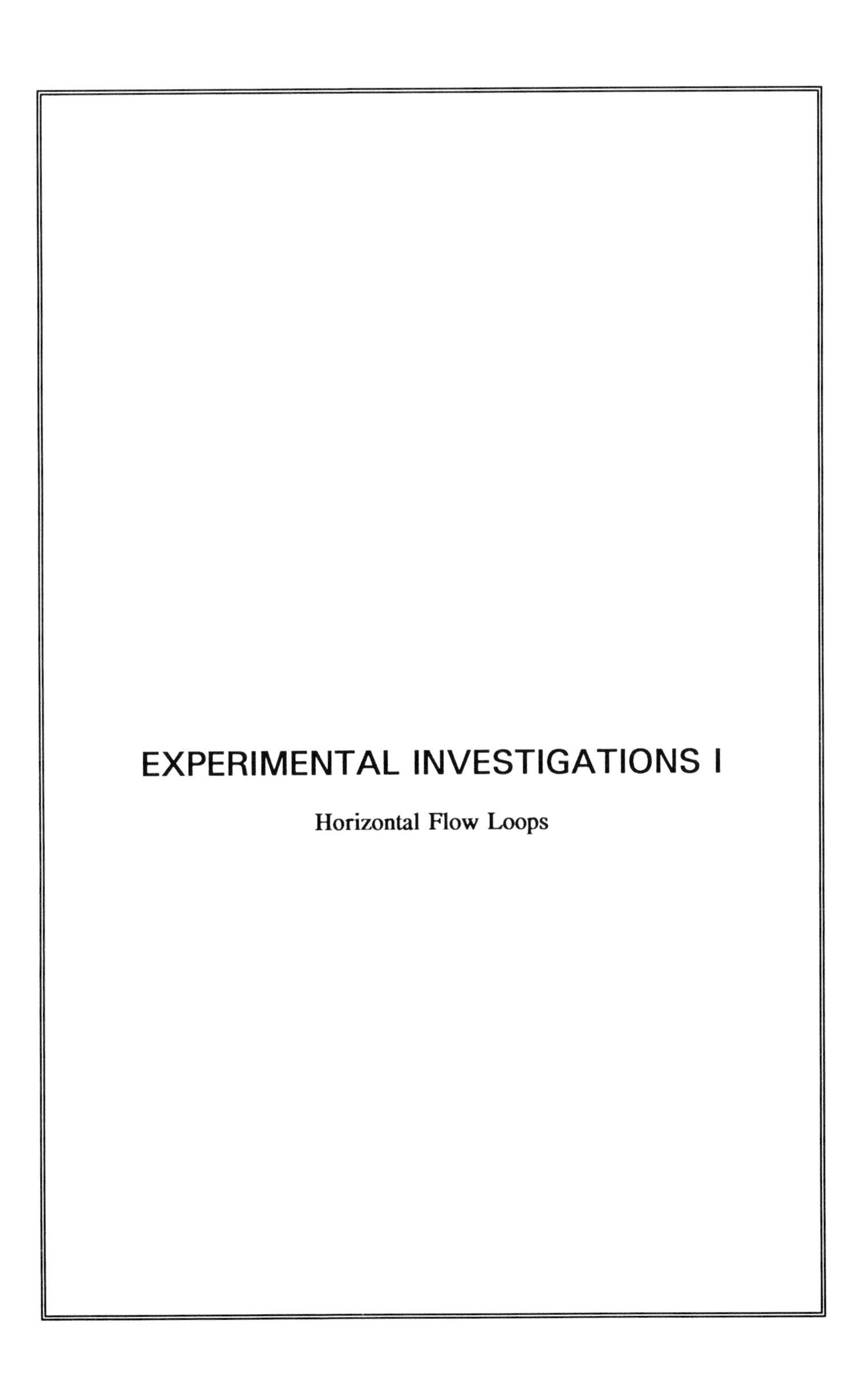

EXPERIMENTAL INVESTIGATIONS I

Horizontal Flow Loops

AN EXPERIMENTAL INVESTIGATION OF THE EFFECTS OF THE WATER PHASE IN THE MULTIPHASE FLOW OF OIL, WATER AND GAS

A.R.W.Hall (AEA Technology, Harwell, UK), G.F.Hewitt and S.A.Fisher (Department of Chemical Engineering, Imperial College, London)

ABSTRACT

Flows of oil, water and gas occur in subsea multiphase pipelines in areas such as the North Sea, where as fields mature, an increasing fraction of water is produced. Due to the interest in such flows, an experimental investigation has been carried out using the High Pressure Multiphase Flow Facility at Imperial College into three-phase flow of air, water and a light lubricating oil (with a viscosity of approximately 50 mPas). The study focussed on three-phase slug flows, and measurements were made of pressure gradient and holdup of oil and water (using quick-closing valves). Visual observation of the flows allowed the determination of which phase formed the continuous phase and whether the liquid flowed as a dispersion or separated into distinct layers. Experimental results are presented and compared to theoretical models.

INTRODUCTION

Three-phase oil-water-gas flows occur in the petroleum recovery industry, since reservoirs of oil and gas almost invariably contain water. Also, as the natural pressure of an oil reservoir decreases with production, water is often injected to boost the pressure. Due to the greater mobility of the water than the oil through the pores of the rock, water is usually found to enter the production wells soon after injection starts. The resulting three-phase mixture may then be required to flow vertically up to a platform, or horizontally along a sub-sea pipeline for transport to central processing facilities. This latter situation is of increasingly common occurrence in North Sea fields where satellite fields are developed around existing production platforms.

Many features of three-phase flows are important in the design of pipelines and ultimately in the economic operation of mature oil fields. In inclined and vertical pipes, the holdup of each phase has an important influence on the hydrostatic component of the pressure gradient and hence the possible production rate. The flow pattern, in particular the distribution of the two liquid phases, has a strong influence on the frictional pressure gradient and on the possible corrosion of the pipeline. Intermittent flows can lead to large fluctuations in pressure and volumetric throughput of liquid, which must be accommodated in the design; the prediction of the region of intermittent flow is therefore of key importance. The flow history of the fluids along a pipeline may also influence the separation of the oil and water phases; clearly, with the limited space available on an offshore platform, this can have an important bearing on the economics of a project.

HIGH PRESSURE MULTIPHASE FLOW FACILITY

2.1 DESCRIPTION OF THE WASP FACILITY

The WASP (Water, Air, Sand, Petroleum) Project in the Department of Chemical Engineering at Imperial College was set up to study flows of importance in the Petroleum Industry. A new high pressure multiphase flow facility was constructed with the following features:

(a) Operation with up to *four* phases, namely air, water, oil and sand (suspended in the water phase).
(b) High pressure operation (up to 50 bar), the air being supplied from the hypersonic wind tunnel air reservoirs in the adjacent Aeronautics Department.
(c) Industrial-scale dimensions, with a test section which has an internal diameter of 3″ (78 mm) and length of 40 m, which may be inclined by a few degrees to the horizontal.
(d) Flow visualisation.
(e) Computer control and data acquisition systems.

The need for a new facility such as this is demonstrated in Figure 1 where the pipe diameters and operating pressure of various experimental facilities are compared with pipelines taken from the UK National Multiphase Flow Database*. It will be seen that research facilities used in published studies of three-phase flows are generally low pressure loops with small diameter test sections, while industrial pipelines are of much larger diameter and have higher operating pressures. The WASP facility therefore bridges this gap. It should be noted that since air has a greater density than natural gas, the 50 bar maximum pressure for WASP corresponds to around 90 bar (to obtain equal densities) for a natural gas system.

The WASP facility is shown schematically in Figure 2. The air supply is drawn from either high pressure air reservoirs or, for low pressure experiments (up to approximately 5 bar) from the site compressed air main. The liquids are taken separately from two pressurised tanks, one for oil, the other for water (or water/sand slurry), flowing under the action of either the applied air pressure or centrifugal pumps.

* This is an ongoing project of AEA Petroleum Services at Harwell

After metering, the three streams are combined in a mixer section; the internal design of this section introduces the three phases as closely as possible to stratified layers to give the shortest and most reliable development of the flow.

From the mixer section, the main test section of the rig continues as a straight, horizontal tube for approximately 40 m. The test section was built of seamless stainless steel tube in sections of length 7 m with tongue-and-groove flanged joints to ensure continuity and concentricity of the bore.

Towards the end of the test section is the visualisation section. This was constructed of a polycarbonate tube of length 0.8 m with a wall thickness of 31 mm, mounted in a steel jacket with window slots. At the end of the test section is the slug catcher vessel, the main function of which is to separate the air and the liquids at full rig pressure. The size of the slug catcher is sufficient to contain several times the volume of the test section and the vessel is fitted with baffles and plates inside to break up the liquid slugs and to dissipate the potentially damaging momentum of the liquid phases. The liquids (oil + water) drain from the bottom of the slug catcher to a 'dump tank' where the oil and water separate under gravity. Water and oil can be pumped back serially through a return line to their respective tanks after a suitable settling time.

2.2 FLOW METERING & INSTRUMENTATION

The water flowrate was measured using a magnetic flow meter while the oil and air flowrates were measured using orifice plates. Pressure gradient was measured by using two Druck PDCR820 general purpose transducers of range 3.5 bar to obtain the pressure drop over a length of 11.10 m (before the quick-closing valves were installed — discussed below) and 14.58 m (after the quick-closing valves were installed). Two individual transducers, mounted flush to the test section, were used in order to avoid the problems of trapped liquid in long lines to the transducers. The difference in the response of the two transducers could be determined accurately by pressurising the test section (without flow) at a series of absolute pressures. After correction for this difference, the error in pressure drop was estimated to be ±2 mbar.

One of the most important parameters in multiphase flow is the *holdup*, or mean fraction of the pipe cross-section which is occupied by each of the phases. For all flow patterns, apart from smooth stratified flow, the prediction of holdup by analytical or numerical solutions is difficult, if not impossible at present. There is therefore a need for experimental measurements of holdup in order to develop, validate and improve empirical prediction methods.

There are several methods for the measurement of holdup in two-phase (gas-liquid) flow which can be extended to three-phase (oil-water-gas) flow. The most readily applicable method is volume averaging by trapping the fluids in a section of pipe, giving a direct measure of the fractions of the phases. Nuclear techniques require more specialist equipment, but give more localised and continuous measurement. In the present work, holdup measurements were carried out using the quick-closing valve method. However, current work on the WASP facility aims at using a dual-energy gamma absorption method to obtain local oil and water fractions, with the quick-closing valve technique providing important calibration data.

The design of quick-closing valve unit chosen was to replace one of the 7 m sections of the main test line of the rig with a new section equipped with a full-bore ball valve at either end. These valves were operated by air actuators which were supplied with compressed air via a single large-bore manifold to ensure simultaneous operation. The air actuators were modified to enable rapid operation and typically the closing time was around 0.5 sec. It is believed that simultaneous operation of the valves is more important than fast operation, and experiments by Hashizume[1] showed no effect on measured void fraction of shut-off time ranging from 0.05 sec to 2 sec. When the valves were closed, the control system was arranged to simultaneously cut power to the pumps, to prevent a surge in pressure.

After operation of the quick-closing valves, the liquids were blown out of the trapped section and collected in tall 4″ diameter glass measuring cylinders from which the individual volumes of oil and of water could be determined.

2.3 OIL SELECTION AND PROPERTIES

One of the main selection criteria for the oil was that it should have a high flashpoint. Two types of oil considered were light lubricating oils (eg Shell Tellus oil) which have flashpoints in the region of 200–220°C, and synthetic phosphate ester fluids whose flashpoints are in the range 230–240°C. The density of phosphate ester fluids is greater than that of water and the Shell Tellus oil was therefore chosen.

It was found that once the oil had been circulated through the test loop, a small quantity of water became stably dispersed in it. Viscosity measurements were therefore carried out for both clean oil and for a representative sample from the test facility. Both clean and used oils were Newtonian; the value of the clean oil viscosity was about 50 mPas at 20°C, while that of the used oil was 11% greater.

The density of the pure oil was given by:

$$\rho_{\mathrm{oil}} = 860.5 + 4.5\left(\frac{24.7 - \mathrm{T}}{9.7}\right), \tag{1}$$

with T in °C, and this density must be corrected for the effect of the suspended water:

$$\rho_{\mathrm{ow}} = (1 - \phi_{\mathrm{w}})\rho_{\mathrm{oil}} + \phi_{\mathrm{w}}\rho_{\mathrm{water}} \tag{2}$$

where ϕ_{w} is the water fraction in the mixture. From density measurements, the stable water fraction was determined to be 7% by volume.

EXPERIMENTAL INVESTIGATION OF THREE-PHASE OIL-WATER-GAS FLOWS

The ranges of experimental parameters in the three-phase oil-water-air experiments were:

Air Superficial Velocity	0.98 - 4.1 m/s
Oil Superficial Velocity	0 - 0.54 m/s
Water Superficial Velocity	0 - 0.83 m/s
Temperature	13 - 24°C
Slug Catcher Pressure	1.10 - 3.36 bar abs
Oil/Water Viscosity Ratio	40 - 60
Oil/Water Density Ratio	0.878

Experimental measurements were taken for approximately 100 points in the three-phase slug flow regime (see below). These results are compared below with models and correlations under the headings of pressure gradient, holdup and flow visualisation.

3.1 PRESSURE GRADIENT

3.1.1 Comparisons with pressure gradient correlations

It was found by Hall[2] that where the oil and water phases were well-mixed, the best agreement with experimental pressure gradient data was given by the Beggs & Brill[3] correlation with a suitable adjustment for the liquid viscosity. Previous workers had taken the effective viscosity as a mean of the oil and water viscosities weighted by volume fraction. This produced curves of the ratio of measured to calculated pressure gradient against water fraction with a characteristic maximum corresponding to the inversion point from a water-continuous to an oil-continuous dispersion. A more accurate method of predicting pressure gradient is to use one of the equations for the viscosity of a dispersion of one liquid in another, for example that due to Brinkman[4]:

$$\mu_{\text{mix}} = \frac{\mu_{\text{cont}}}{(1-\phi)^{2.5}}. \tag{3}$$

Since the largest part of the pressure drop in slug flow is due to the motion of the liquid slugs, where the oil and water are well-mixed, it would be expected that this approach would apply to the present results, where the flow regime in the tests reported was slug flow. Thus, measured pressure gradient was compared to calculated pressure gradient using various correlations and using liquid viscosities calculated both from equation (3) and on the basis of a volume average. Table 1 shows a comparison of the percentage errors in these calculations. The standard deviation is the best indication to the reliability of a correlation, but a simple average is also a useful quantity, since it shows whether a correlation consistently over- or underpredicts. Hence it is clear that the Beggs & Brill correlation is the best of those compared to this data and that most correlations overpredict pressure gradient. Table 2 shows the performance of the Beggs & Brill correlation when compared to the two-phase data points, and also using the linear liquid viscosity calculation.

Pressure Gradient Correlation	Standard Deviation	Average Error %
McAdams[5]	29	-12
Schlichting[6]	37	7
Beggs & Brill[3]	21	18
Dukler[7]	36	81
Friedel[8]	59	59
Lockhart & Martinelli[9]	31	22

Table 1. Errors in pressure gradient calculations from correlations using Brinkman viscosity

Method	Standard Deviation
Air-oil flow	10
Air-water flow	12
Three-phase flow (linear viscosity)	52
Three-phase flow (Brinkman viscosity)	21

Table 2. Errors in pressure gradient calculations using the Beggs & Brill correlation

A number of figures are presented to demonstrate the behaviour of the pressure gradient in three-phase flow. Figure 3 shows a comparison of calculated and measured pressure gradient, using the Beggs & Brill correlation and both the linear and Brinkman liquid viscosities; the superior predictions using the Brinkman viscosity is clear. Figure 4 shows a comparison of pressure gradient calculated from the Beggs & Brill correlation with measured values for both two-phase (air-oil and air-water) and three-phase flows. In Figure 3 it can be seen that the linear viscosity overpredicts pressure gradient in the water-continuous region ($\phi \lesssim 0.6$) while underpredicting in the oil-continuous region.

The reason for the behaviour is illustrated by looking at Figure 5 which shows the effective liquid viscosity as a function of oil fraction, as calculated by the two different viscosity methods.* The linear viscosity is clearly too high in the water-continuous region and far too low in the oil-continuous region.

The inversion point at which the flow changed from a water-continuous to an oil-continuous dispersion was found to be at an oil fraction of about 0.6. Some points with smaller oil fractions were also found to have been oil-continuous: this was determined from both the visual observations and from the large pressure gradients. In some cases, the prediction of pressure gradient in the inversion region was not as good as was found at lower oil fractions: this was thought to be due to the formation of a transient dispersion of one phase in the other, whose properties could neither be described by assuming a water- or an oil-continuous flow.

The average error between measurements and calculations (using the Brinkman viscosity equation) is well within the margins expected when applying the Beggs & Brill correlation to *two-phase* flows, and the Brinkman viscosity correction may therefore be satisfactorily used to explain the effect of a second liquid phase in this system.

3.1.2 Comparison with slug flow models

As well as comparisons with correlations above, the experimental pressure gradient measurements were compared with a model for three-phase slug flow adapted from Dukler & Hubbard[10]. In this model, the pressure gradient in slug flow is formed of two components: a pressure drop associated with the acceleration of liquid from the slow-moving film to the slug body and a pressure drop due to the frictional resistance to the motion of the liquid slug. The accelerational component is related to the liquid density which is taken to be a volume average of the oil and water densities. The frictional component is related to the liquid viscosity, which is given by the Brinkman equation (3). Comparisons were also made between pressure gradients measured for two-phase air-oil and air-water slug flows and the Dukler & Hubbard model. Table 3 shows the errors in the pressure gradient calculations compared to the measurements for the various types of flows.

The most important results are those for water-continuous three-phase flows (the majority of the data points) where it will be seen that there is a considerable improvement of the modified Dukler & Hubbard model over the Beggs & Brill correlation. Figure 6 shows a comparison of calculated against measured pressure gradient for water-continuous three-phase flows using the slug flow model. It is to be expected that a model for a particular type of flow would give more accurate predictions than a general correlation.

---

* Note that there is some variation at a given oil fraction since the operating temperature varied from case to case.

Method	Standard Deviation	Average Error %
air-oil-water flow (water-continuous)	17	+9
air-oil-water flow (oil-continuous)	11	-27
Air-water flow	15	+5
Air-oil flow	9	-14

Table 3. Errors in pressure gradient calculations using the adapted Dukler & Hubbard slug flow model

3.2 HOLDUP

The measurement of holdup presented particular problems due to the intermittent nature of the flow. The ratio of the length of the trapped section (6.78 m) to the length and frequency of slugs was such that trapped liquid volumes could vary enormously, depending on whether or not a liquid slug was trapped. Time delay between activating the control and the valves actually closing, together with the unpredictable arrival of the slugs at the holdup section meant that it was not simple to ensure that consistent samples were trapped.

However, the experimental technique is important for future work and there is currently very little published work on holdup in three-phase slug flow. Early workers (eg Malinowsky[11]) did not published their data, claiming it to be too unreliable. Some recent data is presented by Lahey et al[12]) for a flow in a tube of diameter 0.019 m.

Very generally, the average holdup in slug flow seems to be similar across the range of flow conditions covered in this study. The important feature, therefore, is how the holdup of oil and water vary separately, as a function of the input oil/water ratio. The oil/water ratio (OWR) is related to the oil fraction by:

$$\mathrm{OWR} = \frac{\mathrm{U_o}}{\mathrm{U_w}} = \frac{\phi}{1-\phi} \tag{4}$$

and hence the inversion point of $\phi = 0.6$ corresponds to an oil/water ratio of 1.5.

Figures 7 and 8 show the measurements of holdup of oil and water respectively, plotted against the oil/water ratio, showing the expected tendency for water holdup to decrease and oil holdup to increase with increasing oil/water ratio. Also shown on these figures are the values of holdup calculated assuming that the flow was stratified and the values calculated from the modified Dukler & Hubbard slug flow model. The slug body was assumed to be well-mixed, while the film region between slugs was assumed to be separated. The average holdup over the slug unit (which is the measured quantity in the experiments) is obtained by a volumetric average over the film and slug regions.

The measured water holdup follows the trend of the stratified flow model calculations and agrees very well with the slug flow model. The measured oil holdup agrees less well with the stratified flow model and there is only a slight improvement with the slug flow model. This almost certainly arises from an overestimation of the oil holdup in the film region, which in practice is much smaller than the value obtained from the stratified flow model. This is because for part (or all) of the length of the film, a large proportion of the oil is carried as droplets within the water phase, which leads to a greater oil velocity and smaller oil holdup than would be the case with a separate oil layer. Indeed, it is doubtful whether a smooth oil layer (as assumed in the stratified flow model) was ever formed in the film regions in these experiments. In comparison, smooth oil layers were observed to be formed in the experiments reported by Stapelberg *et al*[13].

Not all the liquid trapped in the holdup section may have been collected in the measuring cylinders, this being a particular problem with experiments in horizontal pipes. However, considerable care was taken and comparison with other experiments (eg Russell *et al*[14]) suggests that 95% of the liquid should have been recovered, and this effect is not likely to explain the discrepancies with prediction.

These measurements of holdup therefore show two important features. Firstly, that the experimental technique is useful, while confirming the view that 'holdup' in slug flow is a difficult quantity to characterise due to the intermittent nature of the flow. Secondly, there are indications that the transport of the liquid phases is complicated. In water-continuous flows, oil was preferentially transported in the liquid slugs, while in oil-continuous flows there appears to be better distribution of the phases. The large viscosity of the oil phase is likely to play an important role in this process.

3.3 FLOW VISUALISATION

As discussed in Section 2, the WASP facility was fitted with a visualisation section at the outlet end of the test section. In the present experiments, the following flows were observed:

- Two-phase air-water slug flow.
- Two-phase air-oil slug flow.
- Three-phase slug flow with water as the continuous phase and with separation of oil and water layers in the film region.
- Three-phase slug flow with water as the continuous phase but with the oil and water were dispersed throughout (no separate layers).
- Three-phase slug flow with oil as the continuous phase. All oil-continuous flows were dispersed throughout.

Figure 9 shows a schematic representation of possible flow patterns in three-phase stratified and slug flows. Flows observed in this study correspond to the second and third diagrams for slug flow. In the case of the "oil and water dispersed" slug flow, either the oil or the water could form the continuous phase depending on the oil/water ratio.

The rate of drainage of the liquid film in water-continuous flows was seen to decrease very markedly as the oil fraction approached the point for inversion to oil-continuous flow. Under some conditions, separation of oil and water layers in the film region was observed intermittently, depending on the interval between slugs.

An additional observation was that in flows where a separate layer appeared, the water layer was undisturbed at the slug front and only became mixed with the oil phase approximately two to four pipe diameters further downstream. This is important in modelling the stratified to slug flow transition. The particular features of the flow pattern observations, namely slug frequency, separated to dispersed transition and stratified to slug flow transition are discussed separately below.

3.3.1 Slug frequency

The slug frequency was calculated by counting the number of slugs in a fixed time interval. Figure 10 shows a plot of slug frequency against the total liquid superficial velocity.

Stapelberg *et al* discussed the correlation of Tronconi[15] for the slug frequency:

$$\omega = \frac{1}{2}\left(1.22\frac{\rho_\ell}{\rho_g}\frac{u_g}{h_g}\right) \tag{5}$$

and suggested that for three-phase flow this should be modified by changing the $\frac{1}{2}$ to $\frac{1}{4}$. This equation does not explicitly consider the liquid velocity but Figure 10 shows that this is an important parameter. Nevertheless, the frequencies were calculated using both Tronconi's equation and Stapelberg's modification; these results are plotted in Figure 11 and the mean errors are given in Table 4.

Method	Factor in Equation 5	Standard Deviation
Tronconi	$\frac{1}{2}$	49
Stapleberg	$\frac{1}{4}$	24

Table 4. Errors in slug frequency calculations

The apparent dependence of slug frequency on the total liquid superficial velocity as illustrated by Figure 10 suggested a comparison should be made with the method of Hill

& Wood[16] which takes into account the liquid velocity:

$$\frac{\omega D}{(u_g - u_\ell)} = 2.74 \frac{\tilde{h}_\ell}{\left(1 - \tilde{h}_\ell\right)} \tag{6}$$

A comparison of the Hill & Wood method with that suggested by Stapelberg is shown in Figure 12, where the ratio of measured to calculated slug frequency is plotted against the total liquid superficial velocity. Hill & Wood performs well, with a mean error of 8%, in the region up to a total liquid superficial velocity of 0.6 m/s, which was the region where a significant variation in slug frequency with liquid velocity was observed. However, at liquid velocities above 0.6 m/s Hill & Wood overpredicts the slug frequency.

3.3.2 Separated-dispersed transition

The transition between slug flows where the oil and water can form separate layers and flows where they are always dispersed can be modelled by considering three quantities. Firstly, the largest drop size in the slug body is calculated assuming that the oil is fully dispersed in the water from the critical Weber number[17]:

$$We_c = \frac{\rho_{cont} \overline{v^2} D_{max}}{\sigma}. \tag{7}$$

Then the terminal rise velocity of these droplets in the film region is calculated from:

$$U_t = \frac{(\rho_w - \rho_o) D_{max}^2 g}{18\mu}. \tag{8}$$

The expected time for separate layers to form can therefore be estimated from this velocity and the liquid height. The time interval between slugs can be calculated from equation (5) and thus a crude transition boundary is given when these two times are equal.

A comparison with this transition is shown in Figure 13 where the measured gas velocity is plotted against the gas velocity calculated for transition (using the liquid velocity and oil fraction from the data points). Where the measured gas velocity is greater than the calculated transition velocity, the flow would be expected to be dispersed, while a smaller measured velocity would imply a flow where separation is expected.

3.3.3 Stratified-slug transition

The transition between stratified and slug flows in three-phase separated systems was discussed by Hall & Hewitt[18]. The linear stability theory presented by Lin & Hanratty[19] for gas-liquid flow was modified for three-phase flow by assuming that the water layer was undisturbed at the point of slug initiation. The comparison of experimental data with the linear stability theory is shown in Figure 14 where the calculated superficial liquid velocity for transition is compared to the measured superficial liquid velocity at fixed superficial gas velocity and water fraction. The greater the difference between the measured velocity and the calculated transition velocity, the more likely the flow to be in the slug flow regime. In fact, in the experiments, all flows were observed to be in the slug flow regime, and Figure 14 shows that this is to be expected. A number of points will be seen to be very close to the transition boundary; it was found that the slug frequency in these cases was much smaller, compared to the other points.

It is interesting to compare the slug frequency with the difference between the measured liquid velocity and the calculated liquid velocity for transition. This is shown in Figure 15 where it will be seen that there is a good correlation between the two quantities. However, this agreement may be true only over this relatively limited range of flowrates.

Linear stability theory was used since the steady-state approach of Taitel & Dukler[20] was found to give transition liquid velocities which were too high in three-phase flows.

3.4 SUMMARY

For three-phase oil-water-air flow, new experimental data has been presented, covering pressure gradients, slug frequency, flow patterns observation and holdup measurement.

The pressure gradient was found to agree reasonably with the Beggs & Brill correlation, after suitable modification of the liquid viscosity to take account of the presence of two liquid phases. Better agreement was obtained when pressure gradient was calculated using a modified form of the Dukler & Hubbard slug flow model.

Measured slug frequency was found to agree with calculations from the Tronconi correlation as adapted for three-phase flow by Stapelberg, and with the method presented by Hill & Wood for liquid superficial velocities up to 0.6m/s. Flow observations for three-phase slug flows were classified according to the distribution of the oil and water phases. For flows where the water formed the continuous phase, a simple criterion was derived for distinguishing flows where separate oil and water layers were observed from flows where the oil and water were dispersed. This was found to agree well with the observations. The transition between stratified and slug flows was predicted using linear stability theory adapted from Lin & Hanratty, and showed that all the flows in the present experiments are in the slug flow regime, as observed.

Finally, holdup of oil and water was measured and compared to a model for slug flow. Good agreement was observed with measurements of water holdup, but not oil holdup. It is suggested that the oil, as dispersed phase, was preferentially transported in the liquid slugs, hence leading to a lower average oil holdup than expected.

ACKNOWLEDGEMENTS

The work described in this paper was supported by the programme of Corporate Investment Research of AEA Technology and the Thermie Programme of the Commission of the European Communities. Published with permission from AEA Technology.

© UKAEA 1993

REFERENCES

[1] **K.Hashizume**, "Flow pattern, void fraction and pressure drop of refrigerant two-phase flow in a horizontal pipe — I. Experimental data," *International Journal of Multiphase Flow*, vol. 9, no. 4, pp. 399–410, 1983.

[2] **A.R.W.Hall**, *Multiphase flow of oil, water and gas in horizontal pipes*. PhD thesis, University of London, 1992.

[3] **H.D.Beggs and J.P.Brill**, "A study of two-phase flow in inclined pipes," *Journal of Petroleum Technology, Transactions*, vol. 255, pp. 607–617, 1973.

[4] **H.C.Brinkman**, "The viscosity of concentrated suspensions and solutions," *Journal of Chemical Physics*, vol. 20, no. 4, p. 571, 1952.

[5] **W.H.McAdams W.K.Woods and L.C.Heroman**, "Vaporization inside horizontal tubes — II Benzene-oil mixtures," *Transactions of the ASME*, vol. 64, pp. 193–200, 1942.

[6] **P.Schlichting**, "The transport of oil-water-gas mixtures in oil field gathering systems," *Erdöl-Erdgas-Zeitschrift*, vol. 86, pp. 235–249, 1971. (In German).

[7] **A.E.Dukler M.Wicks and R.G.Cleveland**, "Pressure drop and hold-up in two-phase flow," *AIChE Journal*, vol. 10, no. 1, pp. 38–51, 1964.

[8] **L.Friedel**, "Improved friction pressure drop correlations for horizontal and vertical two-phase pipe flow," in *European Two-Phase Flow Group Meeting*, no. E2, (Ispra, Italy), June 1979.

[9] **R.W.Lockhart and R.C.Martinelli**, "Proposed correlation of data for isothermal two-phase, two-component flow in pipes," *Chemical Engineering Progress*, vol. 45, no. 1, pp. 39–48, 1949.

[10] **A.E.Dukler and M.G.Hubbard**, "A model for gas-liquid slug flow in horizontal and near horizontal tubes," *Industrial and Engineering Chemistry Fundamentals*, vol. 14, no. 4, pp. 337–347, 1975.

[11] **M.S.Malinowsky**, "An experimental study of oil-water and air-oil-water flowing mixtures in horizontal pipes," Master's thesis, University of Tulsa, 1975.

[12] **R.T.Lahey M.Açikgöz and F.França**, "Global volumetric phase fractions in horizontal three-phase flows," *AIChE Journal*, vol. 38, no. 7, pp. 1049–1058, 1992.

[13] **H.H.Stapelberg F.Dorstewitz M.Nädler and D.Mewes**, "The slug flow of oil, water and gas in horizontal pipelines," in *5th International Conference on Multiphase Production*, (Cannes, France), pp. 527–552, June 1991.

[14] **T.W.F.Russell G.W.Hodgson and G.W.Govier**, "Horizontal pipeline flow of mixtures of oil and water," *Canadian Journal of Chemical Engineering*, vol. 37, pp. 9–17, 1959.

[15] **E.Tronconi**, "Prediction of slug frequency in horizontal two-phase slug flow," *AIChE Journal*, vol. 36, no. 5, pp. 701–708, 1990.

[16] **T.J.Hill & D.G.Wood**, "A new approach to the prediction of slug frequency," *SPE Paper 20629*, 1990.

[17] J.O.Hinze, "Fundamentals of the hydrodynamic mechanism of splitting in dispersion processes," *AIChE Journal*, vol. 1, no. 3, pp. 289–295, 1955.

[18] A.R.W.Hall and G.F.Hewitt, "Effect of the water phase in multiphase flow of oil, water and gas," in *European Two-Phase Flow Group Meeting*, no. F4, (Stockholm, Sweden), June 1992.

[19] P.Y.Lin and T.J.Hanratty, "Prediction of the initiation of slugs with linear stability theory," *International Journal of Multiphase Flow*, vol. 12, no. 1, pp. 79–98, 1986.

[20] Y.Taitel and A.E.Dukler, "A model for predicting flow regime transitions in horizontal and near-horizontal gas-liquid flow," *AIChE Journal*, vol. 22, no. 1, pp. 47–55, 1976.

NOMENCLATURE

Symbol	Description	
g	Acceleration due to gravity	m/s
h_g	Height of gas layer	m
$\tilde{h}_\ell$	Dimensionless equilibrium liquid layer height	-
T	Temperature	°C
U	Superficial velocity	m/s
u	Velocity	m/s
U_t	Terminal velocity	m/s
μ	Viscosity	mPas
ρ	Density	kg/m^3
σ	Surface tension	N/m
ϕ	Volume fraction of dispersed phase	-
ω	Slug frequency	1/s

LIST OF FIGURES

FIGURE 1: TYPICAL OIL-WATER-GAS FLOW CONDITIONS

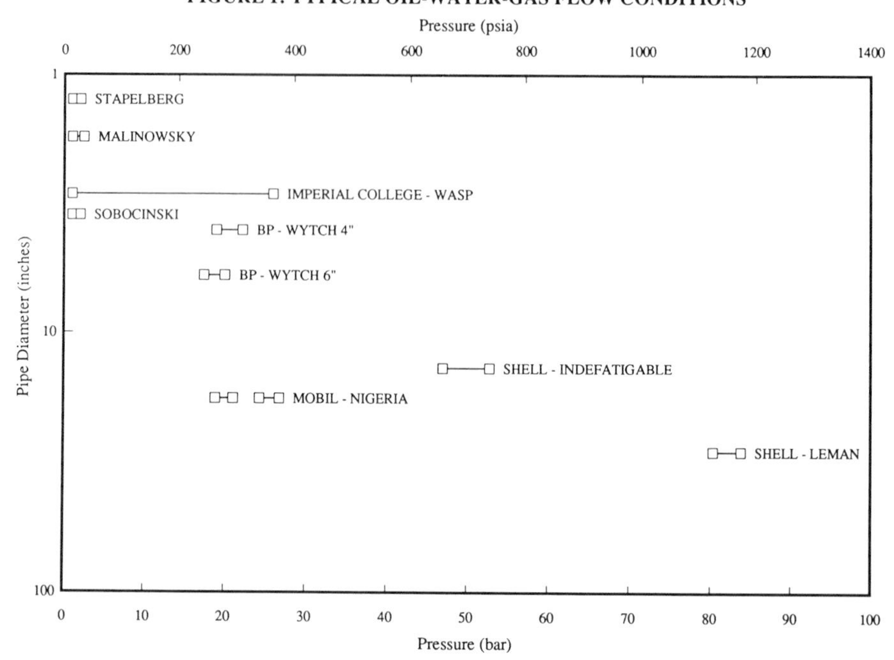

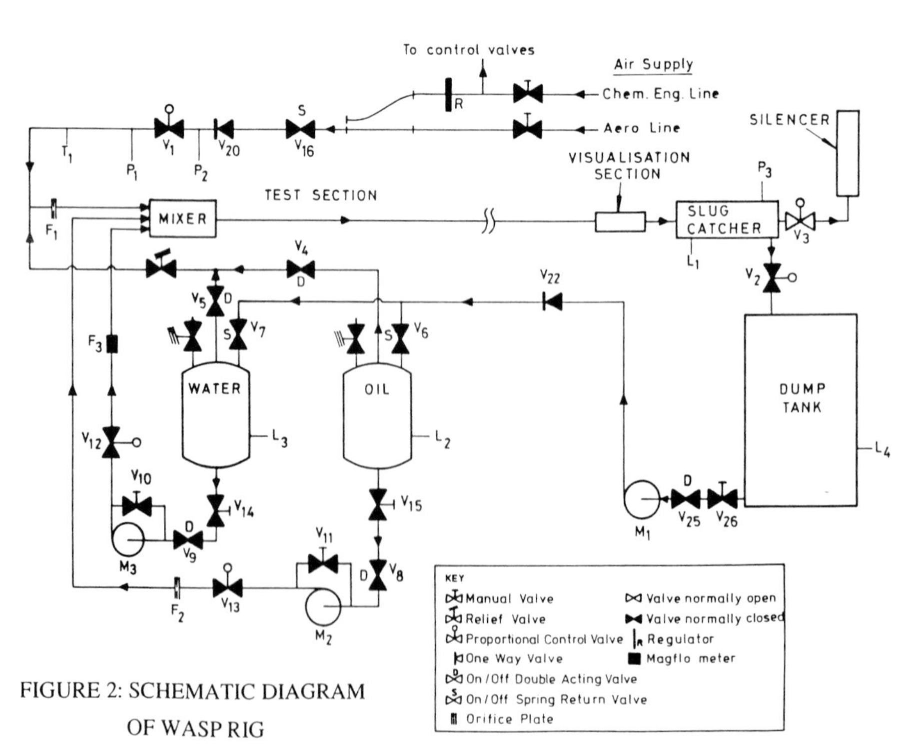

FIGURE 2: SCHEMATIC DIAGRAM OF WASP RIG

FIGURE 3: THREE-PHASE FLOW PRESSURE GRADIENT
(Comparison with Beggs & Brill correlation)

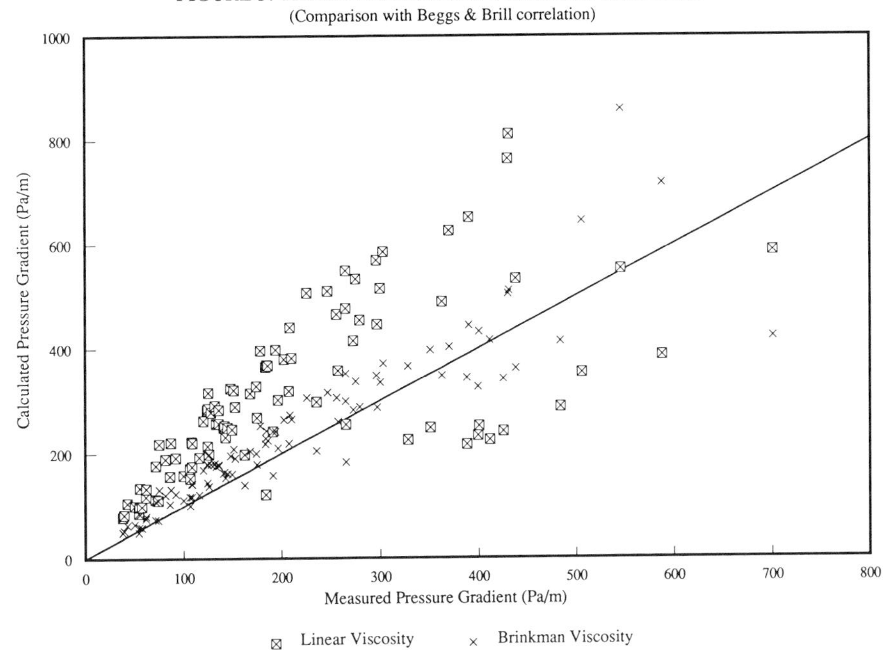

FIGURE 4: THREE-PHASE FLOW PRESSURE GRADIENT
(Comparison with Beggs & Brill correlation)

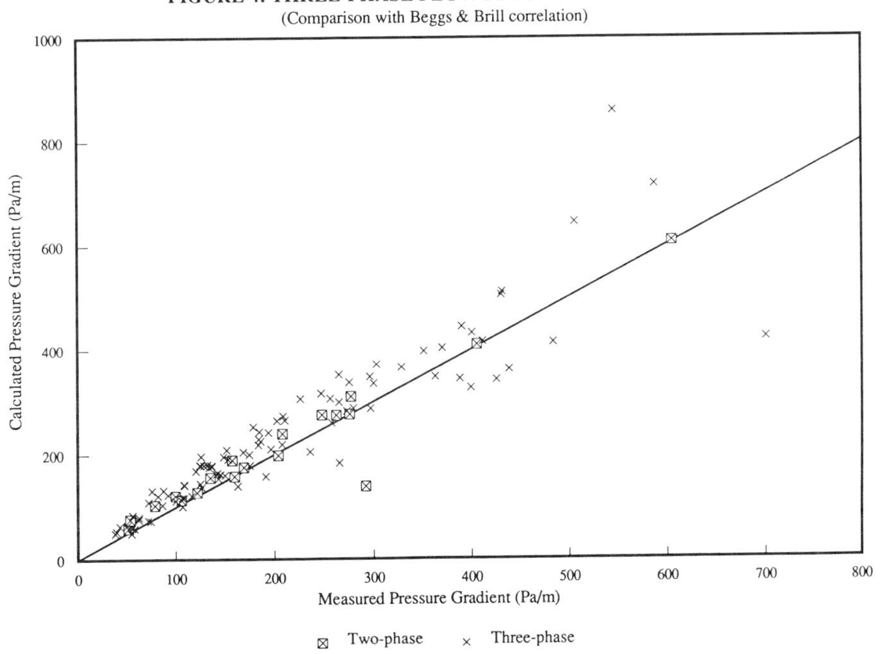

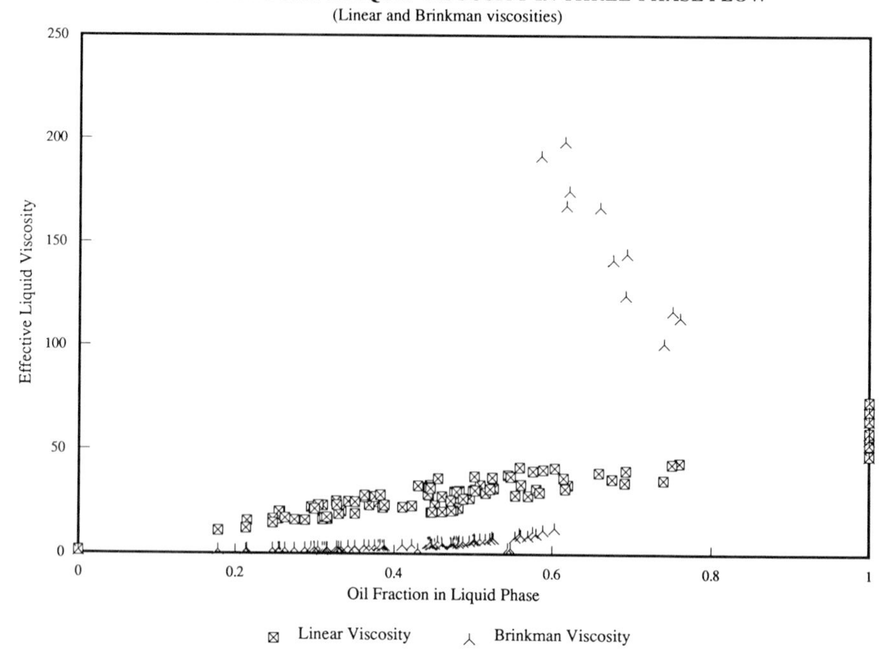
FIGURE 5: EFFECTIVE LIQUID VISCOSITY IN THREE-PHASE FLOW
(Linear and Brinkman viscosities)
Effective Liquid Viscosity
Oil Fraction in Liquid Phase
0
50
100
150
200
250
0.2
0.4
0.6
0.8
1
Linear Viscosity
Brinkman Viscosity

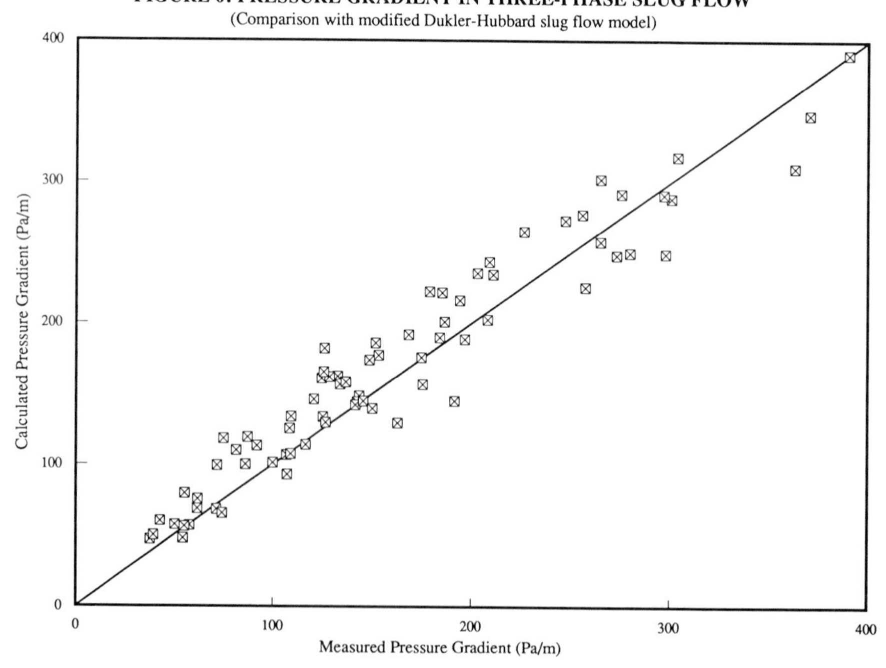
FIGURE 6: PRESSURE GRADIENT IN THREE-PHASE SLUG FLOW
(Comparison with modified Dukler-Hubbard slug flow model)
Calculated Pressure Gradient (Pa/m)
Measured Pressure Gradient (Pa/m)
0
100
200
300
400

FIGURE 7: THREE-PHASE OIL HOLDUP
(Comparison of measurements and models)

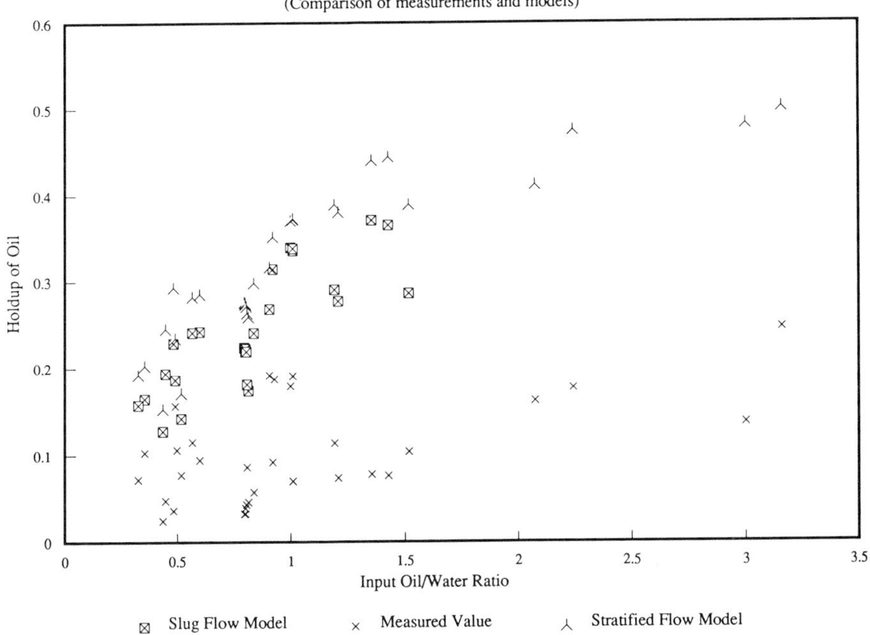

FIGURE 8: THREE-PHASE WATER HOLDUP
(Comparison of measurements with holdup)

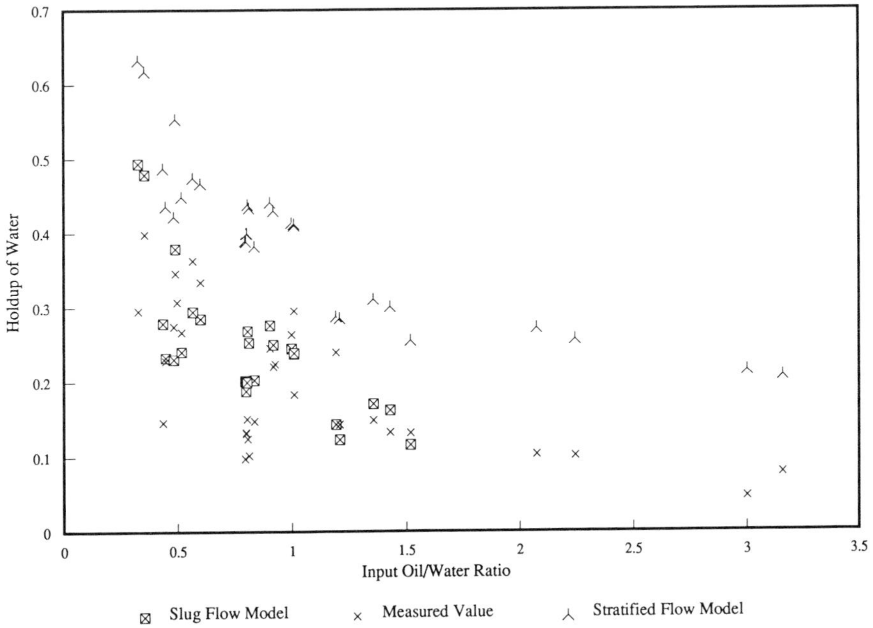

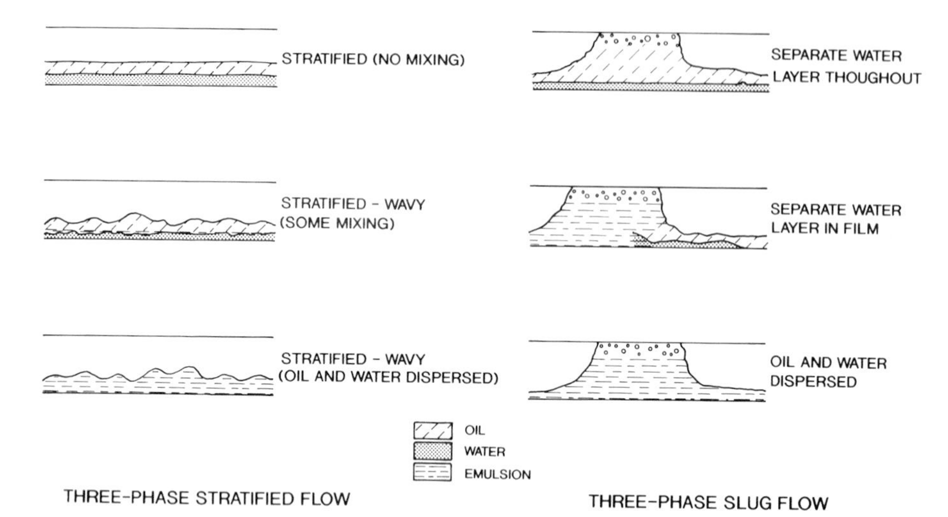

FIGURE 9. THREE-PHASE FLOW PATTERNS

FIGURE 10: THREE-PHASE SLUG FREQUENCY

(Variation with total liquid superficial velocity)

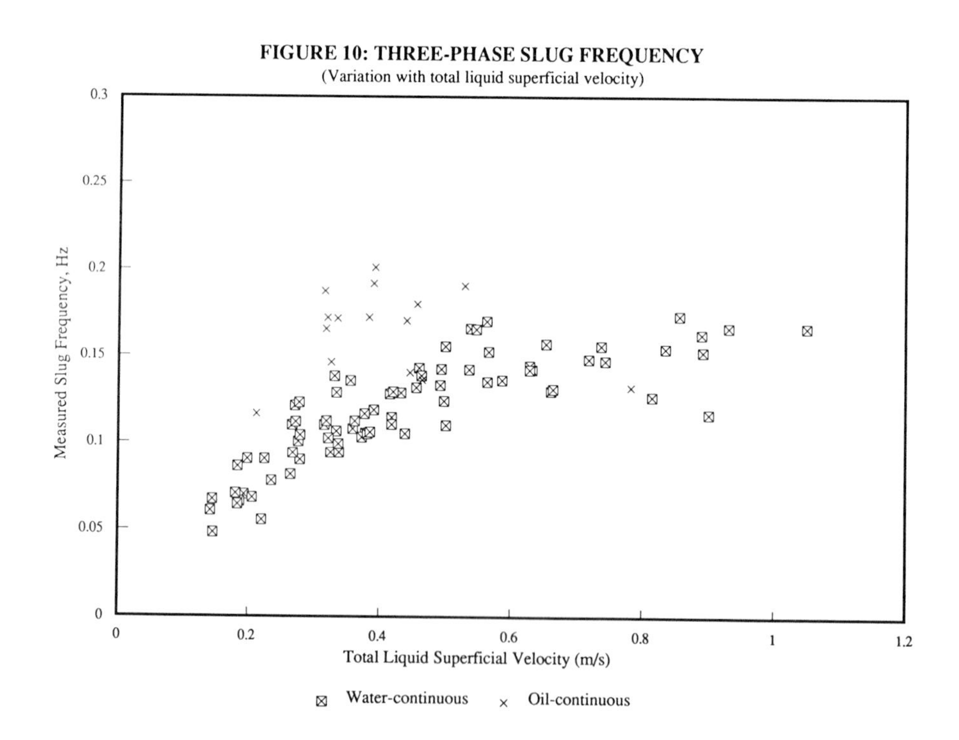

FIGURE 11: THREE-PHASE SLUG FREQUENCY
(Comparison of measurements with correlations)

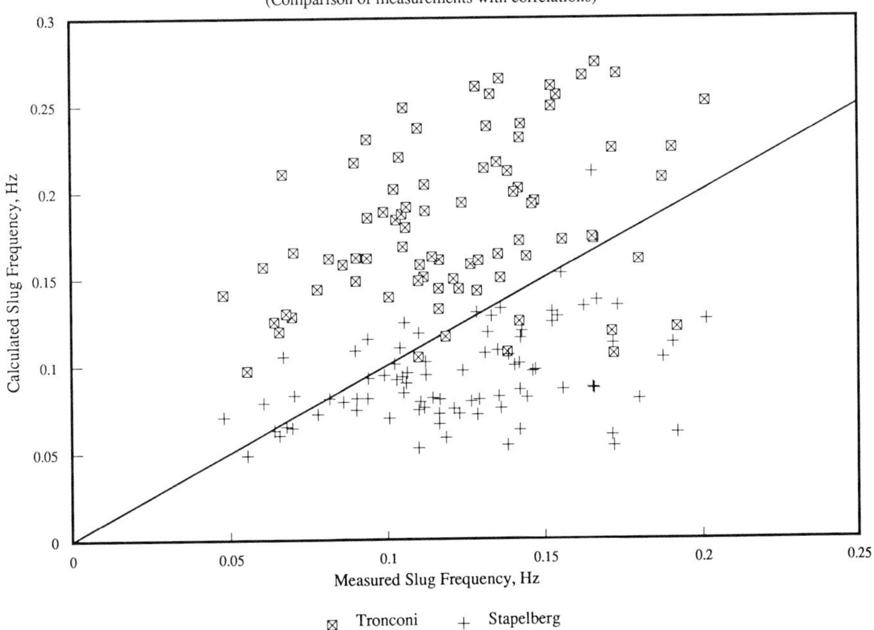

FIGURE 12: THREE-PHASE SLUG FREQUENCY
(Variation with total liquid superficial velocity)

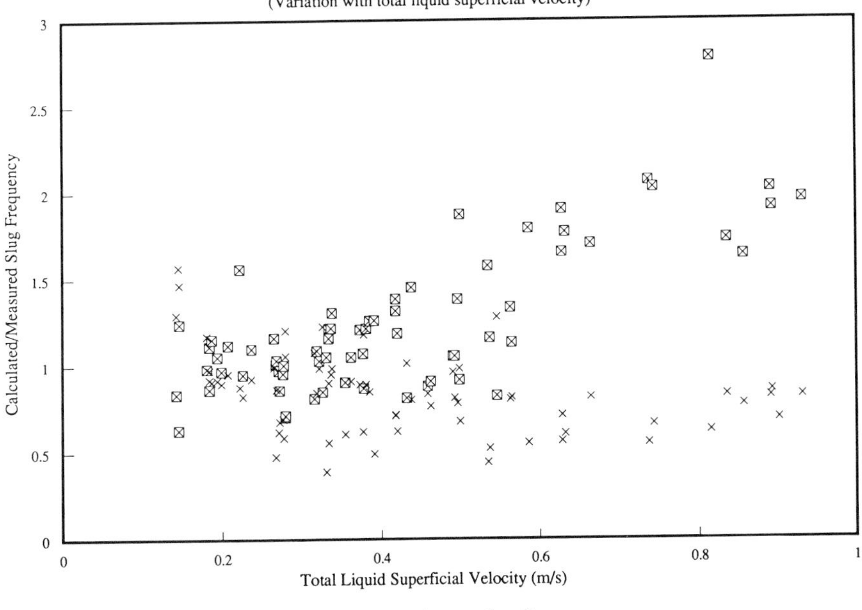

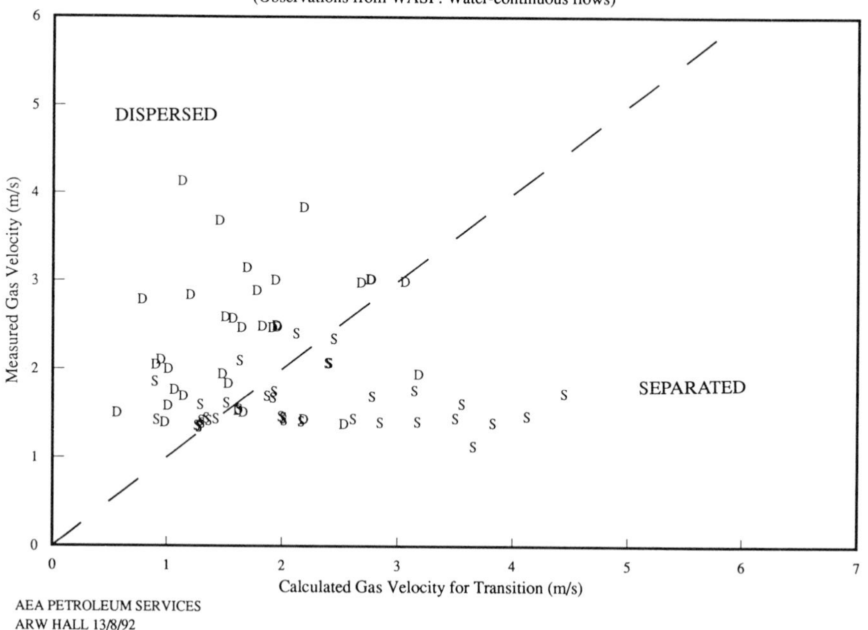
FIGURE 13: PREDICTION OF WATER SEPARATION
(Observations from WASP: Water-continuous flows)
DISPERSED
SEPARATED
Measured Gas Velocity (m/s)
Calculated Gas Velocity for Transition (m/s)
AEA PETROLEUM SERVICES
ARW HALL 13/8/92

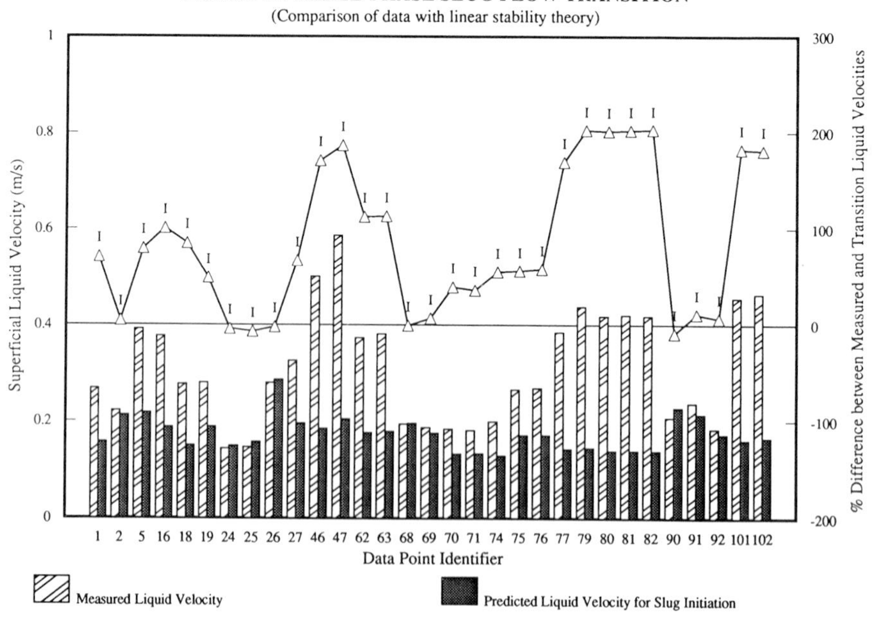
FIGURE 14: THREE-PHASE SLUG FLOW TRANSITION
(Comparison of data with linear stability theory)
Superficial Liquid Velocity (m/s)
% Difference between Measured and Transition Liquid Velocities
Data Point Identifier
Measured Liquid Velocity
Predicted Liquid Velocity for Slug Initiation
% difference between measured and transition liquid velocity

DESIGN OF LARGE-SCALE EXPERIMENTAL FACILITIES FOR MULTIPHASE PRODUCTION SYSTEMS

Hironori Furukawa, Masaru Ihara - Japan National Oil Corporation

Shingo Takao, Kazuo Kohda - NKK Corporation

ABSTRACT

In the process of development of new simulation models or equipment, validation based on the field data is required by the industry. In view of this, Japan National Oil Corporation plans to construct a large-scale, multi-purpose test facilities for multiphase production systems. Basic design of the facilities was completed in accordance with JNOC's policy that the facilities will be open to public, can cope with a variety of research needs, and can be expandable. Because of the purposes of experiment the test facilities will be capable of generating all flow patterns and meaningful transient flow data.

The test facilities were designed so that they would be suitable for the experiment for the following studies;
- Prediction of pressure drop and liquid holdup in horizontal, inclined and hilly-terrain pipes
- Splitting and convergence of multiphase flow at tees
- Effect of changes in diameter of the pipes
- Transient multiphase flow due to changes in flow rates
- Transient multiphase flow due to start-up, shut-down or pigging

The major specifications of the test facilities were determined as follows;
- Test fluids: Gas phase: air and nitrogen
  Liquid phase: water and kerosene
- Test pipelines: 105 mm diameter x 500 m long (1,500 m for future expansion)
  54 mm diameter x 250 m long
- Maximum operating pressure: 980 kPa (3,920 kPa for future expansion)
- Inclined section: 10 m long, 0 - 90 degrees
- Hilly-terrain section: 10 m x 4 slopes, -10 to +10 degrees
- Supply system: Pump: 180 m^3/h x 834 kPa
  Compressor: 2,000 Nm^3/h x 834 kPa
  Separator: 4,000 Nm^3/h (gas), 360 m^3/h (liquid)
  Water tank: 40 m^3
- Instrumentation: Meters for pressure, temperature, liquid holdup to be located at appropriate locations and will have flexibility.

1. INTRODUCTON

In the past studies of gas-oil two-phase flow, the newly developed models were evaluated using the experimental data taken at a laboratory scale test facilities. A few evaluation was carried out using the field data, because it is difficult to obtain from the fields the meaningful data over a wide range of operation. However, the validation of new models or equipment based on the field data is required by the industry. In order to respond to this need, Japan National Oil Corporation (JNOC) has determined to construct a large scale test facilities which can supply the high quality data equivalent to the field data in magnitude.

In 1992 JNOC constructed a test field in Kashiwazaki City in Niigata Prefecture (Kashiwazaki Test Field), which is to be used for various experimental studies of petroleum engineering and geophysics, including drilling and seismic tests. The area of the test field is approximately 240,000 m^2, and it has two wells, two seismic test lines (1,000 m and 500 m) and ten air gun pits. The test facilities for multiphase production systems will also be constructed in the Kashiwazaki Test Field. Fig. 1 shows the location and layout of the Test Field.

Because of the policy of JNOC, the test facilities will be open to public use, and will be designed so that the facilities can cope with a variety of research needs and can be expandable. The background, the design philosophy and the outcome of engineering of the test facilities will be presented in the following sections.

2. REVIEW OF EXISTING FACILITIES

There are a lot of test facilities for experimental studies of multiphase flow in pipes as shown in Table 1.

The Technology Research Center (TRC) of JNOC has a test pipeline of 54 mm (2 inches) diameter and approximately 100 m long, with several injection equipment along the line, which can be used to simulate influx to a horizontal well tubing from the reservoir. The test fluids are air and water, and the maximum operating pressure is 970 kPa. The TRC test facilities were used to collect the data for development of a new model for the pressure distribution along horizontal wells [(1) (2)].

NKK's test pipeline is 105 mm (4 inches) in diameter and approximately 1,400 m long, and has a pig-launcher and a receiver. The test fluids are air and water, and the maximum operating pressure is 690 kPa. This test pipeline was used to collect the data for development of a new simulation program for transient two-phase flow, including the transient caused by pigging the line [(3) (4)].

TUFFP's largest test pipeline is 78 mm (3 inches) in diameter and 420 m long. The test fluids are air-water and air-kerosene, and the maximum operating pressure is 860 kPa. These test facilities were used to collect the data for many experimental studies in multiphase flow, including Zheng's slug flow in hilly-terrain pipelines and Minami's transient model [(5) (6)].

SINTEF has the test facilities with 190 mm (8 inches) in diameter, 450 m long, with the maximum operating pressure of 10,000 kPa. The data collected at the facilities was used to develop the dynamic simulation program "OLGA" [7].

Other oil companies have various types of test facilities for multiphase flow, of which a few examples are listed in Table 1.

3. PURPOSE OF NEW TEST FACILITIES

3.1 Purpose of Study

After some new pieces of equipment or new simulation models have been developed based on laboratory-scale facilities under highly-controlled environment, their validity has to be evaluated using the "field data" before putting into field application. However, it is very difficult to collect from the actual fields the meaningful data over the range wide enough to be used to verify the new equipment or models. Therefore, there is need for the test facilities which is large enough to simulate the field, and whose performance can be highly controlled over a wide range.

In view of the above, the objectives of the usage of the test facilities will be as follows;

- Experiment under high pressure
- Experiment using a long pipe
- Experiment using various types of fluids

At JNOC/TRC the major study objectives of these test facilities have been identified as development of equipment and software for multiphase production systems. In addition, since the test facilities will be open to public use, they will be designed to cope with various experimental studies to be proposed by other research organizations.

3.2 Items to be investigated

In accordance with the above study purposes, the softwares and models to be developed at the test facilities will include the following;

- Prediction of pressure drop and liquid holdup in horizontal, inclined and hilly-terrain pipes
- Splitting and convergence of multiphase flow at tees
- Effect of changes in diameter of the pipes
- Transient multiphase flow due to changes in flow rates
- Transient multiphase flow due to start-up, shut-down or pigging

For development of multiphase production systems, the equipment to be tested and evaluated at the test facilities will include the following items. The test facilities will be designed to have flexibility and space for future expansion to cope with these equipment and others which will be proposed by the outside researchers.

- Two-phase flow pumps
- Multiphase flow metering equipment
- Separators
- Pigging systems

4. DESIGN CONSIDERATION

The new test facilities were designed based on the following technical consideration.

4.1 Test fluids

Mass transfer between the phases is inevitable in multiphase flow. However, it was concluded that the fluids to be used in the test facilities should be the ones without phase changes because of the following reasons;

- Very large pressure difference is required to make phase changes in the test pipeline.
- Temperature of the test pipeline must be controlled accurately.
- It is not practical to achieve the above two conditions in large scale test facilities.
- The effect of phase changes on the models can be included by modifying the models using the PVT data taken from the actual fluids under consideration.

In view of the above, the following fluids were selected;

- Gas phase: air and nitrogen
- Liquid phase: water and kerosene

4.2 Flow pattern

The test pipeline must be capable of generating all of the flow patterns known to exist in pipes, that is; Stratified, Wavy, Intermittent (Slug, Elongated bubble and Churn), Bubbly, Dispersed bubble and Annular mist.

Simulation calculations were carried out to predict the flow patterns in horizontal and inclined pipes for various combinations of gas and liquid superficial velocities, and the required minimum velocities to generate all the flow patterns were determined. Based on them and the pipe diameters, the required minimum gas and liquid flow rates were determined. The mechanistic models by Taitel and Barnea were used for flow pattern prediction[(8) (9) (10)].

In the simulation the pressure at the outlet of the pipeline was set at 245 kPa, and the required pressure at the inlet and the flow pattern were calculated. Table 2 shows that all of the flow patterns can be generated for a 105-mm (4-inch) pipeline with the air flow rate up to 2,000 Nm^3/h and the water flow rate up to 140 m^3/h. In this case the minimum required pressure at the inlet is 834 kPa, and the minimum required pipe length is 250 m.

Flow patterns in inclined pipes were also predicted. Table 3 shows the example calculations for 105-mm (4-inch) inclined section at an inclination angle of 20 degrees. Table 4 shows the example calculations for 105-mm hilly terrain section.

4.3 Transient flow

The test pipeline must be capable of generating transient flow conditions whose characteristics can easily be measured. For example, if the pipeline is too short, the difference in gas flow rates at inlet and outlet will not be able to be distinguished when the transient condition is caused by the change in gas flow rate.

Simulation calculations were carried out to determine the minimum required length of the test pipeline for various patterns of flow rate changes. The computer code for transient two-phase flow, TAPTWO was used for the calculations [(3) (4)]. See the appendix for the outline of TAPTWO. Flow rates of air or water were increased or decreased linearly for the 105-mm pipeline of the length of 250 m, 500 m and 1,000 m respectively.

From simulation studies it became clear that most transient phenomena would be measurable in 250 m pipeline, but in some cases 500 m pipeline is required. Fig. 2 shows the example calculations for 500 m pipeline. The inlet air flow rate was changed linearly from 10 to 2,000 Nm^3/h, and the inlet water flow rate was changed linearly from 180 to 10 m^3/h in 5 min, respectively. The outlet pressure was kept at 245 kPa. The outlet air and water flow rates and inlet pressure caused by these changes were calculated. Fig. 2-a through c indicate that the transient changes in flow rate and pressure are large enough to be measured.

4.4 Line size and length

The diameters of 152, 105 and 54 mm (6, 4 and 2 inches) were considered in this study. In view of collecting the data equivalent to the field data, 152-mm line is desired. However, since the pump and compressor requirements are very large for 152-mm line, the pipe size was determined to be 105 mm, and the space for 152-mm line and related facilities was reserved for the future expansion.

From the above discussions 250 m is enough for most cases, and 500 m is necessary for some of the transient experiments. For future expansion the pipe space for total length of 1,500 m is reserved.

4.5 Pressure

From the discussions of Section 4.2, all the flow patterns can be generated with an inlet pressure of 834 kPa. Therefore, the maximum operating pressure is determined as 980 kPa. However, for future expansion the design pressure of all the facilities is set at 3,920 kPa. For test runs of the newly developed multiphase pumps, the design pressure of a short section of the pipeline downstream of the pump will be 13,700 kPa.

4.6 Inclined section

For experiments of risers and hilly-terrain pipelines, the test pipeline has two inclined sections. One is a section of 10 m, which can be inclined at several inclination angles from horizontal (0 degree) to vertical (90 degrees). The other is the hilly-terrain section with one or two hills with the maximum inclination angle of 10 degrees. (See Table 4.)

5. SPECIFICATIONS OF TEST FACILITIES

Based on the above consideration the major specifications of the test facilities have been designed as follows. Table 5 shows the list of major specifications.

5.1 Supply system and control system

Figures 3 and 4 show the simplified flow diagram and the plot plan of the test facilities, respectively.

The gas phase is compressed, circulated through the test pipeline with the liquid phase, separated, and recirculated. The flow rate at the inlet of the pipeline is controlled by a control valve. The capacity of the compressor was determined to be 2,000 Nm^3/h with the suction pressure of 245 kPa and the discharge pressure of 834 kPa.
The liquid phase is pressurized by a pump, circulated through the test pipeline with the gas phase, separated by the separator, and returns to a tank. The flow rate at the inlet of the pipeline is controlled by a control valve. The capacity of the pump was determined to be 180 m^3/h at a differential head of 106 m (834 kPa for kerosene).

The separator is a horizontal type with a capacity of 4,000 Nm^3/h of gas and 360 m^3/h of liquid. The separator is designed to operate at or above 245 kPa in order to send the liquid to the tank. The capacity of the liquid tank is 40 m^3.

The space for the additional tanks, high pressure pumps and compressors for future expansion, and equipment to be tested is reserved.

5.2 Test pipeline
Figure 5 shows the layout of the test pipeline and approximate location of instrumentation. The test pipeline of 105 mm in diameter and 500 m long, and 54 mm and 250 m are laid at a same level. The minimum radius of the curve is 2 m in order to minimize the effect of curvature. The final portion of 80 m is in a straight line, and includes the inclined sections.

5.3 Instrumentation
Instrumentation and its flexibility is the key to this type of test facilities. As shown in Fig.5, the basic design of instrumentation is as follows.

Pressure taps and bases for temperature gauge are located at the following points;
- Highest point of inclined section
- End point of inclined section
- Immediate upstream and downstream and 100 D upstream and downstream of the last curve
- Immediate upstream of inclined section
- Center of each straight section
- End point of straight section downstream of the inclined section

Transition pieces for liquid holdup sensors and transparent pipe for flow pattern observation are installed at the same location as above.

The liquid holdup sensors will be used to detect the passage of pigs. Instrumentation will be prepared to measure the pressure difference across a pig .

For data collection of transient multiphase flow, flow control system and the separator are designed as follows;
- Linear change of flow rates of gas and liquid will be controlled automatically. The range of flow rate change will be 0 to 100 %, and the duration of change will be adjustable with a minimum of 5 min.
- The change in weight of the separator will be measured by a load cell in order to measure the instantaneous flow rate of liquid from the pipeline.

6. CONCLUSIONS

Basic design of JNOC's large-scale two-phase test facilities was completed. Based on extensive simulation studies, the design work was carried out to satisfy the requirement in view of the purposes of the experiment such as capability of generating all flow patterns and meaningful transient data.

Based on the above consideration the major specifications of the test facilities were determined as follows;

- Test fluids: Gas phase: air and nitrogen
  Liquid phase: water and kerosene
- Test pipelines: 105 mm diameter x 500 m long (1,500 m for future expansion)
  54 mm diameter x 250 m long
- Maximum operating pressure: 980 kPa (3,920 kPa for future expansion)
- Inclined section: 10 m long, 0 - 90 degrees
- Hilly-terrain section: 10 m x 4 slopes, -10 to +10 degrees
- Supply system: Pump: 180 m^3/h x 834 kPa
  Compressor: 2,000 Nm^3/h x 834 kPa
  Separator: 4,000 Nm^3/h (gas), 360 m^3/h (liquid)
  Water tank: 40 m^3
- Instrumentation: Meters for pressure, temperature, liquid holdup to be located at appropriate locations and will have flexibility.

The new test facilities will be open to public, will have capability of coping with a variety of research needs, and will be expandable. When completed, JNOC will be ready to invite the research organizations of all over the world to consider the joint research projects with JNOC utilizing the Kashiwazaki Test Facilities.

7. ACKNOWLEDGEMENT

The authors would like to acknowledge JNOC/TRC for permission to publish this paper.

8. REFERENCES

1. Ihara, M., Yanai, K. and Takao, S. 'Two-Phase Flow in Horizontal Wells', SPE 24493, the 5th Abu Dhabi Petroleum Conference and Exhibition (May, 1992).

2. Ihara, M., Furukawa, H., Takao, S. et al. 'Experimental Investigation on the Interaction between the Fluid Influx from the Reservoir and the Pressure Distribution along a Horizontal Well Configuration', OSEA 92209, the 9th Offshore South East Asia Conference & Exhibition (1-4 December, 1992).

3. Kohda, K., Suzukawa, Y. and Furukawa, H. 'Analysis of Transient Gas-Liquid Two-Phase Flow in Pipelines', J. of Energy Resources Technology, pp.93-101 (June, 1988).

4. Kohda, K., Suzukawa, Y. and Furukawa, H. 'Pigging Analysis of Gas-Liquid Two-Phase Flow in Pipelines', NKK Technical Review, No. 60, pp. 30-36 (1990).

5. Zheng, G.H., Brill, J.P. and Shoham, O. 'An Experimental Study of Two-Phase Flow in Hilly Terrain Pipelines', SPE 24788, the 67th SPE Annual Technical Conference and Exhibition (4-7, Oct., 1992).

6. Minami, K. 'Transient Flow and Pigging Dynamics in Two-Phase Pipelines', Ph.D Dissertation, the University of Tulsa, (1991).

7. Bendiksen, K.H., Malnes, D., Moe, R. et al. 'The Dynamic Two-Fluid Model OLGA: Theory and Application', SPE Production Engineering, pp. 171-180 (May, 1991).

8. Barnea, D. 'Transition from Annular Flow and from Dispersed Bubble Flow - Unified Models for the Whole Range of Pipe Inclinations', Int. J. Multiphase Flow, Vol. 12, No. 5, pp. 733-744 (1986).

9. Barnea, D. 'A Unified Model for Predicting Flow-Pattern Transition for the Whole Range of Pipe Inclinations', Int. J. Multiphase Flow, Vol. 13, No. 1, pp. 1-12 (1987).

10. Taitel, Y. 'Flow Pattern Transition in Two Phase Flow', Int. J. of Heat and Mass Transfer, Vol. 1 (1990).

APPENDIX

Outline of TAPTWO

TAPTWO is a computer program developed to analyze transient two-phase flow in pipelines. It analyzes the transient behavior of two-phase flow caused by changes in flow rates and/or pressures. It also calculates drastic transient flow during and after pigging operation.

TAPTWO can analyze two-phase flow over the entire length of the pipeline, single phase gas flow over the entire length, or the transition from one to the other. Pigging analysis can be executed even in the cases where the slug formed in front of the pig diminishes due to evaporation of liquid or liquid flow-back past the pig.

TABLE 1 EXISTING TEST FACILITEIS

Organization	Size	Inclined Section	Fluid	Pump	Compressor
JNOC/TRC	54 mm (D) 105 m (L)	6 m 0-90 deg. 11 m 0-10 deg.	Water Air	70 m^3/h 100 m (Head)	 170 Nm^3/h 970 kPa
NKK	105 mm (D) 1,436 m(L)	Horizontal	Water Air	65 m^3/h 120 m (Head)	 2,200 Nm^3/h 690 kPa
TUFFP	78 mm (D) 420 m (L)	Partially 0 - 5 deg.	Kerosene Air	45 m^3/h 110 m (Head)	 860 Nm^3/h
SINTEF	190 mm (D) 450 m (L)	54 m 90 deg. 65 m 0 to - 2 deg.	Naphtha Diesel Lubricant Nitrogen	10,000 kPa	 10,000 kPa
TEXACO	152 mm (D)	Horizontal	Crude oil Water Natural gas	133 m^3/h 1,730 kPa	 14,500 Nm^3/h 1,730 kPa
BP	152 mm(D)	-	Water Air	-	
ELF Aquitaine	Vertical and inclined pipes				
AGIP	Onshore multiphase test facilities				

TABLE 2 FLOW PATTERN PREDICTION

(HORIZONTAL, 105 mm ID, 250 m LONG)

GAS FLOW RATE Nm^3/H	LIQUID FLOW RATE m^3/H					
	0.1	1	10	50	100	140
10	STF	STF	IMF	IMF	IMF	DBF
50	STF	STF	IMF	IMF	IMF	DBF
100	STF	STF	IMF	IMF	IMF	DBF
500	STF	WVF	IMF	IMF	IMF	DBF
1000	WVF	WVF	IMF, AMF	IMF	IMF	DBF
2000	AMF	AMF	AMF	IMF	IMF	DBF

STF : STRATIFIED FLOW
WVF : WAVY FLOW
IMF : INTERMITTENT FLOW
DBF : DISPERSED BUBBLE FLOW
AMF : ANNULAR MIST FLOW

TABLE 3 FLOW PATTERN PREDICTION

(INCLINATION ANGLE 20°, 105 mm ID, 10 m LONG)

GAS FLOW RATE Nm^3/H	LIQUID FLOW RATE m^3/H					
	0.1	1	10	50	100	140
10	STF, IMF	STF, IMF	IMF	IMF	IMF	DBF
50	STF, IMF	STF, IMF	IMF	IMF	IMF	DBF
100	STF, IMF	STF, IMF	IMF	IMF	IMF	DBF
500	WVF	WVF, IMF	AMF, IMF	IMF	IMF	DBF
1000	WVF, AMF	WVF, IMF	AMF	IMF	IMF	IMF, DBF
2000	WVF, AMF	AMF	AMF	IMF, AMF	IMF	DBF

STF : STRATIFIED FLOW
WVF : WAVY FLOW
IMF : INTERMITTENT FLOW
DBF : DISPERSED BUBBLE FLOW
AMF : ANNULAR MIST FLOW

TABLE 4 FLOW PATTERN PREDICTION

(HILLY TERRAIN SECTION, 105 mm ID, 10 m LONG × 4)

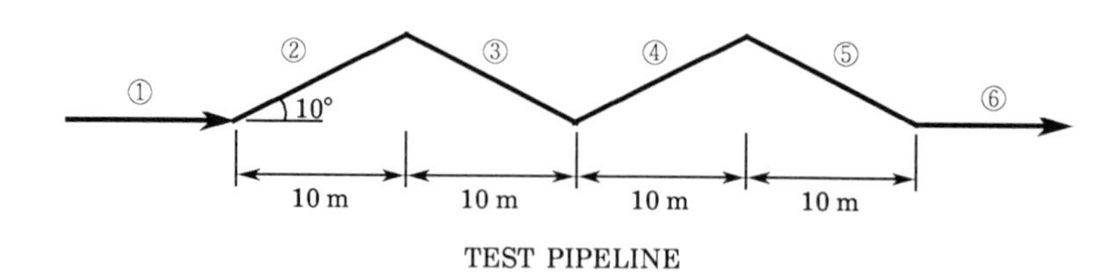

TEST PIPELINE

GAS RATE Nm^3/H	LIQ. RATE m^3/H	FLOW PATTERN ①	②	③	④	⑤	⑥
10	0.1	STF	IMF	WVF	IMF	WVF	STF
100	140	DBF	DBF	DBF	DBF	DBF	DBF
500	10	AMF	IMF	WVF	IMF	WVF	AMF
1000	10	AMF	WVF	AMF	WVF	AMF	AMF
1500	160	DBF	IMF	IMF	IMF	IMF	IMF

STF : STRATIFIED FLOW
WVF : WAVY FLOW
IMF : INTERMITTENT FLOW
DBF : DISPERSED BUBBLE FLOW
AMF : ANNULAR MIST FLOW

TABLE 5 MAJOR SPECIFICATIONS

FLOW CONDITIONS

- TEST FLUIDS : GAS PHASE – AIR AND NITROGEN
  LIQUID PHASE – WATER AND KEROSENE
- FLOW RATE : GAS – MAX. 1,500 NM^3/H
  LIQUID – MAX. 180 M^3/H
- OPERATING PRESSURE : 980 kPa

TEST PIPELINES

- PIPE DIAMETER : 105 MM (4 INCH), 54 MM (2 INCH)
- LENGTH : 500 M FOR 105 MM LINE (1,500 M : FUTURE)
  250 M FOR 54 MM LINE

GAS / LIQUID SUPPLY SYSTEM

PUMP	: 180	M^3/H	×	834 kPa	× 2 UNITS
GAS COMPRESSOR	: 2000	NM^3/H	×	834 kPa	
AIR COMPRESSOR	: 200	NM^3/H	×	390 kPa	
GAS COMPRESSOR COOLER	: 22	M^3	×	7.5 kW	× 2 UNITS
KNOCK OUT DRUM	: 6	M^3	×	3920 kPa	
	0.6	M^3	×	980 kPa	

- SEPARATOR : 10 M^3 × 3920 kPa
  CAPACITY – 4000 NM^3/H (GAS), 360M^3/H (LIQUID)
- TANK : 40 M^3 × 2 UNITS
- COALESSOR : 200 M^3/H, 1.5 M^3 × 980 kPa

TEST SECTIONS

HILLY – TERRAIN SECTION

- PIPE DIAMETER : 105 MM, 54 MM
- LENGTH : 10 M × 4 SLOPES
- MAX. INCLINATION ANGLE : -10 +10 DEG.

INCLINED SECTION

- PIPE DIAMETER : 105 MM, 54 MM
- LENGTH : 10 M × 4
- INCLINATION ANGLE : 0 ~ 90 DEG.

MULTIPHASE PUMP TEST SECTION

- PIPE DIAMETER : 105 MM
- LENGTH : 10 M
- MAX. OPERATING PRESSURE : 13,700 kPa

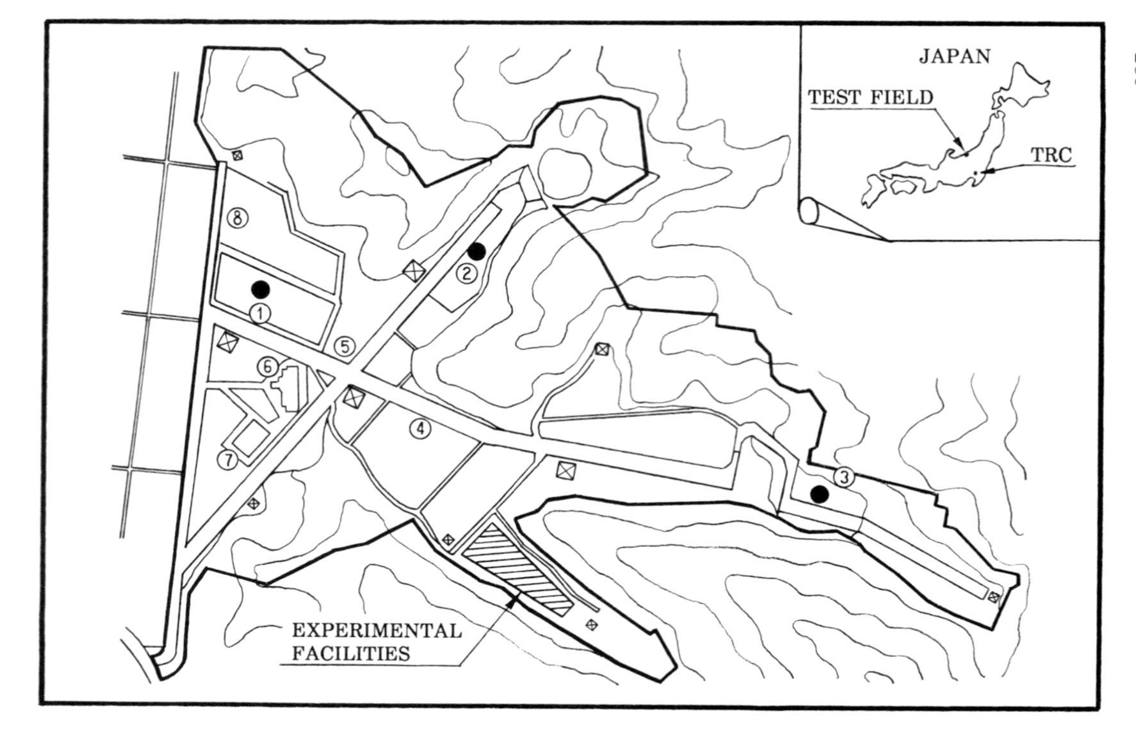

Boundary
Air-gun Pit
① Research Well No.1
② Research Well No.2
③ Research Well No.3 (FUTURE)
④ Seismic Line (950m long)
⑤ Seismic Line (550m long)
⑥ Office
⑦ Warehouse
⑧ Regulating Reservoir

FIG.1 KASHIWAZAKI TEST FIELD

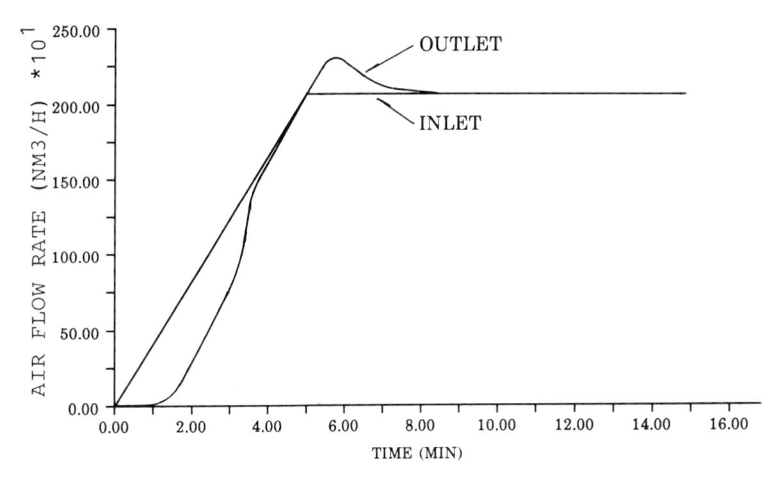

FIG. 2-a TRANSIENT TWO-PHASE FLOW–AIR FLOW RATE

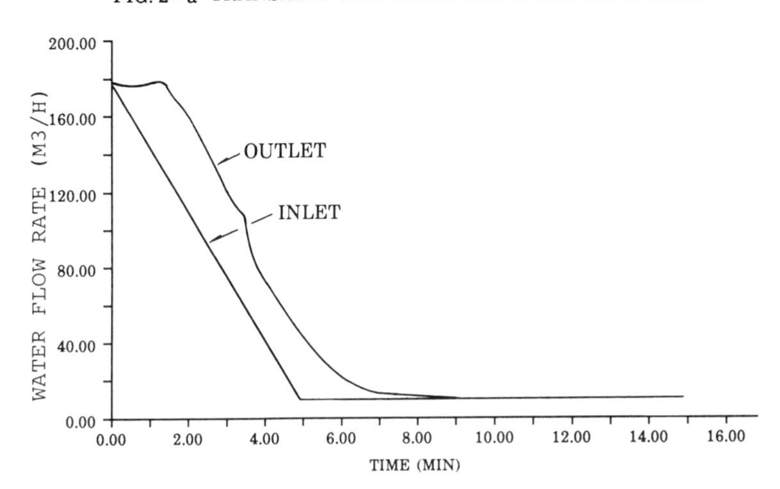

FIG. 2-b TRANSIENT TWO-PHASE FLOW–WATER FLOW RATE

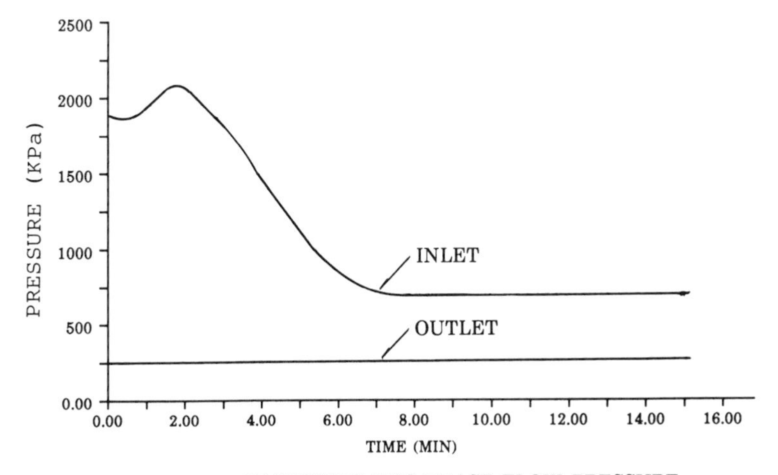

FIG. 2-c TRANSIENT TWO-PHASE FLOW–PRESSURE

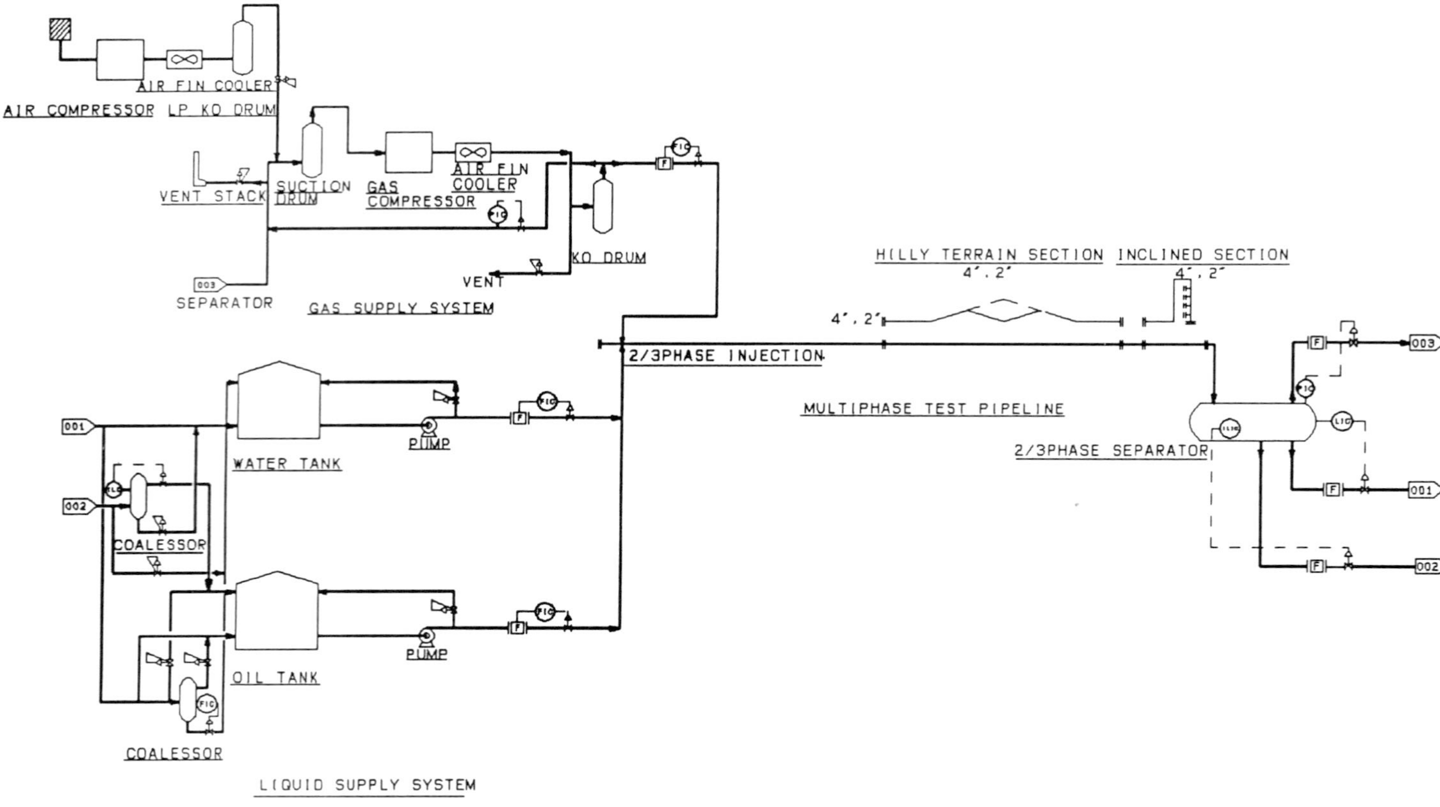

FIG. 3 SIMPLIFIED FLOW DIAGRAM

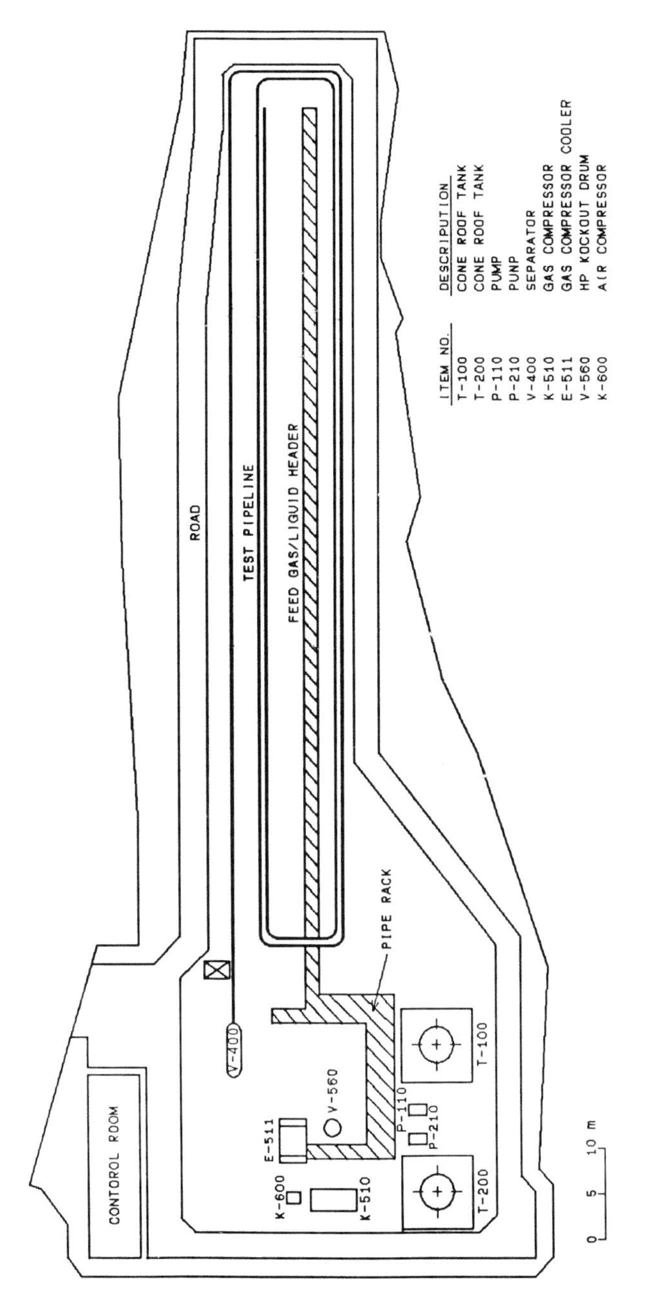

ITEM NO.	DESCRIPTION
T-100	CONE ROOF TANK
T-200	CONE ROOF TANK
P-110	PUMP
P-210	PUNP
V-400	SEPARATOR
K-510	GAS COMPRESSOR
E-511	GAS COMPRESSOR COOLER
V-560	HP KOCKOUT DRUM
K-600	AIR COMPRESSOR

FIG. 4 PLOT PLAN

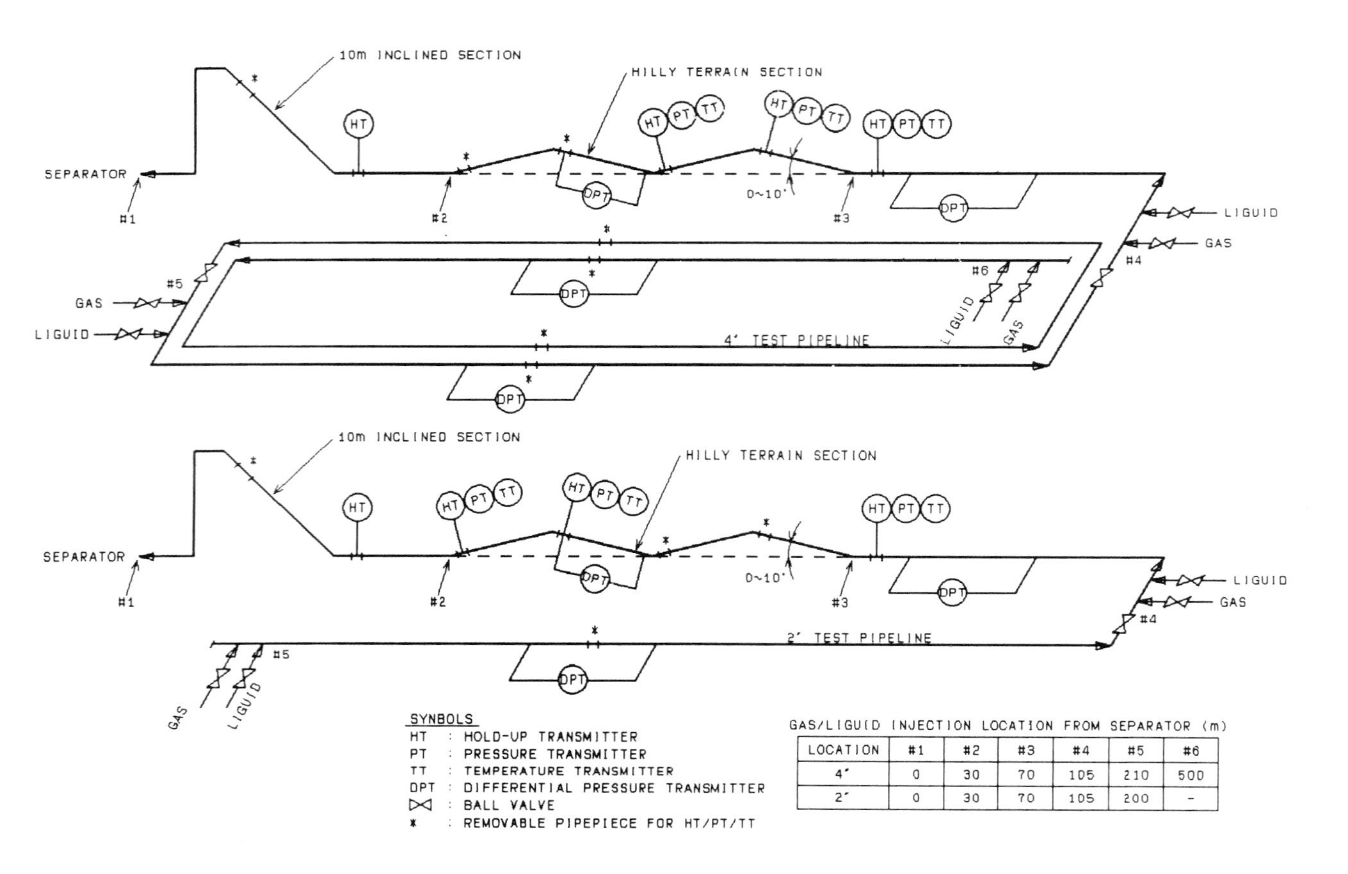

GAS/LIGUID INJECTION LOCATION FROM SEPARATOR (m)

LOCATION	#1	#2	#3	#4	#5	#6
4"	0	30	70	105	210	500
2"	0	30	70	105	200	-

FIG. 5 TEST PIPELINE AND INSTRUMENTATION

Aspects of three-phase flows in gas condensate pipelines

K. Lunde[1], S. Nuland[1] & M. Lingelem[2]

[1]Institutt for energiteknikk, Postboks 40, N-2007 Kjeller, Norway.
[2]Norsk Hydro a.s., Postboks 210, N-1321 Stabekk, Norway.

ABSTRACT

Published data from the North East Frigg gas-condensate pipeline show unexpectedly large accumulation of liquid in the pipeline, and that a significant part of accumulated liquid is water. The data indicate that formation of emulsion does not occur, and emulsification is therefore not the cause of the liquid accumulation.

Three-phase flow experiments, simulating the flow conditions typical of gas-condensate pipelines, were done in IFE's medium pressure, dense gas flow loop. The aim was to simulate the phenomena of accumulation observed in the North East Frigg pipeline. The results from these experiments identified gravitational stratification of the oil and water as the mechanism causing the accumulation of water in the pipeline. When the oil was much lighter than the water, and since the test-section was inclined upwards in the flow direction, then the accumulation of water led to increased liquid holdups in the test-section.

Gas-condensate pipelines may have long inclined stretches, and the condensate density may be significantly lower than that of water. In such a pipeline, water, if produced, will, at low production rates, accumulate and a cause significant increase in the total liquid content of the pipeline, relative to what it would be in a two-phase gas-condensate pipeline.

1. INTRODUCTION

Three-phase flows occur frequently in the operation of oil and gas fields. As the reserves in a field are depleted the water to oil ratio in the well-stream increases . The water may occur naturally, or as a result of water injection aimed at boosting production. Costs associated with removing the water off-shore are very high, making direct transfer to on-shore processing plants of the untreated well-stream economically attractive. Such long-distance multiphase transport requires detailed knowledge of three-phase flows and their potential problems. Here, ability to predict pressure drop, phase fractions, and flow regimes is of primary importance.

Field data from the North East Frigg gas-condensate field show that the presence of water affect the total liquid content in the pipeline. This field is near the end of its productive life and the production is low, with a high water cut. Hansen (1991) estimates the liquid content of the pipeline to be high, and the water-cut of the accumulated liquid to be much greater than indicated by the water-cut in the produced liquid. Available two-phase flow models predict a much lower liquid content, and even accounting for uncertainties in the fluid properties, could not serve to explain this shortcoming of the models.

Most relevant references concentrate on other aspects than phase fractions, and on flow conditions far removed from those of gas-condensate pipe-lines. Many deal, for example, with the transport of very viscous oils, Lonsdale (1971) and Stein (1978), in which the water is added as a means of reducing the pressure drop. There are, however, features of these flows which are of interest also to the study of gas-condensate flows, in particular the occurence and behaviour of oil-water emulsions. Medvedev (1978) describes a model for two-phase flows of oil-water emulsions, which includes effects such as increase in effective viscosity and the effect of drop size on wall friction. Similar increases in viscosity have been reported by several authors, for example Pal & Rhodes (1985), Malinowsky (1975), and Laflin & Oglesby (1976).

An increase in effective viscosity as result of emulsification would in general lead to an increase in the liquid holdup in three-phase flows. This may be especially so for gas-condensate pipelines in which the total liquid flow rate is relatively small, and the reason is a simple geometrical effect. If, in a pipeline of constant gas and liquid flowrate, suddenly a larger force was required to drag the liquid through the pipe due to increased effective viscosity of the liquid, then the gas, in order to provide the extra force, would have to increase its velocity through a reduction in its flow area. If the total liquid holdup in the pipe is small, then even a small change of force will require a relatively large change in gas flow area, and consequently a relatively large increase in the liquid holdup would result.

But it is unlikely that this is the cause of the liquid accumulation reported for the North East Frigg pipeline since no emulsion problems have been reported and since Hansen (1991) from his data inferred a large velocity difference between the condensate and the oil, a situation incompatible with emulsification. That there may be a significant slip velocity between the two liquid phases in three-phase flows was reported by Shean (1976) and Sobocinski *et al.* (1958), who studied such flows in vertical and horizontal pipes respectively. Shean (1976) also reported changes in liquid holdup with input water-cut.

A three-phase flow model in which the motion of the water is coupled to that of the gas and oil through momentum transfer relations was proposed by Fuchs & Nuland (1992). When applied to the North East Frigg case this model predicts a significant slip velocity between the two liquid phases and gives a much better account of the liquid accumulation and its water-cut than did the two-phase models. The North East Frigg data contain large uncertainties. Our study aimed to provide similar data, but from controlled experiments performed in flow conditions typical of gas-condensate pipelines.

2. THE EXPERIMENTS

The experiments reported here were performed in IFE's medium pressure dense gas facility. It is a closed loop flow rig utilising a dense gas at medium pressures to simulate flow conditions in gas-condensate pipelines.

2.1 Why use a dense gas?

Gas-condensate pipelines may be operated at pressures up to hundreds of bars, with corresponding gas densities of hundreds of kg/m^3, and oil densities as low as 600 kg/m^3. At gas-liquid density ratios of this order interfacial phenomena like wave and drop generation will be prominent, even at low velocity differences between the two phases. Both of these factors will affect the transport characteristics of the pipeline in ways which evade simple scaling. Extrapolation of results from experiments performed at atmospheric conditions using air and water is therefore highly questionable.

A particular point of uncertainty is associated with the tendency of atmospheric air-water flows to exhibit slug flow at flow rates of interest to our study. The effect of gas density on slug flow in two-phase flows was investigated by Bendiksen and Espedal [1991], and their conclusions are, in principle, also valid for three-phase flows. They demonstrated that the criterion for the onset of slug flow expressed through Kelvin-Helmholz type instabilities is not a sufficient condition for stable slug flow. An additional condition must be met, namely, that there be enough liquid present in the pipe to sustain the slug once it is formed.

In gas-condensate pipelines, with their high gas density and low liquid flow rate, the liquid transport may be so efficient that stable slug flow is not sustainable. Large waves bridging the pipe may travel as ‘slugs’ for short distances before collapsing, representing a flow-regime between stratified wavy and slug flow. The gas-liquid interface of this regime of proto-slugs will be very rough, with strong generation of liquid drops associated with the proto-slugs, both features which will increase the interfacial friction.

Even if the gas density in IFE's facility is only about 50 kg/m^3 at the maximum system pressure of 8 bara, it is still sufficiently high for many of the phenomena just mentioned to occur. Results obtained in this rig may therefore be used in the prediction of flow characteristics of real pipelines with much greater certainty than results from low gas density systems.

2.2. IFE's medium pressure dense gas facility

This closed loop flow rig consists of a flow section, a return section, a gas-liquid separator, an oil-water separator, pumps and compressors, and a flow rates control system. A more detailed description can be found in Lingelem & Nuland (1993).

The loop has a 20 m long flow section of internal diameter of 0.1 m. Of these 20 m the last 10 m constitute a transparent PVC test-section, enabling visual observation of the flow. The flow section can be lifted to inclinations of 15°.

The gas is circulated by four positive displacement compressors. These can be run separately or together, and in combination with a bypass system they allow the gas superficial velocity in the test-section to be varied continuously from 0 m/s to about 7 m/s. Volume flow rate is measured at the inlet by a Fischer & Porter vortex meter. Measurements of absolute pressure at the inlet and in the test-section allow correction for the expansion of the gas, and the actual volume flow rate in the test section can be calculated. The gas temperature is adjusted after compression by a thermostatically controlled cooling system.

The oil and water are circulated in the loop by centrifugal pumps. The liquid flow rates are quite small, with the superficial velocities for oil and water less than 0.2 m/s and 0.02 m/s respectively. The flow rates are adjusted to the desired values by 'throttling' the pumps using valves, and they are measured using Fischer & Porter rotameters.

The test-section is equipped with systems for measurement of pressure and phase fractions. The absolute pressure in the test-section is monitored by a Honeywell pressure cell. Pressure drop is measured over two lengths of 2 m, situated 4 m apart, by Honeywell dP-cells. The pressure-taps are situated on the top of the test-section, and the pressure-lines are gas-filled.

Phase fractions in two-phase flow are measured by a broad-beam gamma densitometer. In three-phase flows the phase fractions are measured by fast closing valves. The two methods have been checked against each other, and provided certain precautions are taken the agreement is good. For example, it is necessary to allow three-phase flows with very small water flow rates considerable time to develop because of the slow accumulation of water in the test-section. In order to obtain adequate accuracy when used in intermittent flows, more than one sample closing of the valves must be taken, and during the present study three samples were taken for each combination of experimental parameters.

2.3. Experiments performed

Five experiments were performed, an experiment being a series of runs at constant absolute pressure P_{abs}, angle of inclination θ, and liquid to gas input ratio R_{Lin} but with varying water to total liquid input ratio R_{Win} (input water-cut) and gas flow rate U_{sG}. The relevant parameters of the experiments are presented in tables 2.1 and 2.2.

Table 2.1: Invariable experimental parameters

Liquid	ρ (kg/m^3)	μ $(mPas)$	*Gas*	P_{abs} $(bara)$	$\sigma_{gas/oil}$ (N/m) *(Before / After)*	$\sigma_{oil/water}$ (N/m) *(Before / After)*
Water	1000	≈ 1	air	1	0.073 / 0.046	—
D80	800	≈ 2	air	1	0.027 / 0.027	0.017 / 0.010
D80	810	—	SF6	4	0.022 / —	—
D80	996	0.83	FREON	10	0.0154 / —	—

where ρ is the density of the relevant phase, μ its viscosity, and σ its surface tension.

Table 2.2: Variable experimental parameters

Experiment	*I*	*II*	*III*	*IV*	*V*
P_{abs} *(bara)*	8	8	4	4	2.5
θ (°)	2	15	2	2	2
Gas	FREON	FREON	SF6	SF6	FREON
ρ_{gas} (kg/m^3)	50	50	25	25	15
ρ_{oil} (kg/m^3)	960	960	810	810	820
R_{Lin} (%)	2	2	2	0.4	2
R_{Win} (%)	0 - 15	0 - 15	0 - 30	0 - 100	0 - 15
U_{sG} (m/s)	2, 3, 4	2, 3, 4, 6	3, 4, 5, 6	4	4

Two different gases were used: a FREON ($CBrF_3$) and sulphur-hexafluoride (SF_6), both with a density of about 6 kg/m³ at standard pressure and temperature. The oil used for all experiments was Exxol D80. It is a light 'oil', and therefore a good substitute for a gas condensate. The water used was ordinary tap water. In order to make it distinguishable from the oil the water was coloured blue by 'methyl blue', a substance found not to affect the oil.

The gases differed in the extent to which they were absorbed by the oil, which absorption modified its density. The SF6 was only slightly absorbed whereas the absorption of FREON was strongly pressure dependent. Due to this absorption of FREON the oil was quite heavy in experiments I and II and the relative density difference between it and the water was only 4%. By comparison, the relative density difference between the oil and the water for the other experiments was of the order of 20%.

The surface tensions relevant for this study are those of oil-gas, water-gas, and oil-water at the experimental conditions. We have only a few measurements of these, as well as of viscosity, at conditions near those in the experiment. Supplementary measurements were made at atmospheric conditions, and the results of all relevant measurements are presented in table 2.1. To check for changes due to contamination the atmospheric surface tension measurements were repeated after the experiments. They indicate that some contamination occurred, and it is not surprising that it was the water that was affected. It is, ultimately, impracticable to keep a large flow facility like the one used absolutely free of contamination.

3. THE RESULTS

3.1. Qualitative observations

The boundaries between the various flow structures that occurred in these experiments were not very well defined. For example, the flows appeared to be in a kind of transition between a stratified wavy and a slug-flow regime for the entire low to medium range of gas flow rates (U_{sG} < about 5 m/s), and it was impossible to determine exactly the transition point between them. In fact, it seemed questionable to classify many of the observed flow structures in terms of the 'traditional' flow regimes, and for these reasons flow regime maps are not presented here. Instead the observed flow structures have been classified into five groups as follows;

Intermittent: Flows in which the liquid transport through the pipe is clearly unsteady. This group includes slugs, proto-slugs and large waves, and generally occurred at low to intermediate gas flow rates and/or large angles of inclinations.

Intermittent/Drops: Same as above, but now with significant or strong drop generation associated with the intermittent structures. This occurred at large angles of inclination and at high gas flow rates.

Stratified Wavy—Coarse: Flows in which the liquid transport at least seems steady, but which exhibit a very rough gas/liquid interface. This occurred at intermediate gas flow rates at low angles of inclinations.

Stratified Wavy—Coarse/Drops: Same as above, but now with significant drop generation associated with the wave-crests. This occurred at intermediate and high gas flow rates, at mainly large angles of inclination.

Stratified Wavy—Fine/Drops: Flows in which the liquid transport is steady, and the gas/liquid interface is wavy with small waves. There is strong drop generation at the interface, apparently due to waves being blown apart by the gas. This occurred at high gas flow rates at low angles of inclinations.

At low to medium gas flow rates ($U_{sG} < 4$ m/s for $\theta = 2°$, $U_{sG} < 5$ m/s for $\theta = 15°$) the flow was intermittent with fast moving proto-slugs and/or slower moving large waves. Few slugs were longer than 5 to 15 pipe diameters. The gas void fraction in these slugs was very high (10% to 60%) and in many cases one could see gas blowing through the slug, depositing drops ahead of it. At the bottom of the pipe the liquid moved forward only when waves and proto-slugs passed, otherwise it was stagnant or flowed backwards. As gas flow rate increased beyond 3 and 4 m/s the flow became stratified with a very rough interface on which some large waves were still moving. There was significant entrainment into the gas of liquid drops in conjunction with the wave crests, and these and the waves maintained a thin film on the pipe wall above the gas-liquid interface. This film was wavy. At 5 and 6 m/s the flow was stratified with a much finer interface.

Any large waves rearing from the liquid layer at these high gas flow rates were quickly blown apart by the gas. For this reason there was a high concentration of suspended, fast-moving drops in the gas, and it seems likely that they constituted a large proportion of the positive liquid transport through the pipe.

In all of the experiments the water clearly accumulated at the bottom of the pipe. The state of the water varied with input water-cut and gas flow rate such that for small input water-cut and medium to large gas flow rates ($U_{sG} > 3$ m/s) it appeared as a layer of drops at the bottom, while for the higher input water-cut and lower gas flow rate some of the water appeared as a continuous, meandering stream at the bottom of the pipe and some as a layer of drops between the water and the oil. The flow of the water in the meandering stream was always reversed or stagnant. As the gas velocity was increased ever more of the water drops were suspended in the oil until for large gas flow rate ($U_{sG} > 5$ m/s) they all seemed mixed in.

A dominating mechanism for the entrainment of water drops into the oil was flow structures reminiscent of shear instability waves appearing on the oil-water interface. This phenomenon was ubiquitous in all the stratified flow regimes including the slug-bubbles of the intermittent flows, but it also appeared in the intermittent flow regime. Here the water drops were lifted into the oil by shear instability waves occurring under intermittent structures moving on the gas-liquid interface. Once suspended, the drops would stay in the oil and be transported with it for some time, even at low gas flow rates. It appears that the only positive transport of water in these flows was through drops suspended in the oil.

In experiments I and II the water had no observable effect on the structure of the gas-liquid interface. In experiment III, the addition of water had the remarkable effect of stabilising the flow at the lowest gas flow rate from intermittent flow to stratified wavy, in spite of an increased liquid level in the pipe. The two liquid phases were well separated, and the water, occurring as a layer of drops under the oil, constituted roughly 50% of the total liquid holdup. This layer of water drops presumably acted as an energy sink and damped the waves generated at the oil-gas interface, such that they did not grow into slugs. As mentioned above, it was very clear how a passing wave on the oil-gas interface entrained and carried along with it large amounts of water drops.

In many runs the oil and the water were thoroughly mixed. But on using the fast-closing valves it became clear that even if the liquids appeared well mixed they separated completely within a few tens of seconds. In fact, only in one experiment, IV, did the mixture appear as a single homogenous liquid. The visible transition from two 'well mixed' but separate liquids to a homogenous mixture was not sudden, but occurred between water input rates of 40% to 70%. Visually, the transition manifested itself through the mixture becoming very viscous, and again this was easily seen on using the fast valves, when the stopped fluid would only just run down the 2° incline! This was not a very stable emulsion, as it separated within 3 to 8 minutes.

In this experiment the water had a significant effect on the gas-liquid interface. On the surface of the very viscous mixture some peculiar and steep waves occured. They appeared perfectly two-dimensional, spanning the pipe, and had a height of 30 to 40 mm, significantly larger than the average liquid depth on which they were moving. The speed of the waves through the pipe was low; in fact, they looked like 'static' Kelvin-Helmholz waves. They gave the impression of large waves on a shallow liquid wanting to break yet incapable of this destruction due to the high viscosity of the liquid on which they were moving. These large and steep waves made the interface very rough.

We have no methods for determining the exact structure of the emulsion that was formed. Careful observations of the mixture through the transparent test-section gave the following conclusions. During all the runs, including those with a water-cut of 80% in the holdup, the oil was the continuous phase. There were no signs of inversion of the emulsion when the water-cut in the input was increased from 0% to 90%, either visually or in the quantitative results. Thus, a better descriptive term for the final and viscous mixture in which the water was the dominant but dispersed phase and the oil constituted the continuous film is 'water-oil foam'.

3.2. Quantitative results

The results, that is phase fractions, water-cut in holdup, pressure gradients and flow structure, from all five experiments in Table 2 are presented in Tables 3a-e. The results, except those relating to flow structure, are also presented graphically, and plots of total liquid holdup as function of input water-cut R_{Win} are presented in figures 1a to 5a. Corresponding plots of water-cut in holdup (that is, the ratio of water in holdup to total liquid holdup) against input water-cut are shown in figure 1b to 5b. Some key observations relating to the phase fractions results follow:

(i) The liquid holdup exhibited the familiar dependency on gas flow rate, namely, a decrease with increasing gas flow rate.

(ii) In accordance with the visual observations, the quantitative results indicate that there was significant water accumulation in the test-section for all the experiments.

(iii) The gas flow rate was the parameter that induced the clearest effect on the accumulation of water. The accumulation was highest for the lowest gas flow rates and then decreased with increasing gas flow rate. There was also a tendency for the water accumulation to be strongest in the experiments with the largest relative density difference between the oil and the water, figure 3b.

(iv) The accumulation of water caused significant increase in the total liquid holdup only in the experiments where the relative density difference between the oil and the water was large, that is, experiments III, IV, and V, figures 4a, 5a, and 3a (P_{abs} = 2.5 bara) respectively.

(v) There was no evidence in the phase fraction measurements, figure 5a, of the emulsification observed visually during experiment IV, other than a generally high level of liquid holdup. There was a change in sign of the oil water slip velocity, figure 5b. From lagging the oil (accumulating in the test-section) the water changed to leading (the oil now accumulating) at an input water-cut between 60% to 70%.

Plots of pressure gradients for all experiments except V appear in figures 6 and 7. The results from experiments I and II are presented in fig. 6 as functions of input water-cut. The results from experiments III and IV are presented in figure 7. Its format, with pressure gradient plotted as a function of liquid holdup, enables comparison of results from two and three-phase experiments performed under slightly different conditions. Results from three-phase flows (experiment III) are represented by filled symbols, from two-phase gas-oil flows by open, unconnected symbols, and from two-phase gas-water flows by open, connected symbols. The dashed lines represent the results from the three-phase flows of experiment IV.

Some key observations relating to the pressure gradient results follow:

(i) The pressure gradients exhibited the familiar dependency on gas flow rate, exhibiting a minimum as the flow passed from slugging through proto-slugging to stratified wavy, where it increased with increasing gas flow rate.

(ii) In experiment II the pressure gradient increased slightly with input water-cut, while in I, the pressure gradient seemed unaffected by input water-cut, figure 6.

(iii) There was some increase in the pressure gradient as function of input water-cut for all gas flow rates of experiment III, figure 7. A significant observation is, however, that the changes exhibited by two-phase flows at similar liquid holdups and gas flow rates are of the same order and trend. Strictly speaking, the changes seen in experiment III can therefore not be said to be caused by three-phase effects.

(iv) There was a dramatic effect on the pressure gradient as function of input water-cut in experiment IV. These changes were clearly caused by three-phase effects and were associated with the emulsification observed visually.

3.3. Discussion of results

Our results indicate that there is a simple mechanism of gravitational stratification at the base of the liquid accumulation observed in the North East Frigg pipeline. In our experiments very stable stratifications occurred even when the water was dispersed as drops in a continuous oil phase and the relative density difference between the oil and the water was as low as 4%. It should be kept in mind that very stable stratifications occur in the natural environment at even smaller relative density differences than this. The natural effect of the stratification was for the water to collect at the bottom of the pipe where the transport velocity was very small, even negative! In that situation the water required a very large flow area for its conveyance, hence the accumulation of water. This is an example of a 'distribution slip velocity' known from other multiphase flow situations.

In the cases of small relative density difference between the oil and the water the accumulated water simply displaced the oil but did not add much to the weight of the liquid in the cross-section. When the relative density difference between the oil and the water was large the accumulation increased the total weight of the liquids, and since the liquid flowed up an inclined pipe extra force was required to drag it through the pipe. This force was provided by the gas increasing its velocity through decreasing its flow area, hence the increased total liquid holdup in these cases.

A point of interest here is that this mechanism of water accumulation will only lead to increased holdups in flows up inclined pipes. In horizontal pipes, or in cases of very small relative density difference between the oil and the water, the water will simply displace the oil from the bottom of the pipe. In doing so it will improve the transport rate of the oil, not because the oil 'slides' on top of a less viscous substance, water, as referred to in the Introduction, but because the oil is 'standing on the shoulders' of the water.

In the experiments in which the liquids did not form emulsions there were no significant three-phase effects on the pressure gradients.

It came as a surprise that this oil-water system formed a kind of emulsion, as was the case in experiment IV. Whether a 'real' emulsion, or formed as a result of some contamination of the two liquids, is impossible to say; but the situation provided an example of the effect of emulsification on both liquid holdup and pressure drop. This effect was to increase both, far more than would be expected from the simple water accumulation mechanism just described. It is therefore likely that an additional mechanism was at play, namely, the very large increase in the effective viscosity that resulted from the emulsification of the oil and water.

When seen in conjunction, the water-cut and pressure gradient measurements from experiment IV provide a more accurate estimate of the point of transition to emulsion in this flow. Visually it was estimated to take place between input water-cuts of 40% and 70%, but figures 5 and 7 indicate that the transition point occurred at about 60%. This point has been circled in figure 7.

4. CONCLUSIONS

There was a strong accumulation of water in the test-section as a result of gravitational stratification of the oil and the water. Gas-condensate pipelines will be prone to such stratification, and hence accumulation of water, because of their relatively low liquid flow rates and because condensate and water do not easily form stable emulsions.

In the cases of large relative density differences between the oil and the water, the accumulation of water led to a significant increase in total liquid holdup in the test-section. In gas-condensate pipelines, where the condensate density may be significantly lower than that of water, the accumulation of water will, depending on the pipeline profile, lead to a significant increase in total liquid content of the pipeline relative to that in a similar two-phase condensate pipeline. The very high liquid content reported by Hansen (1991) for the North East Frigg pipeline therefore seems quite plausible.

There were no significant three-phase effects on the pressure gradients measured, except in the one experiment in which the oil and water formed a kind of emulsion, and here the effect was one of dramatic increase.

5. REFERENCES

BENDIKSEN, K. & ESPEDAL, M. 1992 Onset of slugging in horizontal gas-liquid pipe flow. *Int. J. of Multiphase Flow*, **18**, No. 2, pp. 237-247.

FUCHS, P. & NULAND S. 1992 Three-phase modelling is a must. *Multiphase Transportation III - Present Application & Future Trends*, Røros, Norway.

HANSEN, T.E. 1991 The North East Frigg full scale multiphase flow test. *5th International conference on multi-phase production.* Cannes, France.

LAFLIN, G.C. & OGLESBY, K.D. 1976 An experimental study on the effect of flowrate, water fraction and gas-liquid ratio on air-oil-water flow in horizontal pipes. *BSc thesis,* University of Tulsa, USA.

LINGELEM, M.N. & NULAND, S.N. 1993 Two-phase flows in inclined pipes at high gas density. *12th Int. conf. on Offshore Mechanics and Arctic Engineering,* June 1993, Glasgow, Scotland.

LONSDALE, J.T. 1971 Drag reduction for oil water mixtures in pipe. *MSc thesis* Purdue University, USA.

MALINOWSKY, M.S. 1975 An experimental study of oil-water and air-oil-water flowing mixtures in horizontal pipes. *MSc thesis,* University of Tulsa, USA.

MEDVEDEV, V.F. 1978 Turbulent flow of concentrated emulsions in pipes. *Journal of applied mech. and physics,* **10**, No.6.

PAL, R. & RHODES, E. 1985 A novel viscosity correlation for non-Newtonian concentrated emulsions. *J. of Coll. and Interf. Sc.*, **107**, No. 2.

SHEAN, A.R. 1976 Pressure drop and phase fraction in oil-water-air vertical pipe flow. *MSc thesis,* Massachusetts Institute of Technology, USA.

STEIN, M.H. 1978 Concentric annular oil-water flow. *PhD thesis,* Purdue University, USA.

Table 3a: Results from experiment I of Table 2.2.

Superficial gas vel. (m/s)	Superficial liquid vel. (m/s)	Total liquid holdup (fraction)	Oil holdup (fraction)	Water holdup (fraction)	Water-cut in holdup (%)	Pressure gradient (Pa/m)	Flow structure (—)
Water-cut in input: 0% of total liquid input							
2.0	0.04	0.225	0.225	0.000	0	92	I
3.1	0.06	0.159	0.159	0.000	0	102	I
4.1	0.08	0.095	0.095	0.000	0	145	SW-C
Water-cut in input: 5% of total liquid input							
2.1	0.04	0.225	0.144	0.081	36	96	I
3.1	0.06	0.162	0.112	0.050	31	126	I
4.1	0.08	0.100	0.088	0.011	11	148	SW-C
5.2	0.10	0.067	0.061	0.006	9	272	SW-C/D
6.3	0.12	0.057	0.050	0.007	12	396	SW-F/D
Water-cut in input: 10% of total liquid input							
2.1	0.04	0.230	0.133	0.097	42	99	I
3.1	0.06	0.142	0.105	0.037	26	111	I
3.7	0.08	0.092	0.075	0.018	20	171	SW-C
5.2	0.10	0.075	0.063	0.013	17	256	SW-C/D
6.3	0.12	0.063	0.053	0.010	16	403	SW-F/D
Water-cut in input: 15% of total liquid input							
2.0	0.04	0.206	0.136	0.070	34	99	I
3.1	0.06	0.131	0.094	0.037	28	118	I
4.1	0.08	0.102	0.076	0.025	25	168	SW-C

Legend for flow structure symbols used in tables 3a-e:

I	= Intermittent (Slugs/proto-slugs/large roll-waves)
I/D	= Intermittent with significant amount of drops
SW-C	= Stratified Wavy-Coarse
SW-C/D	= Stratified Wavy-Coarse with significant amount of drops
SW-F/D	= Stratified Wavy-Fine with significant amount of drops

See section 3.1 for a more detailed description of these flow structures.

Table 3b: Results from experiment II of Table 2.2.

Superficial gas vel.	Superficial liquid vel.	Total liquid holdup	Oil holdup	Water holdup	Water-cut in holdup	Pressure gradient	Flow structure
(m/s)	(m/s)	(fraction)	(fraction)	(fraction)	(%)	(Pa/m)	(—)
Water-cut in input: 0% of total liquid input							
2.1	0.04	0.235	0.235	0.000	0	548	I
3.1	0.06	0.132	0.132	0.000	0	422	I
4.2	0.08	0.086	0.086	0.000	0	428	I/D
Water-cut in input: 5% of total liquid input							
2.1	0.04	0.215	0.128	0.086	40	607	I
3.1	0.06	0.156	0.1	0.056	36	548	I
4.1	0.08	0.096	0.084	0.011	11	453	I/D
6.1	0.12	0.063	0.057	0.007	11	572	SW-C/D
Water-cut in input: 10% of total liquid input							
2.1	0.04	0.216	0.138	0.077	36	584	I
3.1	0.06	0.140	0.102	0.039	28	492	I
4.1	0.08	0.102	0.081	0.021	21	480	I/D
6.2	0.12	0.067	0.055	0.012	18	613	SW-C/D
Water-cut in input: 15% of total liquid input							
2.1	0.04	0.206	0.115	0.091	44	609	I
3.1	0.06	0.146	0.097	0.049	34	514	I
4.1	0.08	0.110	0.078	0.032	29	508	I/D
6.2	0.12	0.068	0.052	0.016	24	626	SW-C/D

Table 3c: Results from experiment IV of Table 2.2.

Superficial gas velocity = 4.0 m/s Superficial liquid velocity = 0.016 m/s						
Water-cut in input	Total liquid holdup	Oil holdup	Water holdup	Water-cut in holdup	Pressure gradient	Flow structure
(%)	(fraction)	(fraction)	(fraction)	(%)	(Pa/m)	(—)
0	0.058	0.058	0.000	0	74	SW-C/D
10	0.083	0.066	0.017	20	82	SW-C/D
20	0.093	0.059	0.033	35	90	SW-C/D
30	0.106	0.059	0.047	44	119	SW-C/D
40	0.116	0.056	0.060	52	125	SW-C
50	0.129	0.055	0.074	57	116	SW-C
50	0.127	0.055	0.072	57	112	SW-C
60	0.146	0.054	0.092	63	137	SW-C
70	0.162	0.053	0.108	67	199	SW-C
80	0.160	0.044	0.116	73	210	SW-C
90	0.166	0.033	0.133	80	322	SW-C
100	0.066	0.000	0.066	100	81	SW-C

Table 3d: Results from experiment III of Table 2.2.

Water-cut in input	Total liquid holdup	Oil holdup	Water holdup	Water-cut in holdup	Pressure gradient	Flow structure
(%)	(fraction)	(fraction)	(fraction)	(%)	(Pa/m)	(—)
Superficial gas velocity = 3.2 m/s Superficial liquid velocity = 0.06 m/s						
0	0.19	0.19	0	0	84	I
5	0.223	0.144	0.087	39	120	I
10	0.237	0.132	0.105	44	106	SW-C
15	0.244	0.123	0.121	50	115	SW-C
20	0.245	0.121	0.125	51	115	SW-C
25	0.249	0.119	0.131	53	122	SW-C
30	0.248	0.115	0.134	54	119	SW-C
Superficial gas velocity = 4.1 m/s Superficial liquid velocity = 0.08 m/s						
0	0.127	0.127	0	0	118	SW-C/D
5	0.159	0.133	0.026	16	130	SW-C/D
10	0.175	0.125	0.050	29	136	SW-C/D
15	0.181	0.112	0.069	38	142	SW-C/D
20	0.193	0.108	0.085	44	157	SW-C
Superficial gas velocity = 5.1 m/s Superficial liquid velocity = 0.1 m/s						
0	0.102	0.102	0	0	164	SW-F/D
5	0.114	0.104	0.010	9	167	SW-F/D
10	0.124	0.103	0.021	17	173	SW-F/D
15	0.130	0.098	0.032	25	187	SW-F/D
Superficial gas velocity = 6.3 m/s Superficial liquid velocity = 0.12 m/s						
0	0.087	0.087	0	0	235	SW-F/D
5	0.093	0.089	0.004	4	245	SW-F/D
10	0.098	0.084	0.014	14	250	SW-F/D
15	0.106	0.085	0.021	20	257	SW-F/D

Table 3e: Results from experiment V of Table 2.2.

Superficial gas velocity = 4.0 m/s Superficial liquid velocity = 0.08 m/s					
Water-cut in input	Total liquid holdup	Oil holdup	Water holdup	Water-cut in holdup	Flow structure
(%)	(fraction)	(fraction)	(fraction)	(%)	(—)
0	0.166	0.166	0	0	SW-C/D
5	0.186	0.161	0.025	13	SW-C/D
10	0.197	0.151	0.047	24	SW-C/D
15	0.206	0.14	0.066	32	SW-C/D

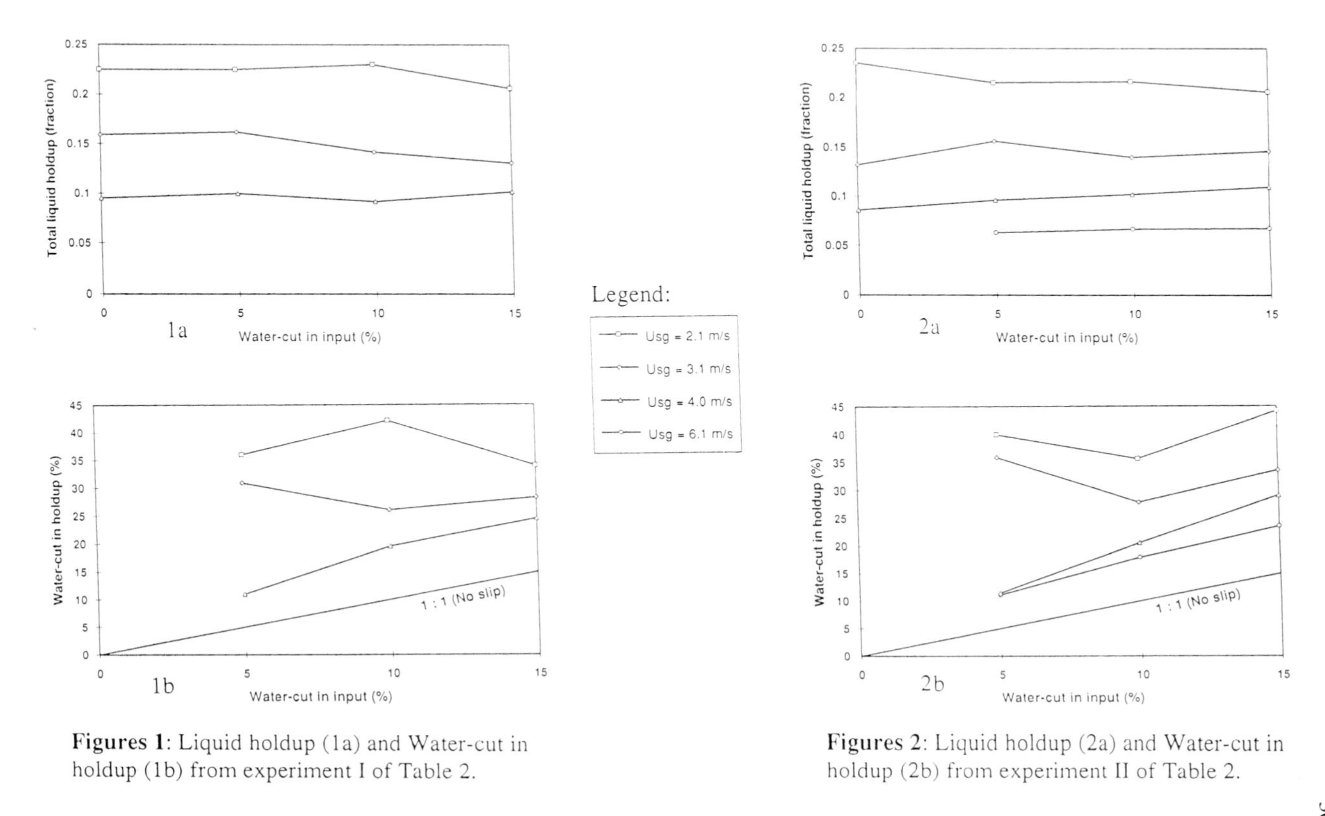

Figures 1: Liquid holdup (1a) and Water-cut in holdup (1b) from experiment I of Table 2.

Figures 2: Liquid holdup (2a) and Water-cut in holdup (2b) from experiment II of Table 2.

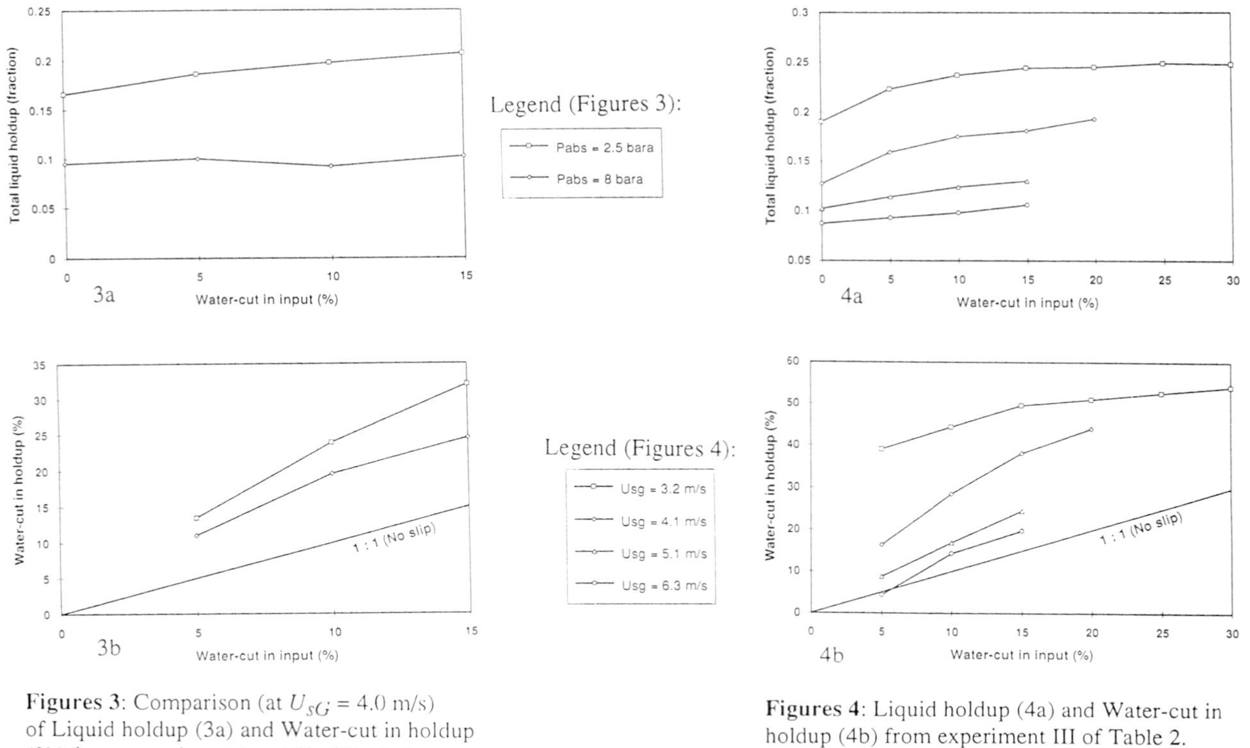

Figures 3: Comparison (at $U_{SG} = 4.0$ m/s) of Liquid holdup (3a) and Water-cut in holdup (3b) from experiment I and V of Table 2.

Figures 4: Liquid holdup (4a) and Water-cut in holdup (4b) from experiment III of Table 2.

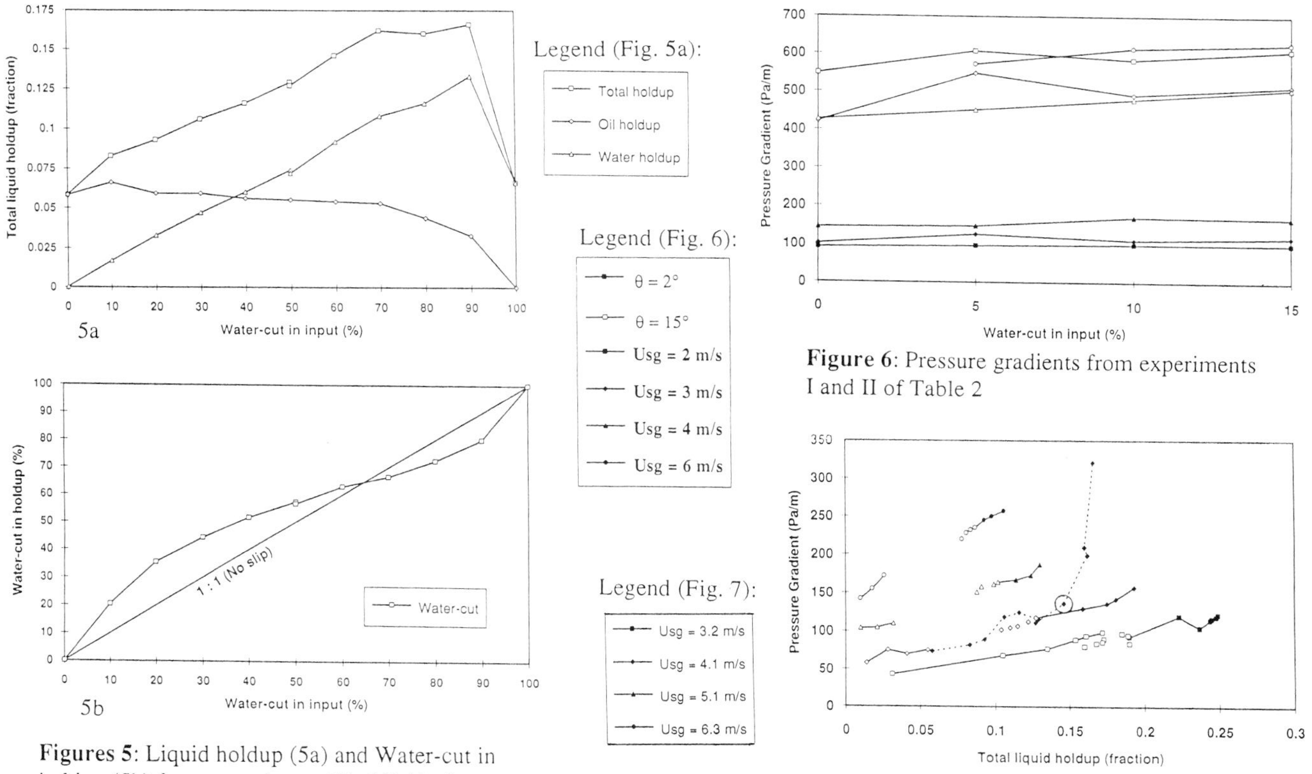

Figures 5: Liquid holdup (5a) and Water-cut in holdup (5b) from experiment IV of Table 2.

Figure 6: Pressure gradients from experiments I and II of Table 2

Figure 7: Pressure gradients from two and three-phase flow experiments (III, IV). See text for details.

THE EFFECT OF GAS VELOCITY ON GLYCOL DISTRIBUTION AND TOP-OF-THE-LINE CORROSION IN LARGE WET GAS PIPELINES

Rolf Nyborg, Arne Dugstad and Liv Lunde
Institutt for energiteknikk, P.O. Box 40, N-2007 Kjeller, Norway

ABSTRACT

Glycol can be used to prevent hydrate formation and reduce corrosion in subsea wet gas pipelines. Transport of glycol from a liquid phase trapped under a hydrocarbon layer to the top of the line in a large wet gas pipeline has been studied in a three-phase flow loop. The small amounts of glycol in the gas phase were measured with a radioactive tracer technique using ^{14}C-labeled glycol. The amount of glycol in droplets in the gas increased markedly with the gas velocity in the pipeline. These droplets provide the gas phase with a glycol source which can keep the gas phase saturated with glycol vapour. This can make a significant contribution to the glycol content of the liquid condensing in the top of the line. Top-of-the-line corrosion was studied in a one phase gas flow loop with wet gas containing glycol and CO_2. The corrosion rate increased with CO_2 partial pressure, gas flow rate and condensation rate. A large reduction in corrosion rate in the top of the line was observed when glycol was added to the gas.

INTRODUCTION

Unprocessed gas from offshore gas fields with large water depth will in the future be transported over long distances in large multiphase pipelines for processing onshore or at existing platforms. As the gas in the pipeline is gradually cooled, water and hydrocarbons condense out. The condensed water phase is corrosive and can attack both the top of the line and the bottom of the line. Glycol is often added in order to prevent hydrate formation, and glycol addition will also reduce corrosion in the pipeline[1,2,3]. Normally monoethylene glycol (MEG) or diethylene glycol (DEG) is used. If the water/glycol mixture is trapped under the hydrocarbon condensate in the bottom of the line, the gas phase and the condensing water can become almost free from glycol after some distance[1]. If the glycol is going to have an effect on corrosion in the top of the line, glycol must then be transported from the liquid water/glycol phase in the bottom of the pipeline to the gas phase as glycol vapour or as droplets. The transport of glycol as droplets to the gas phase has been studied in a large three-phase flow loop with 100 mm test section diameter under flow conditions similar to those expected during operation of large wet gas pipelines. The corrosion rate of pipeline steel in the top of the line has been studied in a one phase gas flow loop with wet gas containing glycol and CO_2.

GLYCOL TRANSPORT STUDIES

Three-Phase Flow Loop Experiments

The glycol transport studies were performed in a large diameter three-phase flow loop in order to obtain quantitative data on mass transport of glycol from a liquid water/glycol phase trapped under a hydrocarbon condensate layer to the gas phase. The corrosion rate in the top of the line depends among other factors on the amount of water and glycol condensing on the pipe wall.

The high-density gas freon 13B1 was used as the gas phase, and the experiments were performed with a pressure of 9 bar in the loop. The droplet generation rate is assumed to be proportional to the density of the gas when the gas flow rate is constant. Based on the difference in density experiments with freon at 9 bar then correspond to methane at 85 bar, typical of large gas pipelines. The loop was operated as a closed system and emissions of freon to the atmosphere were kept at an absolute minimum. In subsequent experiments in this flow loop freon has been replaced with another heavy gas, sulphur hexafluoride. The two liquid phases used in the loop were a diethylene glycol/water mixture with a ratio of 2:1 and a hydrocarbon phase consisting of Shellsol D70, which has physical properties comparable to a hydrocarbon condensate. The hydrocarbon to glycol/water ratio was 6:1 in all experiments with hydrocarbons present.

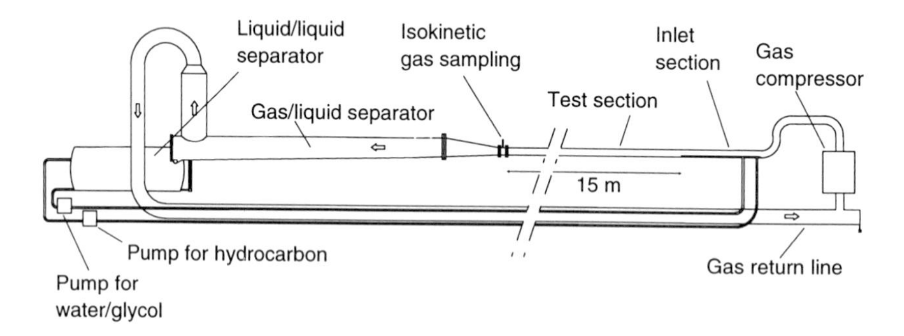

FIGURE 1. Three-phase test loop used for glycol transport studies. Test section diameter 100 mm and length 15 m.

The three-phase flow loop used in the experiments includes a 15 m long horizontal test section with inner diameter 100 mm. A schematic overview of the loop is shown in Figure 1. In the inlet section, a horizontal steel plate separates the liquid inlet from the gas inlet, as shown in Figure 2. The two liquids are pumped in under the plate in order to obtain a smooth liquid flow and prevent formation of emulsion where the gas and liquids meet. Stable three-phase flow is developed during the 15 m long test section, and at the end of the test section samples of the gas phase with droplets can be taken out with an isokinetic probe as shown in Figure 3. This probe consists of a pitot tube in the centre of the 100 mm pipe, a differential pressure transmitter and a sampling container. The flow through the pitot tube is adjusted in order to obtain zero pressure drop between the pitot tube and the pipe. This assures that the samples of gas and droplets are taken out with the same velocity as the flow in the test section, and that the droplet fraction through the isokinetic probe is equal to the droplet fraction in the test section. During gas sampling, the gas flows through the pitot tube and into a horizontal gas

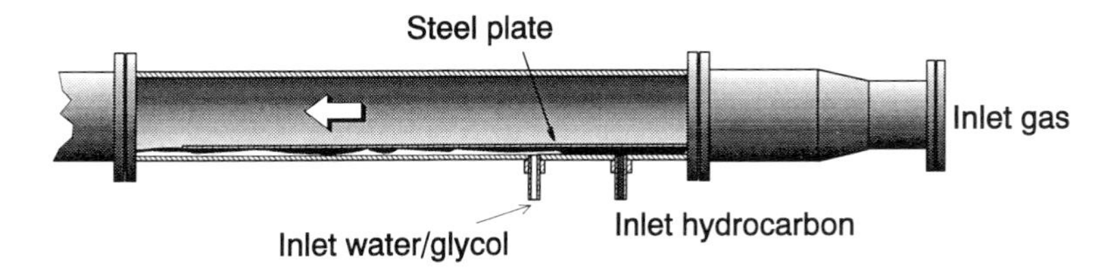

FIGURE 2. Inlet section in three-phase loop.

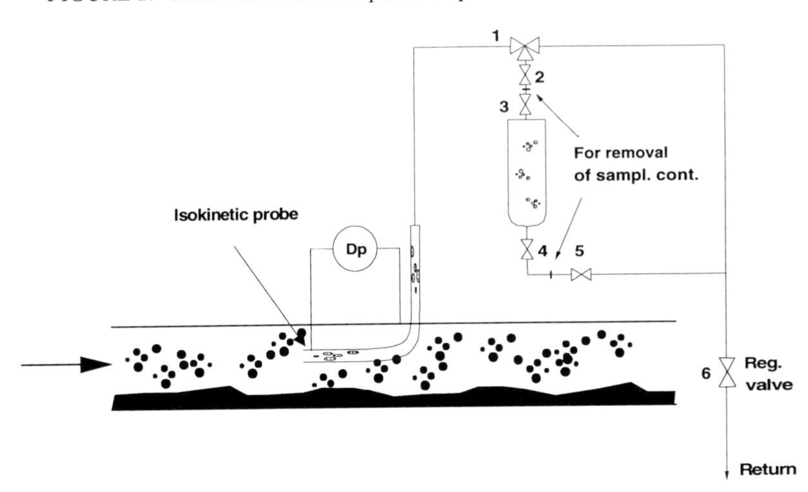

FIGURE 3. Section for isokinetic sampling of gas with droplets.

sampling container, where liquid droplets are accumulated because the gas flow rate is several times smaller through the sampling container. Afterwards the sampling container is removed for extraction of the liquid sample.

After the test section in the loop, the flow goes into a gas/liquid separator consisting of a 5 m long pipe with 300 mm inner diameter, see Figure 1. In this section the droplets in the gas phase fall down and are accumulated together with the rest of the liquids in the bottom of the pipe. After this the liquids go to a liquid/liquid separator and the gas goes through a return line to the compressors. The gas feeding and measuring system consists of four parallel compressors, a gas cooler and a Vortex flowmeter. The hydrocarbon and glycol/water feeding and measuring systems include separate pumps fed from the liquid/liquid separator and rotameters.

The loop is designed for operation at 10 bar pressure. All main components are made of stainless steel. The first experiments were done with a transparent test section in order to make visual observations of the flow at ambient temperature. The transparent test section was not certified for elevated temperature, and was replaced with a stainless steel test section before experiments with measurement of droplet content were performed at 40 °C. In the experiments with visual observation of the flow a blue colour was added to the glycol/water phase in order to be able to distinguish between the two liquid phases. This gave very good results when a combination of visual observation, video recording and photographs was used to characterize the flow. Before the experiments with measurements of glycol content in droplets in the gas phase were performed, the glycol/water solution was replaced with a similar solution without colour additives, but with the ^{14}C-labeled diethylene glycol added.

The physical properties of the three fluids and the geometry of the flow loop were used in computer simulations with the OLGA program in order to obtain test conditions representative for the flow in large diameter pipelines for subsea transport of wet gas over long distances. This includes superficial gas velocities of 4 - 6 m/s, superficial liquid velocities of 0.005 to 0.01 m/s and a total liquid volume fraction in the pipeline between 0.5 and 1.5 %.

The measurement of glycol content in droplets in the gas phase was based on a radioactive tracer technique. Small amounts of ^{14}C-labeled diethylene glycol were added to the glycol/water mixture used in the loop. The small glycol content in the droplets trapped in the gas sampling container was extracted with a larger volume of nonactivated glycol added to the sampling container. The ^{14}C content in the liquid in the container was measured with a sensitive scintillation counting equipment, and this gave a direct measure of the glycol content in the droplets in the gas phase. This method made it possible to detect glycol contents as low as 10^{-4} ml in the sampling container, corresponding to a glycol content in the droplets down to 10^{-9} volume fraction in the test loop.

Glycol Content in Droplets in the Gas Phase

A series of flow loop experiments were performed with superficial gas velocities between 1.2 and 6.4 m/s and superficial liquid velocities between 0.005 and 0.02 m/s. Visual observation of the flow showed that the flow regime was wavy stratified with some water dispersed in the hydrocarbon phase at the lowest gas velocities of 1.2-2 m/s. At a gas velocity of 4.2 m/s the flow was still wavy stratified, but more wavy and with the liquid film creeping upwards. At a high gas velocity of 6.4 m/s a transition towards annular flow with droplets was observed, and the water was very dispersed in the hydrocarbon phase.

The results from the measurements of glycol content in droplets in the gas phase are shown in Figure 4. The results are given as volume fraction of liquid glycol in droplets in the gas phase. The total droplet fraction is much higher than this, since the droplets will consist mostly of hydrocarbon and water. When a hydrocarbon phase is present, the glycol droplet content is low at 1.2 and 4.2 m/s gas velocity, but increases markedly from 4.2 to 6.4 m/s, as shown by the filled points in the figure. In the experiments without a hydrocarbon phase, the glycol droplet content is higher because the glycol/water droplet formation is not hindered by a hydrocarbon phase on top of the glycol/water phase. These experiments are shown with open points in the figure.

The results show that the presence of a hydrocarbon layer strongly reduces the glycol droplet content in the gas at gas velocities larger than 4 m/s. For instance, for a gas velocity of 6.4 m/s and a superficial liquid velocity of 0.01 m/s the glycol content is 20 times higher in the experiments without hydrocarbon than in the experiments with hydrocarbons. The same tendency is also seen at 4.2 m/s gas velocity, while all measurements are very low at 1.2 m/s gas velocity. The presence of hydrocarbons reduces the glycol droplet formation not only under wavy stratified flow, but also when the water/glycol phase is dispersed in the hydrocarbon condensate phase.

The increase in glycol droplet content from 4.2 to 6.4 m/s gas velocity is very pronounced. This is a typical range for flow velocities in large wet gas pipelines. The results indicate that an effect of glycol on top-of-the-line corrosion will be more pronounced at high flowrates. This is interesting as the corrosion experiments described below showed that the top-of-the-line corrosion rate will be very low as long as glycol is present in the gas phase.

The glycol contents were measured in a 0.1 m diameter test loop, and the applicability of the results in a large diameter pipeline can be discussed because distribution of droplets may be different in a large pipeline. The results give a good indication of glycol transport to the gas phase by droplet formation in wet gas pipelines with glycol addition, but the results do not tell how these droplets will be transported vertically in the large pipeline or whether they will actually hit the upper wall. However, the results show that the droplet formation is large enough to provide the gas phase with a glycol source which can keep the gas phase saturated with glycol vapour. The glycol droplet content in the gas phase is always at least ten times higher than the glycol content corresponding to the vapour pressure of glycol, as indicated by the horizontal broken line in Figure 4. When condensation at the top is slow, this can make a significant contribution to the glycol content of the water phase condensing in the top of the line, even when a hydrocarbon layer covers the glycol in the bottom.

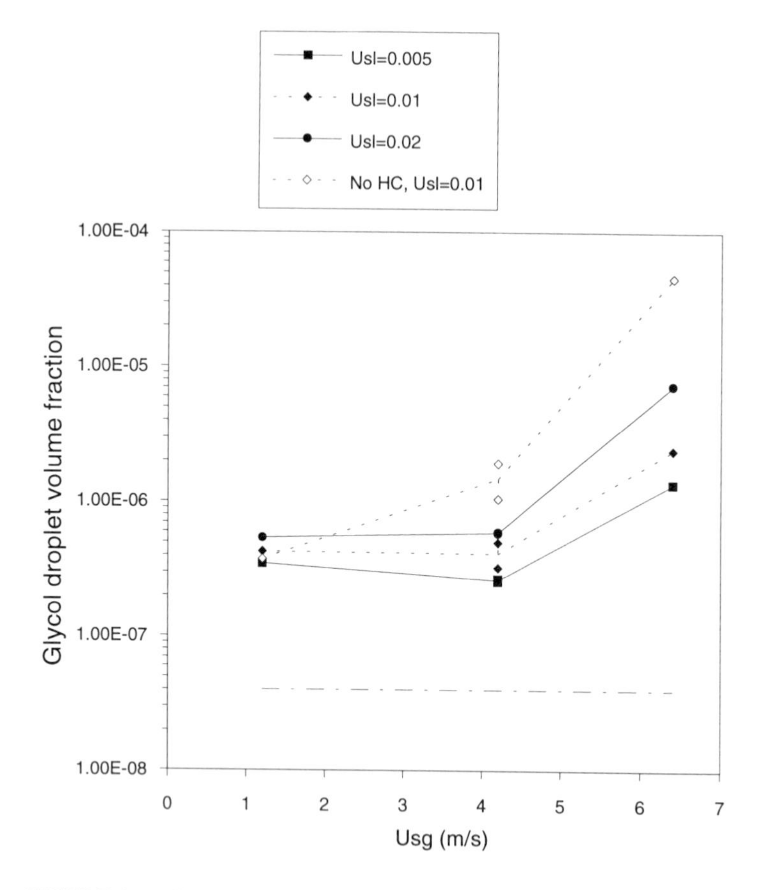

FIGURE 4. Volume fraction of glycol in droplets as function of superficial gas velocity at different superficial liquid velocities. The horizontal broken line indicates the glycol volume fraction calculated from the vapour pressure.

Previous calculations of the distribution of glycol in large wet gas pipelines have indicated that the condensing water can be almost free from glycol after only a few km in a pipeline with stratified flow if the contact between the glycol/water phase in the bottom and the gas phase is completely blocked by a hydrocarbon condensate layer on top of the glycol/water phase[1]. In this situation top-of-the-line corrosion might be a problem under certain conditions. The present experiments have shown that the generation of droplets containing glycol will be large enough to keep the gas phase saturated with glycol under flow conditions typical for large wet gas pipelines.

TOP-OF-THE-LINE CORROSION STUDIES

One-Phase Flow Loop Experiments with Condensation

The corrosion experiments were performed in a stainless steel loop with 15 mm inner diameter. A schematic overview of the loop is shown in Figure 5. The loop can be used for both one-phase water experiments, one-phase gas experiments and bubble and mist flow experiments. In the present dewing corrosion experiments the circulating gas was bubbled through the separator, which was half-filled with a mixture of monoethylene glycol (MEG) and distilled water. This ensures that a tempered gas saturated with water vapour and glycol enters the test sections. The gas phase was a mixture of CO_2 and argon with total pressure 16 bar. Maximum gas flowrate is 8 m/s. Corrosion rates at two different flowrates are obtained simultaneously as one part of the gas circuit is branched in two. Details in the experimental technique have been published previously [4].

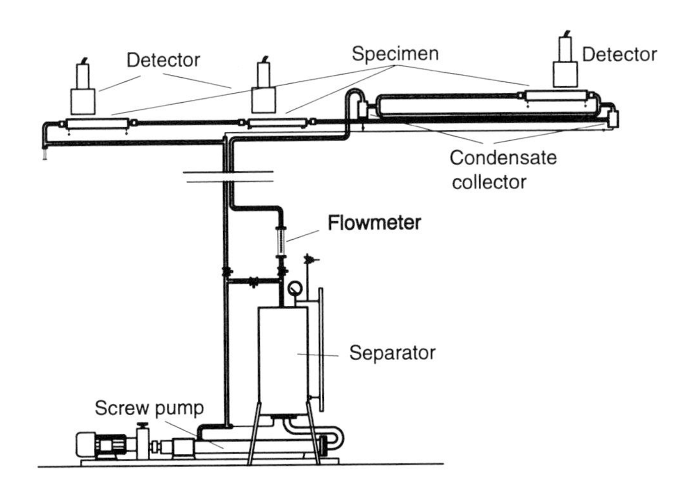

FIGURE 5. Gas flow loop used for top-of-the-line corrosion studies.

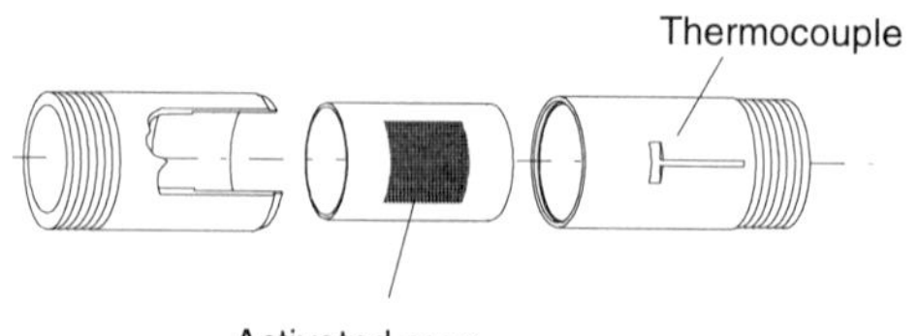

FIGURE 6. *Test tube with activated area.*

The specimens are normalized St 35 carbon steel tubes with the same diameter as the stainless steel tubes and mounted flush with the loop wall. The test tube is equipped with 70 cm long cooling jackets. The temperature difference between the surface of the test tube and the flowing gas can be varied between 0 and 20 K with an accuracy of 0.1 K. The experiments were performed with a temperature of 40 °C in the gas.

The corrosion rate is measured with a radioactive technique based on neutron activated specimens[5]. The detectors are mounted close to the cooling jacket. The neutron activated tubular specimen is installed flush in the test tube as illustrated in Figure 6. The two parts are electron beam welded together, giving a very smooth flow channel. By activating only a region at the top of the tube it is possible to differentiate between attack in the top of the tube (between 10 o'clock and 2 o'clock positions) and the bottom of the tube. The sensitivity in the measurement of metal loss due to corrosion is 0.2 μm.

The radioactive technique does not discriminate between uncorroded metal and corrosion products. If corrosion products accumulate on the surface, the corrosion rate becomes underestimated when films are growing in thickness. This is not a problem at steady state corrosion where film is removed at the same rate as it is formed. In the present experiments only a few percent of the corroded metal remained on the surface, and at steady state corrosion the rate determining mechanism is believed to be film removal. Thus, the corrosion film does not influence the measurements much.

Dewing corrosion is strongly influenced by the condensation rate, which depends on the subcooling[6,7]. In previous experiments with large subcooling, the condensation rate was estimated by measuring the heat transfer between the flowing gas and the tube wall[6,7]. This method is not very accurate and could not be applied for the low condensation rates in the present experiments. Therefore, a new technique for condensation rate measurements based on radioactive iron carbonate scale was developed[4,8]. A thin layer of radioactive $FeCO_3$ was precipitated on the inner wall of a stainless steel tube. This tube was mounted in the loop and cooled in the same way as the test tubes. The water condensing on the pipe wall dissolves the iron carbonate, and the amount of $FeCO_3$ which can be dissolved in the condensing water is limited by the solubility of iron carbonate[8]. The amount of radioactive $FeCO_3$ dissolved and removed from the surface then gives a measure of the amount of condensed water.

Corrosion Rates During Condensation

Top-of-the-line corrosion experiments were performed at 40 °C with 1 and 4 K subcooling and 0-50 % monoethylene glycol. Two flowrates were tested simultaneously, 3 and 6 m/s in experiments without glycol and 1.5 and 3 m/s when glycol was present. The reduction in flow was necessary to avoid glycol foam formation[4].

Corrosion rates at different flowrates and subcooling temperatures are shown as a function of CO_2 partial pressure in Figure 7. The corrosion rate increased with CO_2 partial pressure, flowrate and subcooling. The corrosion rate was relatively low in all experiments, and the presence of glycol in the gas reduces the corrosion rate further. Only one experiment with glycol resulted in a corrosion rate above the detection limit.

After the exposure the specimens were covered with a black corrosion film consisting primarily of FeC_3 and $FeCO_3$. The film was soft and adhered poorly, and corresponds to a few per cent of the total material loss. Similar films develop in liquid flow under the same CO_2 partial pressure and temperature[8] when the water is saturated with Fe^{2+}. The presence of the film shows that the supply of water to the specimens by dewing is so slow that the corrosion reaction saturates the water on the specimen surface with Fe^{2+}.

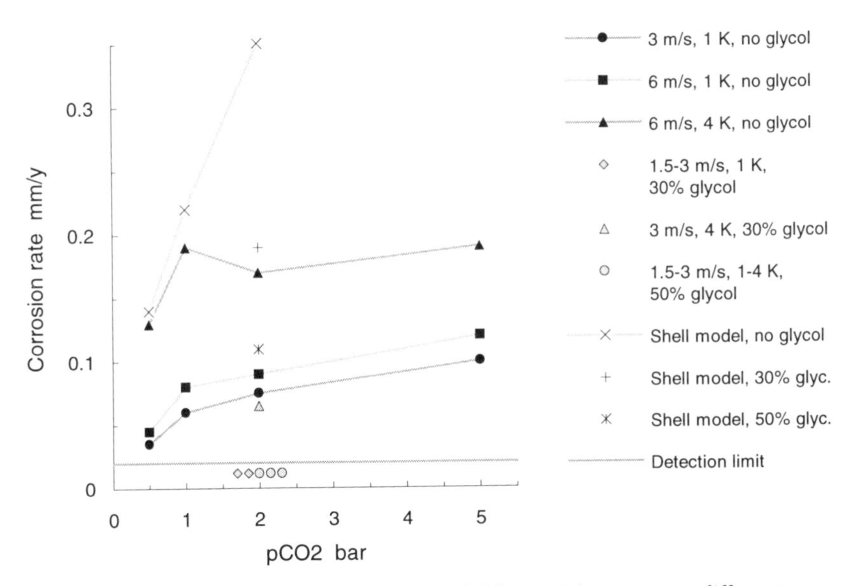

FIGURE 7. Corrosion rate as function of CO_2 partial pressure at different flowrates and subcooling. Only one experiment with glycol gave a corrosion rate above the detection limit.

When steel is exposed to a liquid phase of pure condensed water with only CO_2 added, a very high initial corrosion rate is obtained at 40 °C. However, when the water supply is limited (slow renewal), the corrosion products (Fe^{2+} and CO_3^{2-}) accumulate in the water and the pH increases until $FeCO_3$ precipitates out. The accumulation of corrosion products in the water and the precipitation of $FeCO_3$ reduces the corrosion rate until $FeCO_3$ is transported away from the surface at the same rate as it is formed. The corrosion rate therefore corresponds to the rate at which $FeCO_3$ is removed by the condensate[6,7]. Under these conditions the corrosion rate is proportional to the condensation rate and the concentration of dissolved iron carbonate.

Prediction of Dewing Corrosion

The maximum amount of $FeCO_3$ which can be dissolved is dependent on the type of corrosion film formed. When steel corrodes without formation of protective films, corrosion products are released as ions directly to the water[8]. As the precipitation rate of $FeCO_3$ is slow, a high degree of supersaturation can be obtained. The amount of corrosion products which can be removed from the corroding surface can therefore be much higher than that calculated from thermodynamic solubility data. However, when a dense, protective $FeCO_3$ film is formed the corrosion rate is slower and the amount of corrosion products in the water is reduced until the film starts to dissolve. The dissolution rate is then determined by the solubility of $FeCO_3$ and the condensation rate.

The experiments with measurement of condensation rate showed that the condensation rate increased with increasing subcooling rate. At 3 m/s gas flow and 1 K temperature difference between the gas and the pipe wall the condensation rate was measured to 0.36 ml/(m^{2}s), and the calculated corrosion rate assuming $FeCO_3$ saturated water was 0.074 mm/year. This is close to the corrosion rate measured to 0.06 mm/year under the same testing conditions. When the subcooling was increased to 4 K the condensation rate increased to 0.63 ml/(m^{2}s) and the calculated corrosion rate was almost doubled. No corrosion experiments were carried out under these conditions, but a doubling in corrosion rate was obtained at 6 m/s flowrate when the subcooling temperature was increased from 1 to 4 K.

The effect of CO_2 partial pressure, flowrate and subcooling rate shown in Figure 7 is in agreement with the model described. The generally low corrosion rates indicate that protective films have been formed. When the CO_2 partial pressure was increased from 0.5 to 5 bar, the corrosion rate increased between 50 and 100 % and the $FeCO_3$ solubility increased by 50 %. Increasing the subcooling from 1 to 4 K increased the corrosion rate 2 - 3 times while the condensation rate increased 1.5 - 2 times. The higher increase in corrosion rate compared to the condensation rate is probably due to less protective films and switching to a supersaturation mechanism.

One explanation for the reduction in corrosion rate observed when glycol is present can be reduction in the dew point and the condensation rate. Glycol injection before the gas enters the pipeline will reduce the water vapour content in the gas entering the pipeline. Glycol injection will also reduce the condensation rate of water because the dew point is depressed. The lower condensation rate reduces both the film dissolution rate and the probability for a change of the corrosion rate determining mechanism to the supersaturation type. A reduction in the corrosion rate has been reported when glycol is mixed with the water, but a large reduction was only found when the liquid phase contains more than about 50 % glycol[1,2,3]. At the moment it is not clear whether the reduction in the condensation rate or the diluting effect of glycol in the condensate is most important for the overall effect of glycol injection on top-of-the-line corrosion.

The corrosion rate predictions calculated with the Shell model[1] for CO_2 corrosion have been included in Figure 7 for comparison. The calculations are based on a condensation rate of 0.25 ml/(m^{2}s), which is regarded as conservative. The model gives corrosion rates higher than those measured, and the Shell model seems to be conservative under these conditions. The present experiments have shown relatively low corrosion rates in a wet gas with condensation, and the presence of glycol in the gas reduces the corrosion rate further. This indicates that top-of-the-line corrosion will probably not be a large problem in practice when the temperature is reasonably low.

CONCLUSIONS

The glycol transport studies give a good indication of the glycol droplet formation in large wet gas pipelines. The presence of a hydrocarbon layer reduces the glycol droplet content in the gas for flow conditions typical for large wet gas pipelines. The glycol content in the gas is low for 1.2 and 4.2 m/s gas velocity when a liquid hydrocarbon layer is present, but increases markedly from 4.2 to 6.4 m/s. The droplet formation is large enough to provide the gas phase with a glycol source which can keep the gas phase saturated with glycol vapour. These droplets can make a significant contribution to the glycol content of the water condensing at the top, even when a hydrocarbon layer covers the glycol in the bottom.

The maximum corrosion rate in the top of gas pipelines with condensation of water on the pipe wall is limited by the condensation rate and the supersaturation of iron carbonate in the water. The corrosion rate increases with increasing temperature difference between the gas and the steel wall, with increasing CO_2 partial pressure and with increasing gas flow rate due to increased condensation rate. Addition of 50 % glycol in the liquid phase reduced the corrosion rate in the top of the line to very low values.

Corrosion films with protective properties were formed at 1 K subcooling, and the corrosion rate was limited by the condensation rate and thermodynamic solubility of $FeCO_3$. At higher subcooling less protective films were formed and the corrosion rate was governed by the condensation rate and a steady state supersaturation level.

The low corrosion rate in the top of the line and the presence of glycol in droplets in the gas phase indicates that top-of-the-line corrosion should not be a problem in practice in large wet gas pipelines with glycol injection when the temperature is reasonably low.

ACKNOWLEDGEMENT

The experiments described in this paper were carried out in two research projects financed by A/S Norske Shell. Their support and permission to publish this paper is gratefully acknowledged. The authors also wish to thank Geir Sæther and Trygve Furuseth for their contribution in performing the experiments.

REFERENCES

1. C. de Waard, U. Lotz, D.E. Milliams, "Predictive Model for CO_2 Corrosion Engineering in Wet Natural Gas Pipelines", CORROSION/91, Paper No. 577, (Houston, TX: NACE, 1991).
2. L. van Bodegom, K. van Gelder, M.K.F. Paksa, L. van Raam, "Effect of Glycol and Methanol on CO_2 Corrosion of Carbon Steel", CORROSION/87, Paper No. 55, (Houston, TX: NACE, 1987).
3. L. van Bodegom, K. van Gelder, J.A.M. Spaninks, M.J.J. Simon Thomas, "Control of CO_2 Corrosion in Wet Gas Lines by Injection of Glycol", CORROSION/88, Paper No. 187, (Houston, TX: NACE, 1988).
4. R. Nyborg, A. Dugstad, L. Lunde, "Top-of-the-line Corrosion and Distribution of Glycol in a Large Wet Gas Pipeline", CORROSION/93, Paper No. 77, (Houston, TX: NACE, 1993).
5. A. Dugstad, K. Videm, "Radioactive Techniques for Corrosion Monitoring", CORROSION/89, Paper No. 159, (Houston, TX: NACE, 1989).
6. S. Olsen, A. Dugstad, "Corrosion under Dewing Conditions", CORROSION/91, Paper No. 472, (Houston, TX: NACE, 1991).
7. A. Dugstad, L. Lunde, S. Olsen, "Dewing Corrosion - A Possible Problem Area for Pipelines Transporting Unprocessed Gas over Long Distances", 5th International Conference on Multiphase Production, (Cranfield, UK: BHR Group, 1991).
8. A. Dugstad, "The Importance of $FeCO_3$ Supersaturation on the CO_2 Corrosion of Carbon Steels". CORROSION/92, Paper No. 14, (Houston, TX: NACE, 1992).

EXPERIMENTAL INVESTIGATIONS II

Vertical Flow Loops

Experimental simulation of entrainment behind Taylor-bubbles in vertical slugflow.

R. Delfos, M. Alexandersen, R.V.A. Oliemans
Laboratory for Aero- and Hydrodynamics, Delft University of Technology
Rotterdamseweg 145, 2628 AL Delft, the Netherlands

In the understanding of vertical slug flow there are some processes that are still only vaguely understood. Most important in these is the gas transport from the Taylor bubble to the next slug cylinder. In slug flow modelling more knowledge of this process is very important for predicting the voidfraction in the liquid slugs. This work describes a type of experiment in which we can get more insight in the basic phenomena that occur.

1. Physical background.

Two-phase flow conditions are very common in all kind of technical environments. Slugflow is a basic flowpattern occurring under condition where the volume flowrates of the two phases are of the same order of magnitude. It is not restricted to gas-liquid flow only, but is also observed in liquid-liquid, liquid-particles and gas-particles flow. This work concentrates on gas-liquid slugflow in relatively large vertical pipes, as for instance met in oil production or in process equipment.

Slug flow is seen as an intermittent flowpattern (figure 1). It is formed by large, nearly the tube diameter D filling, bubbles rising through liquid, often highly loaded with small bubbles, called liquid slugs. The bubbles are called Taylor- or slug bubbles. Usually In modelling this pattern is taken as periodic, with Taylor bubbles of length L, and slug lengths of length L_s. See for instance Fernandes et.al. (1983).

In vertical concurrent gas-liquid flow the Taylor bubbles are rising much faster (u_b) than the mixture velocity u_m. The liquid that is shed from the slug above is falling along the Taylor bubble and impinges the highly turbulent top of the next liquid slug, thereby tearing small bubbles from the large one. At impact the film has thickness e, and velocity u_f.

The voidfraction in the wake (α_w) is relatively high, because of the bubbles being torn in. What further happens with the small bubbles in the wake is dependent on the tube-diameter (Taitel et.al. 1980). Namely, the rise velocity of bubbles that are small compared to the tube does not dependent on it's diameter. However, the velocity of Taylor bubbles depends via a square root on the tube diameter. For tubes smaller than a certain value the small bubbles in the wake will overtake the leading Taylor bubble and re-coalesce with it. In these flows the void fraction α in the slug is relatively low. For much larger tubes the Taylor bubble rises much faster, so part of the small bubbles from the wake are left behind and will flow into the slug to be overtaken by the next Taylor bubble. Here the void fraction α in the slug is much higher. The critical tube diameter is a very often used one in laboratories, namely 5cm for air/water systems. This discrepancy however disappears for higher mixture velocities. This makes scaling up to larger diameters very difficult. To have a system where under all circumstances the small bubbles are transported downward relative to the Taylor bubble we have a tube diameter of 10cm.

The existing models for slugflow are based on many submodels, for instance the rise velocities of single bubbles. An excellent review on these was recently given by Fabre & Liné (1992). Another model is needed to describe by some physical mechanism the gas transferrate from Taylor bubble to slug cylinder, in order to calculate the gas distribution over bubble and slug cylinder.

The mechanism is explained by Fernandes et.al. by assuming that gas is torn downward with the falling wall-film. The amount of loss is calculated by assuming that all gas having a negative velocity with respect to the bubble tail, is entrained into the wake. The results of this model only agree, with measurements, for a small range of parameters (Fernandes et.al. 1983; Mao & Dukler 1989; van Hout et.al. 1992). It is suspected that the physics of this model is only partly correct.

In their model for horizontal flow Andreussi & Bendiksen (1989) assume that gas starts to be entrained if the liquid velocity entering the tail exceeds some minimum; the onset velocity u_{on}. The entrainment is modelled as being proportional to the real velocity minus u_{on}. Furthermore, the onset velocity was not derived from theory, but had to be correlated from experimental data (see Measurements).

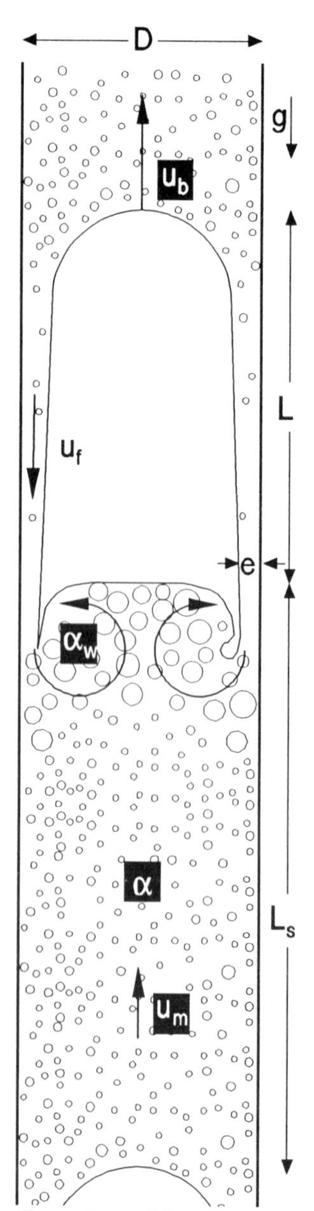

figure 1. model

2. Experiment.

In the experimental situation with upflowing liquid and gas, it is always very complicated to do measurements because the flow is intermittent. Because the pattern is passing by, you can only measure a very short time at one position, or take averages over a large amount of slugs. This requires a measurement technique which discriminates between the different areas in the flowpattern. In experiments where void fractions have been measured (van Hout et.al. 1992, Mao & Dukler 1989), these were found by averaging, although the lengths of the individual bubbles are different. Now we think the gas transferrate and the voidfraction are dependant on fluid impact velocity, and therefore Taylor bubble length, so we ought to study the fine structured zone just behind single Taylor bubbles for a longer time.

As a simulation of the real situation with upflowing liquid we constructed an experimental set-up to study a single Taylor bubble and the phenomena around it. The bubble is held stationary by flowing water downward with an appropriate flowrate. Two tube diameters above the bubble we manipulate the velocity profile with specially shaped honeycomb structures, such that the profiles become representative for different simulated upward flow velocities. Through the centre of this honeycomb air is continuously blown in via a small tube, forming a Taylor-like bubble. The fluid flowing around the bubble forms a wall-bound free-falling film. At the bottom it impacts the turbulent, often foamy, top of the following liquid cylinder, entraining gas bubbles and creating a wake zone. The pattern is in a fixed position according to measuring instruments, which is an advantage to normal slugflow.

It is clear that this situation is not exactly a Galileo transform of the simulated situation, because our boundaries, the tube walls, are not moving downward as they should. So the simulation is only valid when the influence of the tube walls is small. This will especially not be the case in the falling film, which starts as a free-falling film, but develops into a fully turbulent boundary layer within about 1 metre (see Measurements). Here the limit velocity is imposed by the wall and the flowrate, and not by the entrance velocity field. In our experiment the amount of water flowing down is however the same as in the simulated upward flow, but differences might occur.

The idea of this experiment was originally inspired by a short note by Davidson and Kirk (1969). They influenced the velocity profile by an array of drinking straws glued together. This was however not to simulate slugflow in itself, but to stabilize the bubble to be studied. In our experiment we do not want especially to study the bubble itself but in particular the phenomena around it.

Our experiments are about the same as those done in Toulouse by the group of Fabre (Riiser et al. 1992). They also study a stationary wake behind a bubble. There is however an important difference between the experiments. In their case the 'bubble' is formed behind a massive cylinder placed in the centre of the tube. The water flows in through an annulus around this cylinder, whereas the air is blown in through a hole in the middle of it. Both water and air velocity can be varied freely within a certain range. In our experiment the water velocity is fixed for each situation because we have to keep a bubble at a fixed position in the tube, and a change in velocity would lead to rising or falling. Moreover we have adapted our velocity profile to simulate a real slug flow situation. Our entrance profile just upstream of the bubble has only small scale turbulence, instead of being a turbulent 'wall jet'. Especially for shorter bubbles, where the film is not fully developed, differences can be expected.

3. Description of the experimental set-up.

The flow conduit.

The experimental set-up is shown in figure 2. It consists of a closed water flow-loop which is driven by an adjustable pump. The liquid flowrate is measured by means of an electro-magnetic flowmeter. The top of the conduit is an open (atmospheric pressure) vessel; separator 1. From here the water flows down through a honeycomb flow straightener for swirl extinction into the inlet section, where the fluid velocity profile is formed and air is blown in. The measurements are done in the test section, which is an acrylic rectangular box filled with refractive index matched water to minimize refraction of light, with the tube section passing flush through it. This makes it possible to use of optical techniques like photography to determine gas distribution, or Laser Doppler Anemometry to determine fluid velocities. The pressure drop can be measured from pressure taps. The conduit continues with the mixture flowing into separator 2. Water is drawn from the bottom by the pump. The air descended from the water in the separator is not released from the system, but is used for the inlet section passing through Rotameters to measure the flowrate. When well-functioning the air conduit is as the water conduit a closed system, so that the flowrates through the flowmeter and through the testsection are equal. To control the system the water level in the vessel is continuously measured.

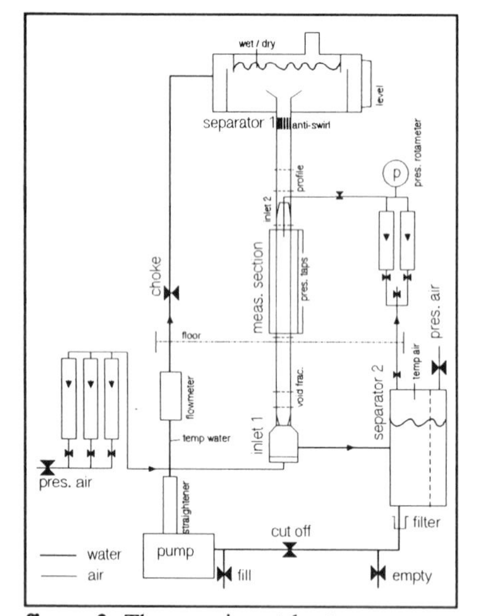

figure 2. The experimental set-up.

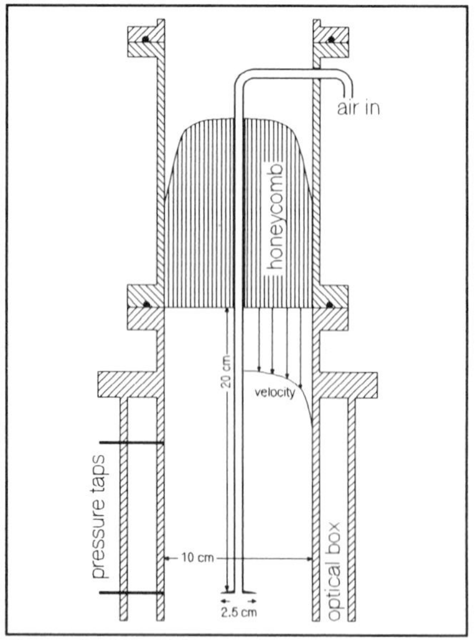

figure 3. The inlet section.

The inlet section.

The inlet section, shown in figure 3, is based on the idea that the fluid velocity profile and the mixture flowrate have much influence on the gas transfer process. A fine-sized honeycomb structure is placed in the downflowing liquid to manipulate the velocity profile. The honeycomb material can be seen as a set of closely-packed drinking straws. The assumption is that the pressure drop over a straw is equal for all of them. This assumption is valid when the pressure drop over the structure as a whole is large compared to $\frac{1}{2}\rho V^2$, the variation in dynamic pressure over the cross section. The velocity profile through the straws is thereby roughly the inverse of the length profile.

By some straightforward calculations we have designed a shape for the profiles, based on the following:
We want to simulate a Taylor bubble during slug flow. The mixture velocity, u_m, is the total velocity of gas and liquid. The time-averaged (mixture)velocity profile u(r) is assumed to be that for fully developed homogeneous turbulent flow far downstream from the Taylor bubble: it is described with sufficient accuracy for this purpose by the eighth-power expression:

$$u(r) = 1.25u_m \left(1 - \left(\frac{r}{R}\right)^8\right) \tag{1}$$

A Taylor bubble in a turbulent pipe flow has the well-known Nicklin rise velocity u_{TB}:

$$u_{Tb} = u_{b0} + u_c \; ; \quad u_{b0} = 0.35\sqrt{g'D} \; (Dumitrescu) \; ; \quad g' = \frac{g(\rho_l - \rho_g)}{\rho_l} \tag{2}$$

The first term on the righthand side is the rise velocity of a single bubble in stagnant fluid. The second term expresses the influence of the mixture velocity u_m. The Taylor bubble rises with respect to the centreline-velocity u_{cl} which for fully turbulent flow exceeds the mixture velocity by about 25%. The velocity in a frame where the bubble is not moving becomes:

$$u_i(r) = u(r) - u_{Tb} = -1.25u_m \cdot \left(\frac{r}{R}\right)^8 - 0.35\sqrt{g'D}$$

$$u_{ls} = 0.25u_m + 0.35\sqrt{g'D} \tag{3}$$

This $u_i(r)$ is the velocity profile to be arranged. The superficial or averaged fluid velocity u_{ls} determines the flowrate. With a fixed diameter (100mm) the mixture velocity, which is the total flowrate of liquid and gas, is the only free parameter. Before choosing this parameter we define the mixture Froude numbers Fr_m, which is important for determination of flowpattern transition:

$$Fr_m = \frac{u_m}{\sqrt{g'D}} = 0.35\frac{u_m}{u_{b0}} \tag{4}$$

We look at three special cases for Fr_m, which is proportional to the ratio u_m/u_{b0}, and thus determines the shape of the honeycomb.

(i) The mixture velocity is zero. This simulates the rise of a single bubble in stagnant fluid. Both mixture and honeycomb profile are flat.

(ii) Fr_m is equal to 1. This is a typical situation for the onset of slugflow at moderate Froude numbers, where both liquid and gas velocity are not too high. The velocity at the wall is about twice the central velocity.

(iii) Fr_m is much higher than 1. This is for the situation where the transitions between the flowpatterns are very close together. The velocity maximum at the wall is very sharp. To have the honeycomb mechanically stable enough Fr_m is chosen 3.

Especially the last case is very promising, because there we simulate flow with high Froude numbers, which always is very difficult to achieve because of the high entrance lengths.

Downstream of the honeycomb we see in fig 3. the air inlet, which is a 5mm i.d. tube passing through the centre of the honeycomb. It terminates with a 25mm diameter endplate in the shape of Dumitrescu's bubble. The tube length is 200mm downstream of the honeycomb, which we chose as a compromise. We wanted to minimize the interaction between the honeycomb (where streamlines are essentially axial) and the streamlines flowing radially around the blunt surface of the bubble. On the other hand, the further downstream of the bubble, the more likely for the velocity profile to get unstable.

4. First observations.

In our first experiments, reported here, we have concentrated on the first condition; i.e. the simulation of a single bubble rising in stagnant liquid. Here the velocity profile we had to create is completely flat. We arranged this by placing a fine copper gauze in the flow. We measured the dynamic pressure profile by the differential pressure between a total pressure tube and a wall pressure hole. The velocity profile we derived from this was seen to be flat within 5% from the centre to only 4mm from the wall.

The bubble surface in the nose region is not as smooth as in the situation where it is rising freely, but has some large-scale disturbances. Partly these are imposed by the turbulent flow above and partly by the endplate the bubble is suspended from, which is not really a free surface and thus creates a vortex sheet, which might create some instability.

The theoretical fluid velocity is not the only one giving a stable bubble. We could lower it by about 10% giving a more bluff bubble with a more slowly oscillating nose before it became unstable and started to rise along the air inlet tube. When we flow water beyond the theoretical value the surface becomes smoother but the bubble nose gets sharper, until about 15% above the theoretical value the bubble shape becomes visibly unrealistic for a free bubble. The optimum flowrate giving the best form according to the shape calculated by Dumitrescu with the smoothest interface is about 10% larger than the theoretical one, a fact that we cannot explain by now.

When we look at the bottom of the bubble we can see the phenomena we are interested in. If we blow in air at a small flowrate, starting from zero bubble length, the entrainment of air behind the growing bubble is seen to be zero. Only after the bubble reaches a length of about 50cm, and therefore is a real Taylor bubble, it starts to shed air in the form of small bubbles being torn from it. This process is not continuous, but is seen to be strongly intermittent, i.e. when a first bubble detaches from the Taylor bubble, a small 'cloud' of bubbles is entrained. The bubbles stay relatively long in the recirculating zone (which we will call the wake further on), typically 10cm long, just behind the bubble. Part of the bubbles coalesce with the original Taylor bubble, the rest is swept away with the bulk fluid flow.

When we increase the bubble length further the intermittency in entrainment seems to decrease. The recirculating zone is now continuously fed with small bubbles and becomes less transparent by the higher voidfraction. The bubbles under the wake zone seem to be distributed quite uniformly in both the axial and the radial direction. When we further increase the bubble length (and thus the gas flowrate) the wake becomes so densely packed with bubbles that it seems foamy. The length of the wake increases little with bubble length.

We have also made video pictures from the nose and the wake. Because of the long exposure time (we used normal video without a shutter) the small bubbles are pictured as streak lines, from which a velocity field can be deduced. In the experiments to come we will use a pulsed light to get sharper pictures, making it possible to get information with a more accurate correlation technique. What we already have seen from analyzing the video frame by frame, was that the flow field in the wake is often far from the idealised situation in fig. 1. The toroidal vortex, which we observed, is only axisymmetric when the entrainment is low. When the gas transferrate is higher it is rather sloshing to all sides of the tube, which we have also observed via the quite unstable surface. When the entrainment is not too high we even saw a few times a kind of resonance between air being entrained, and the surface wobbling, where it's amplitude was a few cm. This could be an important mechanism for onset of entrainment, which we will further investigate in the near future.

When the bubble gets longer than about 70cm the film is seen to become turbulent in a very intermittent, burst-like manner. Possibly the bursts are formed by a three dimensional instability of the large waves on the film. When such a burst runs into the wake zone, we observed the entrainment being highly intensified.

5. Measurements.

In the experiments we keep the water flowrate at the theoretical value within 1%. We measure the air flowrate with calibrated Rota-meters giving an absolute accuracy of 2%. The bubble that grows is seen to become quite stable in length within a typical time of a few times L/u_g. On a much shorter timescale, a few seconds, the length varies a few centimetres caused by the intermittency of the entrainment process. Therefore we determine the length by reading a rule, averaging the surface by the eye. The length is mostly not stable on longer timescales. Especially at higher gas flowrates there is always a kind of overshoot. A bubble growing from zero gets very large because the wake is not sufficiently developed to transfer the high flowrate to the slug cylinder. Only after a while the length starts to converge very slowly to a final value. Often we have to wait for more than half an hour before a real stable length L develops. The shrink after the start-up is often more than 30cm.

The observed length L increases with the air flowrate u_{gs}, as is shown in fig. 4. Even at the lowest gas flowrates tested a bubble grew about 50cm long. This seems to be a lower limit on entrainment, although now and then from bubbles shorter than this a few small bubbles could escape, but this 'flowrate' is an order of magnitude smaller than the measured ones. For the flowrates tested the length is seen to increase almost linearly with gas flow. A linear regression could be made giving a very accurate fit. This correlation could be used as a submodel, for instance to calculate the void fraction in the slug cylinder. It is not advisable to use this correlation in the reversed manner, because the bubble length might be a poor measure on gas flowrate. Even more important is that our simulation is for mixture velocity zero, so not a real slug flow.

Now we compare our results with those found by Riiser et.al. (1992), part of which are shown in fig. 5. They did measurements for different liquid flowrates, but here only one is depicted. The length of their cavity increases with gas flowrate, in a linear relationship just like ours.

A real deviation is seen at the lowest flowrates. In our experiment at some length the entrainment just stops while in their experiment from a certain point the straight line deviates from the higher rates and smoothly crosses zero. The difference must be sought in the entrance of the liquid. In their experiment the liquid enters as a very turbulent wall jet, whereas in our case we have a free falling film with only very fine-scale turbulence, which becomes highly turbulent after (as observed) about 6 diameters.

So qualitatively the results are quite similar. The difference in the parameters is partly because the tube diameter (5cm) is only half of ours. Moreover, their water flowrate is not the Dumitrescu value for stagnant liquid, but it is about 75% higher. Therefore we can not compare the results quantitatively with ours at the moment. Their flowrate is however the one we will simulate in experiments with u_m being $3.u_b$; i.e. case (iii) for high Froude numbers.

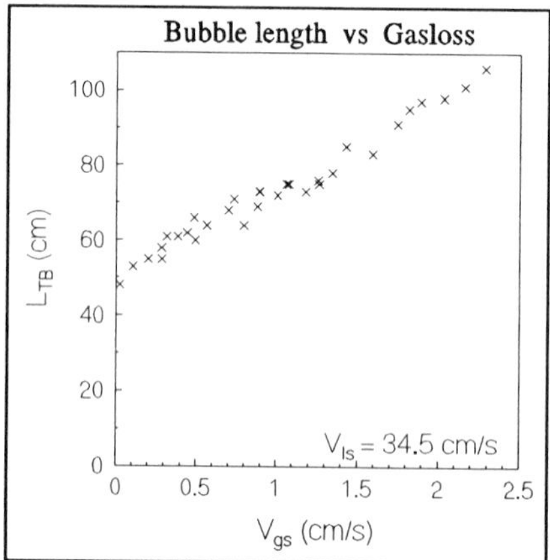

figure 4. Simulated situation: a bubble rising in stagnant liquid.

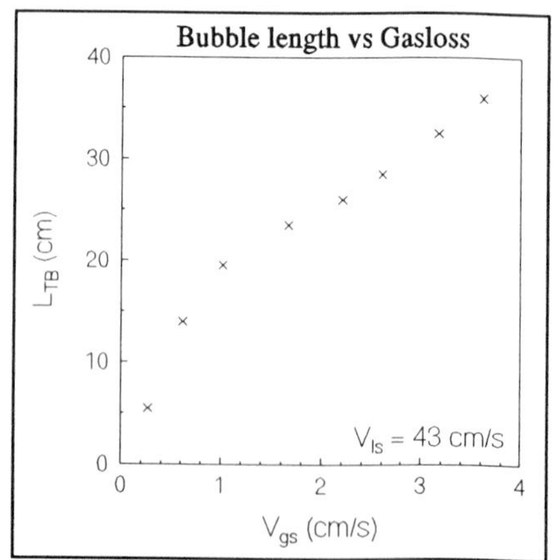

Figure 5. Measurements derived from Riiser et.al. (1990).

Riiser et.al. describe the onset of entrainment to be dependant on turbulent properties. They calculate the friction velocity u_{wl} and film thickness e at entering position with a simple model assuming a free falling film with a constant friction factor with the wall. They find a terminal velocity u_{fT} and a film developing length L_D using the Blasius friction law:

$$u_{fT} = \sqrt[3]{\frac{gDu_{ls}}{2f_w}} \qquad L_D = u_{fT}^2/g \tag{5}$$

$$= 3.1m/s \qquad = 0.97m \quad \textit{in our experiment}$$

With these they define the friction Weber number We^* as $\rho u_{wl}^2 e/\sigma$. With their measurements they find We^*_{crit}, the value at onset, to be about 0.4. Using this model, we find from our measurements the entrainment to start at a somewhat higher value: $We^*_{crit} = 0.87$. This is not exactly the same, but at least the magnitude is the same.

We also can compare our onset of entrainment velocity u_{on} with the correlation by Andreussi & Bendiksen (1989). They proposed the following correlation:

$$u_{on} = 2.6\sqrt{gD}(1-2(D/.025)^2) = 2.25 m/s \qquad \textbf{(6)}$$

With the model by Riiser we find for our u_{on} (assuming fully turbulent flow) a value of 2.5 m/s. Although correlated for horizontal flow, the agreement for our vertical flow is good.

The film model predicts that our film is fully developed after about 1 metre length. So for Taylor bubbles longer than this the entrainment should not continue to increase the way it does. We conclude that our film velocity increases further than expected. The same effect was reported by Mao & Dukler (1989).

We can calculate the gas fraction (α) in the liquid cylinder from the superficial gas velocity u_{gs} as injected in the inlet section and the superficial liquid velocity u_{ls} by using a drift-flux model (Wallis, 1969). It is however not possible to compare these values with those given by the existing slug models, because the situation we simulated up to now is the one for mixture velocity zero. This is why we left those results out by now.

6. Conclusions.

Experiments are being done to get more insight into the hydrodynamics of the wake zone behind a Taylor bubble. In our experiment it is possible to keep one Taylor bubble in a stationary position by flowing liquid downward against the bubble and blowing air into it to compensate for the air entrained into the liquid at the turbulent tail end. This set-up gives us a quite unique kind of experiment which is a simulation of a real slug flow. Although the simulation is not perfect, it enables us to do some measurements that would be nearly impossible in the upflowing experiment. Until now we have measured the length of the bubble that develops by blowing in a variable gas flowrate. The first results of these are:

- For bubble lengths small enough, no air is entrained. This length is about 50cm. The calculated film velocity and thickness give a critical Weber number that is about twice the value found by Riiser et.at. The onset velocity is only 10% lower than the value given by Andreussi's correlation.
- For bubbles longer than the entrainment length we found the entrainment being about linear with bubble length, similar to the results by Riiser et.al.

We realise that our experiments are still in a start-up phase, and that these measurements are only the first reliable data found. In future experiments we will concentrate on simulating real velocity profiles, and on some special phenomena as they occurred during the experiments.

7. Literature.

Andreussi, P. and Bendiksen, K. **An investigation of void fraction in liquid slugs for horizontal and inclined gas-liquid pipe flow.** Int. J. Multiphase Flow Vol. 15, pp. 937-946 (1989)

Davidson, J.F. and Kirk, F.A. **Holding a bubble fixed by downward flow.** Chem. Eng. Sci. Vol. 24 pp. 1529-1530 (1969)

Fabre, J. and Liné, A. **Modelling of two-phase slug flow.** Ann. Rev. Fluid Mech. 24, pp. 21-46 (1992)

Fernandes, R.C, Semiat, R. and Dukler, A.E. **Hydrodynamic model for gas-liquid slug flow in vertical tubes.** AIChE Journal, Vol. 29, no 6, pp. 981-989 (1983)

Mao, Z.-S. and Dukler, A.E. **An experimental study of gas-liquid slug flow.** Exp. Fluids 8, pp. 169-182 (1989)

Riiser, K, Fabre, J. and Suzanne, C. **Gas Entrainment at the Rear of a Taylor-Bubble.** Paper presented at European Two-Phase Flow Group Meeting, Stockholm, june 1-3 1992

Taitel, Y, Barnea, D. and Dukler, A.E. **Modelling flow pattern transitions for steady upward gas-liquid flow in vertical tubes.** AIChE J. Vol 26, no 3, pp. 345-354 (1980)

Van Hout, R, Shemer, L. and Barnea, D. **Spatial distribution of void fraction within a liquid slug and some other related slug parameters.** Int. J. Multiphase Flow Vol. 18, No 6, pp. 831-845 (1992)

Wallis, G.B. **One-dimensional two-phase flow.** McGraw-Hill (1969)

The Characterization of the Flow Reversal Phenomenon in Vertical Concurrent Annular Flow

Guimarães*, P. E., Pedras, M. J., França**, F. A.
Department of Energy
FEM / UNICAMP
13.081 - Campinas - SP - BRAZIL
(*) Currently at PETROBRAS (**) Author for correspondence

INTRODUCTION

The gas-well load-up and the film reversal of upward concurrent annular flows in vertical production columns are closely related phenomena. The progressive reduction of the pressure, in the gas formation results in the decrease of the gas flow rate and velocity inside the column. If there is a flow of water released from the formation or the condensation of gas on the inner wall of the pipe, characterizing an annular flow, the drag on this liquid film also reduces and the film gets thicker. The onset of flow reversal of the liquid film takes place when, for a certain flow condition, the film "hangs" in the pipe. A further and small decrease of the gas flow rate gives rise to a falling film. The liquid accumulates in the bottom of the column increasing the hydrostatic pressure and leading to the well load-up.

Procedures resulting from flooding analysis have been extended to the prediction of the onset of flow reversal, besides the lack of agreement that one can observe when considering practical results. The similarity that exists between the flow reversal of upward annular flows and the flooding phenomenon in counter current flows can explain this fact. The flooding in annular counter current flows is a process that takes place when the gas flow rate is increased and the falling film reverts to an upward moving film. This process is driven by different mechanisms, which depend on physical properties of the fluids, system dimensions and flow operational conditions. Some of these mechanisms are: wave motion or interface disturbance, liquid bridging, transition to churning flow and liquid entrainment in the gas core. They mark the onset of flooding. The flooding is completed when the down flow ceases and the upward moving film shows a more regular wave pattern. In systems where the liquid is introduced in a mid-section of the pipe, the climbing and falling films observed during the flooding also occurs during the process of flow reversal. The direct visualization of the flow also indicates a change in the structure and in the waviness of the interface. Sudden changes in the pressure drop, typical of flooding, are manifested during the flow reversal of concurrent annular flows.

There are not in the literature too many papers dedicated to the flow reversal experimentation and modeling in concurrent annular flows. Wallis (1962) analyzing the transition from flooding to upward flow proposed a value for the non-dimensional superficial gas velocity at which zero down flow occurred, $J_G^*=0.775$, where

$$J_G^* = J_G \cdot \rho_G \cdot \left[D \cdot g \cdot \left(\rho_L - \rho_G \right) \right]^{-1/2} \tag{1}$$

Experimental data on flow reversal was published by Hewitt et al (1965), who verified the development of a film that hung on the pipe inner wall. This happened for J_G^* ranging from 0.7 to 0.9.

Pushkina and Sorokin (1966) reported data on flooding and flow reversal in pipes ranging from 6 mm to 309 mm ID. Based on dimensional analysis their results for air-water systems at low liquid rates can be presented in terms of the Kutateladze criterion,

$$V_G = K \cdot \left[\frac{g \cdot \sigma \cdot \left(\rho_L - \rho_G \right)}{\rho_G^2} \right]^{1/4} \tag{2}$$

where K = 3.2. For a 27 mm pipe it gives $V_G = 15.6$ m/s. They mentioned that the onset of flooding and flow reversal, for practical perspective, took place at constant V_G, the gas velocity.

Focusing on the gas well load-up Turner et al (1969) proposed two models to explain the liquid removal from the production column. Field data corresponding to 66 combinations of flow parameters were compared to the results of a flow film approach, and 106 combinations of flow parameters to the results of a drop lifting model. According their analysis the drop lifting model showed a better agreement with the data. The model lead to the minimum gas velocity that continuously remove liquid from the column. The result was similar to the Kutateladze criterion with the constant increased in 17%:

$$V_G = 3.705 \cdot \left[\frac{g \cdot \sigma \cdot \left(\rho_L - \rho_G \right)}{\rho_G^2} \right]^{1/4} \tag{3}$$

Dukler and Smith (1977) applied Taitel and Dukler's (1976) model for flow regime transition in upward flow to discuss the hanging film region during the flooding process in counter-current flows. According their proposal, the gas velocity that sustains a hanging film is the minimum velocity required to suspend an entrained average drop in the core flow:

$$V_G = \left(\frac{4\cdot\kappa}{3\cdot C_D}\right)^{1/4} \cdot \left[\frac{g\cdot\sigma\cdot(\rho_L-\rho_G)}{\rho_G^2}\right]^{1/4} \tag{4}$$

where $\kappa = 12$ and $C_d = 0.44$.

Asahi (1978) investigated theoretically flow reversal problems of one-dimensional adiabatic annular two-phase flows without phase change. From a separated flow model he obtained the characteristic equation

$$F(\alpha) = \alpha - \frac{1}{\Psi} + \frac{4\cdot\tau_I^*}{\sqrt{\alpha\cdot\Psi}} - 4\cdot\tau_W^* = 0 \tag{5}$$

expressed in terms of the void fraction, α, the dimensionless shear stress at the interface and at the wall, τ_I^* and τ_W^*, and Ψ, a function of fluid properties, gas and liquid dimensionless superficial velocities and the liquid entrainment. Using a classical correlation for the interfacial friction factor, developed for smooth and rough annular flows, he concluded that the flow reversal takes place for a given liquid flow rate and liquid entrainment as the lowest gas flow rate for which the characteristic equation has a root in (0.8, $1/\Psi$).

Taitel et al (1982) presented a model that was claimed to explain the flooding and flow reversal on the basis of a film mechanism. Balancing the forces that act in a steady laminar film they showed that the prediction of the flooding and flow reversal process comes from the solution of the equations given below:

$$\tau_I = \frac{J_L^*}{2\cdot\delta^{*2}} + \frac{2\cdot\delta^*}{3} \tag{6}$$

$$\tau_I = f_I \cdot \frac{J_G^{*2}}{2\cdot(1-2\cdot\delta^*)^4} \tag{7}$$

The system has multiple solutions and the search for the physically realistic one resulted from two criteria, stability analysis and limiting flow imposed by kinematic flow theory. The two criteria lead to similar results.

Turner's minimum gas velocity was verified by Coleman et al (1991) for low pressure gas wells (up to 3500 kPa). They found a somewhat lower value for the constant in Eq. (2), 3.087.

The discrepancies in the results motivated this study. It was undertaken to clarify the flow behavior and measure the flow macroscopic parameters in vertical annular co-current flows at the onset of the flow reversal. A test set-up was built allowing the flow visualization and the measurement of the mean film thickness, the dispersion rate, the pressure and the pressure drop, the phase indicator function and, after calculation, the mean cross-sectional void fraction. The working fluids in this first approach were air and water. The measurements focused on the up flow film at the onset of flow reversal, but flooding data was also taken to verify the mentioned deviation in flow macroscopic parameters. The results were compared to previously published data and results from models. The apparatus and instrumentation have been improved and tests with higher viscosity liquids and larger pipes have been scheduled for the near future.

EXPERIMENTAL APPARATUS

The tests were performed in a simulator, 3.2 meters high. The test column and the collecting tanks were made of transparent Plexiglas. The column was a 27 mm ID pipe. The air entered the pipe through an injector at the bottom of the pipe; the water, through the porous wall of a small pipe, located 2.3 meters from the air injector. The inlet air and water flow rates were measured with carefully calibrated flow meters. Fig. (1) shows a schematics of the apparatus.

Collecting tanks were used to measure the liquid film flow rate in upward and downward flow. The flow rate of the entrained liquid was calculated taking the difference between the inlet liquid flow rate and the liquid film flow rate. Variable reluctance transducers and signal conditioners were used to measure the pressure and the pressure drop, points marked as A in Fig. (1).

The average film thickness was measured by processing the output signal of an electrical sensor. This sensor consisted of a probe and the circuitry, both described by Guimarães (1992). The measuring station is marked as B in Fig. (1). Displacing the probe across the liquid film, from the pipe wall to the core flow, one was able to measure the electrical continuity between the probe tip and the pipe wall. The signal was acquired and processed, and the phase indicator function was obtained for 20 positions along this path. The probability density function of the phase indicator in the liquid film region (from the pipe wall to the crest of the highest waves) was then calculated and the film thickness was assumed to correspond to 50% probability of the phase occurrence.

The entrained liquid was measured by taking the difference between the input liquid flow and the liquid flow rate in the film. The liquid flowing as film on the pipe wall was collected in tanks at the bottom and at the top of the test section and measured by weight. Having the entrained rate and the mean film thickness one was able to calculate the mean void fraction.

Tests were carried for constant liquid input and variable gas rates, from gas flow rates which are representative of rough annular flows down to the onset of flow reversal. Eight liquid flow rates were applied: $J_L = 0.022$ m/s, $J_L = 0.029$ m/s, $J_L = 0.035$ m/s, $J_L = 0.042$ m/s, $J_L = 0.050$ m/s, $J_L = 0.056$ m/s, $J_L = 0.064$ m/s, $J_L = 0.072$ m/s.

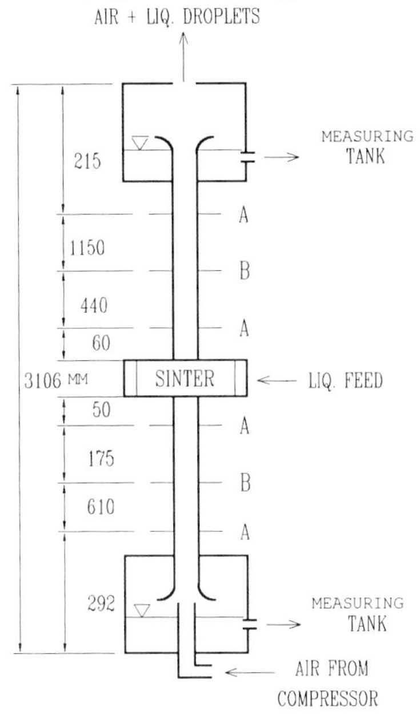

Figure (1) - Schematics of the apparatus. A - pressure measuring station. B - film thickness measuring station.

Tests to track the onset of flooding at the same liquid flow rates were also performed. The development of an incipient climbing film above the liquid injector was the rule which defined the onset of flooding. In such a way, the mentioned similarity between the flooding and flow reversal could be verified and the deviation in the flow macroscopic parameters could be measured.

RESULTS AND ANALYSIS

Fig. (2) shows the results of the flooding experiments in terms of the gas and liquid superficial velocities. The data published by others are shown for comparison, as well as the two limiting conditions proposed by Wallis (1962). In terms of non-dimensional superficial velocities, $J_G^{*1/2} \times J_L^{*1/2}$, they correlate as linear equations, with coefficients C = 0.8 and C = 1.0. The linear fit which better correlates the experimental data has a coefficient of 0.84.

Given the physical particularities of the set-up and the criterion adopted to mark the onset of flooding one can say that the results agree well.

To mark the onset of the flow reversal in co-current annular flows the rule adopted was the development of a liquid film with no net flow in the lower section of the pipe. For a constant input liquid rate, the gas flow was reduced until the hanging film was formed

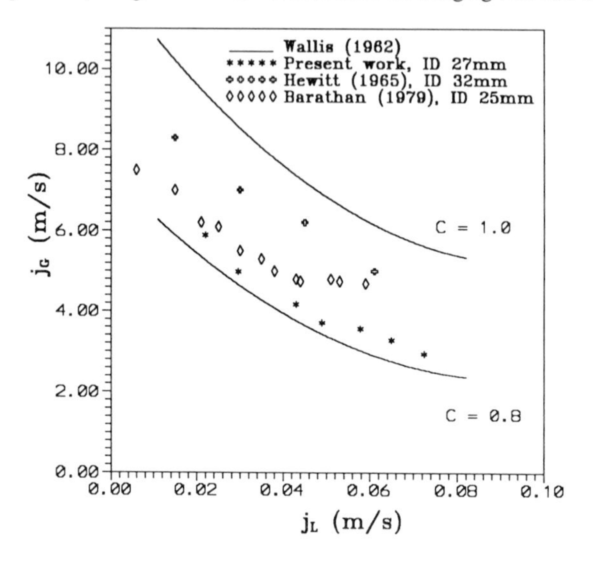

Figure (2) - Flooding data.

bellow the liquid feed. Sometimes, probably due to irregularities in the liquid injector, a localized down flow formed in the pipe. A further and small decrease in the gas flow resulted in a hanging film uniformly distributed on the inner pipe wall. In these experiments, for J_L ranging from 0.022 m/s to 0.072 m/s, the hanging film developed at a constant gas flow rate, given by J_G = 10.4 m/s or , J_G^* = 0.67. This corresponds to an average gas velocity V_G =11.2 m/s.

The velocity suggested by Pushkina and Sorokin (1966) is 15,6 m/s and Turner's et al (1969) correlation predicts V_G = 18,1 m/s. The suspended drop model used by Dukler and Smith (1977) gives the minimum velocity V_G = 12.0 m/s and Coleman's (1991) suggestion results in V_G = 15.0 m/s. The superficial gas velocity proposed by Wallis (1962) is higher than the measured one. Table (1) presents these results.

Figure (3) compares the experimental results of this study with the data published by Hewitt (1965) and the analytical solution proposed by Taitel et al (1982). The results show that the onset of flow reversal takes place at a lower gas flow rate compared to that which marks the onset of flooding. The deviation grows as the liquid flow rate increases. Taking this in account one concludes that the use of procedures resulting from flooding analysis do not correctely predict the onset of flooding or the load up in gas wells.

	V_G (m/s)	J_G^*
This study	**11.2 +**	**0.67 +**
Pushkina and Sorokin (1966)	**15.6**	**-**
Turner et al (xx)	**18.1**	**-**
Dukler and Smith (1977)	**12.0**	**-**
Coleman (xx)	**15.0**	**-**
Wallis (1962)	**-**	**0.775**

(+) average

Table (1) - Gas velocity and superficial velocity at the onset of flow reversal (development of hanging film)

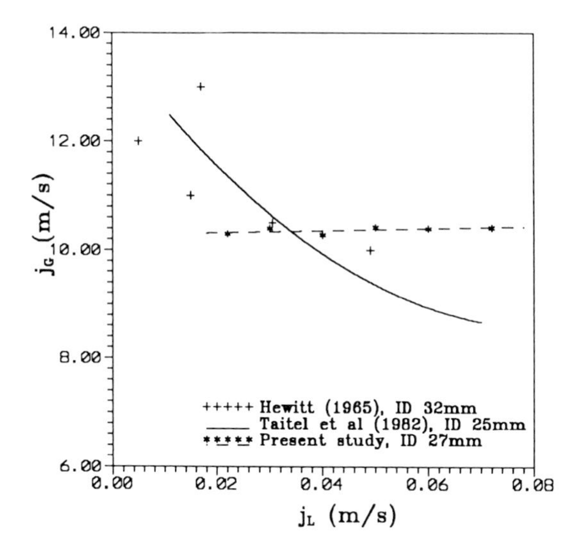

Figure (3) - Flow reversal data

For the particular system used in this study the mean gas velocity calculated by the suspended drop model is in good agreement with the measured one at the onset of flow reversal. Although, the direct observation of the flow indicates that the suspended drop model does not represent the physics of the phenomenon. Even when the gas flow rate is slightly reduced below the critical value, and the hanging film evolves down slowly, no deposition is observed ahead of it. One concludes that the down moving film clearly does not result from the deposition of entrained droplets below the liquid injector. Furthermore, the criterion does not reveal other features of the flow that can be useful as indicators of the onset of flow reversal and the well load-up.

If the entrainment is not the mechanism which drives the development of a hanging film, film mechanisms must be considered. Taking this in account, the flow behavior and the main variables of the upward moving film were analyzed, in a search for a tool to predict the load-up. The one-dimensional separated flow momentum equation, applied to an adiabatic and incompressible upward co-current annular flow writes:
for the core flow (gas plus liquid droplets),

$$W_G \cdot (V_G + \Delta V_G) + W_E \cdot (V_E + \Delta V_E) - W_G \cdot V_G - W_E \cdot V_E = \\ = -(A_G + A_E) \cdot \rho_C \cdot g \cdot \Delta z + (A_G + A_E) \cdot \Delta p_C - \pi \cdot D_I \cdot \tau_I \cdot \Delta z \tag{8}$$

and for the film flow,

$$W_{LF} \cdot (V_{LF} + \Delta V_{LF}) - W_{LF} \cdot V_{LF} = \\ = -A_{LF} \cdot \rho_L \cdot g \cdot \Delta z + A_{LF} \cdot \Delta p_{LF} + \pi \cdot (D_I \cdot \tau_I \pm D \cdot \tau_W) \cdot \Delta z \tag{9}$$

In the above equations W is the mass flow rate, V represents the velocity, A is the area, ρ is the density, g is the acceleration of gravity, z represents the axial distance, p is the pressure, D is the pipe perimeter and τ is the shear stress. The subscript G applies to the gas, E to the entrained liquid, C to the core flow (gas plus liquid droplets), I to the interface, FL to the liquid film and W to the pipe wall.

Further assumptions considered that the liquid droplets travel at the same velocity as the gas in the core. Moreover, considering the experimental results on entrainment and pressure drop, the acceleration term in Eqs. (7) and (8) is small compared to the gravity or to the interfacial and wall shear forces. If the pressure can be assumed uniform across the pipe area and the pressure drop terms are eliminated from the above equations, one obtains:

$$\left[1 - \left(1 - \frac{2\delta}{D}\right)^2\right] \cdot (\rho_L - \rho_G) \cdot g = \frac{4}{D}\left[\frac{\tau_I}{\left(1 - \frac{2\delta}{D}\right)} \pm \tau_W\right] \tag{10}$$

Eq. (10) balances the forces which are representative in annular upward concurrent flows. The double sign for the wall shear term relates to the film transition. The analysis of the experimental data shows that the wall shear stress switches sign as the gas flow rate decreases toward the flow reversal condition. This indicates that for these low gas rates the liquid close to the wall flows downward even for a net upward directed liquid flow.The same was previously observed by Zabaras et al (1986), who reported the transition mechanism in the film. Their results point out that, after the film transition takes place, the gravitational force exceeds the interfacial one, the difference decreasing for increasing liquid flow rates. During the film transition the state which corresponds to a zero wall shear stress was refereed by Moalen-Maron and Dukler (1984) as the "U state". They analytically deduced it by starting from the equation for steady, laminar, one-dimensional motion of an incompressible Newtonian fluid. Comparison with experimental results showed better agreement for low liquid rates.

As the flow condition changes toward the formation of a hanging film, characterizing the onset of the flow reversal, the up moving film grows exponentially as shows Fig. (4). This results in the increase of the gravitational term in Eq. (10).

Not only the behavior of the film thickness but also the pressure drop dictates the trend of the interfacial shear term. Considering the liquid flow rates tested, the pressure drop decreases until the non-dimensional gas velocity is within the range $1.05 \leq J_G^* \leq 1.25$. For lower gas flow rates the tendency is opposite and the pressure drop increases as J_G^* decreases. This is shown in Fig. (5)

The film transition point can be found if the wall shear term is set to zero in Eq. (10):

$$\left[1-\left(1-\frac{2\delta}{D}\right)^2\right]\cdot(\rho_L-\rho_G)\cdot g=\frac{4}{D}\frac{\tau_I}{\left(1-\frac{2\delta}{D}\right)} \tag{11}$$

The interfacial shear stress is defined by

$$\tau_I \equiv \frac{1}{2}\cdot f_I\cdot\rho_G\cdot\frac{J_G^2}{\left(1-\frac{2\delta}{D}\right)^4} \tag{12}$$

where f_I is the friction factor coefficient. The interfacial shear stress can be calculated from a force balance applied to the core flow:

$$\tau_I=\left(-\rho_C\cdot g+\frac{\Delta p}{\Delta z}\right)\frac{4}{D\left(1-\frac{2\delta}{D}\right)} \tag{13}$$

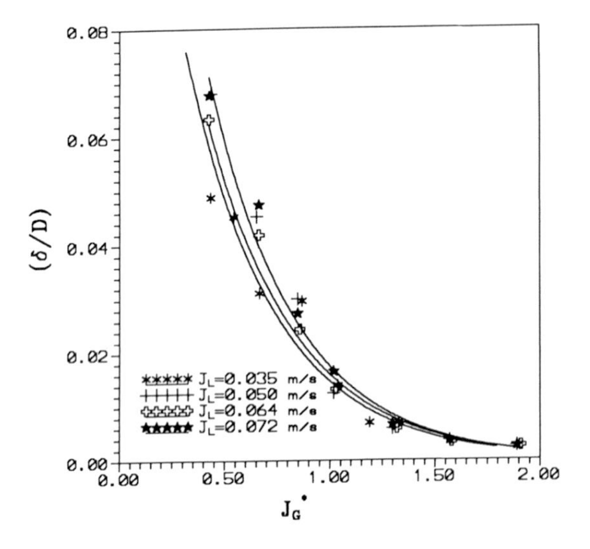

Figure (4) - The film thickness vs. the non-dimensional superficial velocity. Solid lines represent exponential fit.

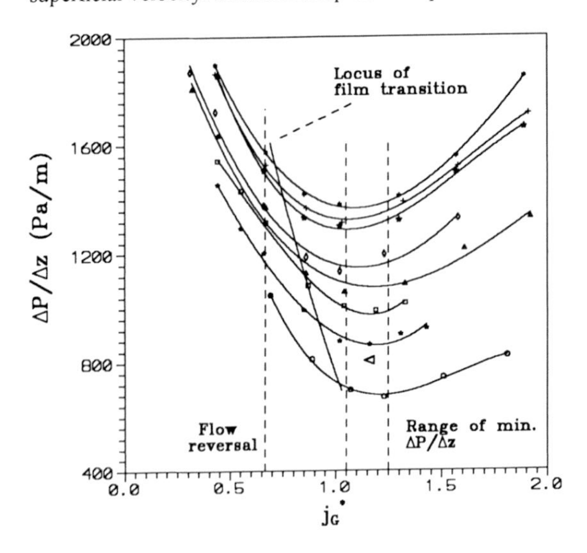

Figure (5) - Pressure drop curves. J_L increases from the lower curve to the upper one.

Equating Eqs. (12) and (13) one gets the interfacial friction factor f_I. Fig (6) shows the plot of f_I against (δ/D). The figure also shows the data obtained by Zabaras et al (1986) and the correlation proposed by Barathan (1978), both focused on transitional annular flows. For comparison Fig. (6) also includes the well known Wallis' (1969) correlation, proposed for developed rough annular flows. This equation clearly under predicts f_I.for flow condition close to the onset of flow reversal. The spread in the data is related to a dependence on J_L as recently discussed by Pedras (1993). If one considers a fit to the data they correlate as:

$$f_I = 0.005 + 3.6 \cdot 10^{-2} \cdot \frac{\delta}{D} \cdot \mathrm{Re}_L^{-0.59} \tag{14}$$

The solution of Eqs. (11), (12) and (13), in terms of the non-dimensional superficial gas velocity, J_G^*, and the relative film thickness, (δ/D), having J_L as a parameter, is depicted in Fig. (7). The locus of the film transition is shown as a solid line, which is also the locus of the maximum J_G^* for a given J_L. If the actual J_G^* is higher than this maximum, the film is moving upward and a negative wall shear stress term in equation (10) must be preserved . If the actual J_G^* is lower, the film transition has already happened and the wall shear stress term in equation (10) is positive. As a consequence, only the solid line represents the physical system under study.

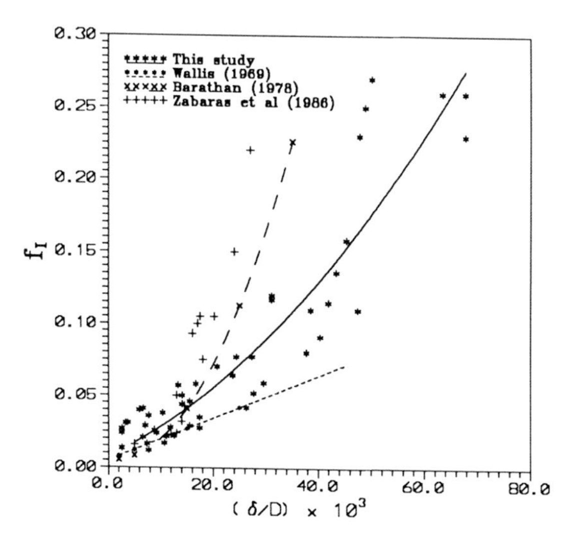

Figure (6) - Experimental friction factor vs. the dimensioless film thickness

Fig. (5) also shows the curve representing the film transition. It tends exponentially to the flow reversal line as the pressure drop and the liquid flow rate increase. Since the flow transition is a condition for the onset of the flow reversal, this tendency leads one to speculate that, in apparatus where the liquid is injected through the porous wall of a short pipe, entrance effects induce the flow mechanisms which regulate the flow reversal.

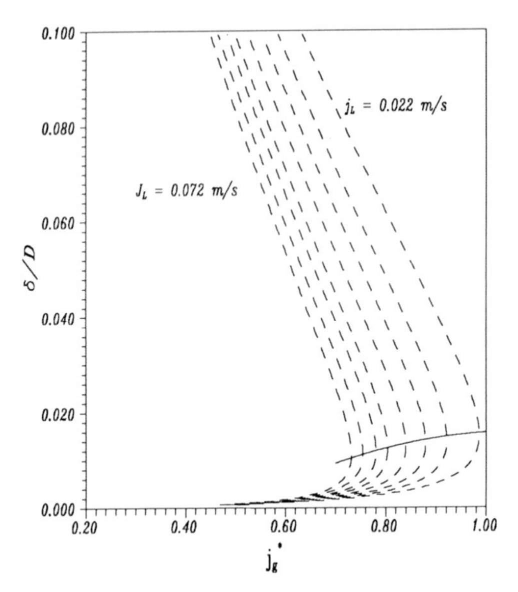

Figure (7) - The locus of the film transition

As the liquid rate and the force resulting from the pressure drop increase, the entrance effects on the gravitational, interfacial or acceleration terms are minimized. Then, the solution for the film transition approximates the flow reversal line. Nevertheless, the film transition model predicts with a good accuracy, specially as the liquid flow rate increases, the onset of the flow reversal in concurrent upward annular flows. Its application improves and generalizes the prediction of the onset of the flow reversal and the load-up in gas wells. Its limitation relies on the use of a friction factor correlation specifically developed for the system under operation.

CONCLUSION

An apparatus was built to measure the flow operational conditions at the onset of flow reversal in vertical pipe. The aim was the anticipation of the load-up of gas wells. It has shown that the mean gas velocity calculated by the suspended drop model is in good agreement with the measured values at the onset of flow reversal. Although, the direct observation of the flow indicates that the suspended drop model does not represent the physics of the phenomenon. The film transition phenomenon that takes place in the liquid film was then considered. The locus of the film transition occurs when the non-dimensional superficial gas velocity, J_G*, is within the range limited by the minimum pressure drop region and the flow reversal line. It predicts with a good accuracy the flow reversal for the higher liquid rates.

Acknowledgment - The authors gratefully acknowledge the financial support of PETROBRÁS

List of Simbols

A ≡ cross sectional area
Cd ≡ drag coefficient
D ≡ pipe diameter
f ≡ friction factor
g ≡ acceleration of gravity
J ≡ superficial velocity
K ≡ Kutateladze number
p ≡ pressure
V ≡ actual velocity
W ≡ mass flow rate
z ≡ axial distance
α ≡ mean void fraction
δ ≡
Δ ≡
σ ≡
τ ≡
ρ ≡

Re ≡ Reynolds number

Subscripts

C ≡ core flow (gas + droplets)
E ≡ entrained liquid
G ≡ gas
I ≡ interface
L ≡ liquid
LF ≡ liquid film
W ≡ wall

(*) ≡ dimensionless variable ($\delta^*=\delta/D$)

References

Asahi, Y. 1978 Flooding and Flow Reversal of Two-Phase Annular Flow, Nuclear Eng. and Design, 47, 251-256.
Barathan, D., Wallis, G.B. and Richter, H.J. 1978 Air-Water Counter-Current Annular Flow in Vertical Tubes, EPRI Report NP-786.
Coleman, S.B.,Clay, H.B., McCurdy, D.G. and Norris III, H.L. 1991 Understanding Gas-Well Load-Up Behavior, J. Petroleum Technology, March.
Dukler, A.E. and Smith, L. 1977 Two-phase Interactions in Counter-Current Flow Studies of the Flooding Mechanism, Annual Rep., NUREG/CR-0617.
Guimarães, P.E. 1991 The Flow Reversal of Two-Phase Vertical Concurrent Annular Flows, Master Thesis, Univ. Estadual de Campinas-UNICAMP, Brazil.
Hewiit, G.F., Lacey, P.M.C. and Nicholls, B. 1965 Transition in Film Flow in a Vertical Tube, UKAEA Report AERE-4614
Moalen-Maron, D. and Dukler, A.E. 1984 Flooding and Upward Film Flow in Vertical Tubes-II, Int. J. Multiphase Flow, 10 (5), 599-621.
Pushkina, O.L. and Sorokin, Y.L. 1969 Breakdown of Liquid Film Motion in Vertical Tubes, Heat Transfer Soviet Research 1, 56-64.
Pedras, M.H.J. 1993 Interfacial Friction in Transitional Annular Flows (in portuguese), Master Thesis, Univ. Estadual de Campinas- UNICAMP, Brazil.
Taitel, Y. and Dukler, A.E. 1976 Flow Regime Transitions for Vertical Upward Gas-Liquid Flow: An Approach through Physical Modelling, Report #1, ERDA E(10-1)-1571.+
Taitel, Y., Barnea, D. and Dukler, A.E. 1982 A Film Model for the Prediction of Flooding and Flow Reversal for Gas-Liquid Flow in Vertical Tubes, Int. J. Multiphase Flow, 8 (1), 1-10.
Turner, R.G., Hubbard, M.G. and Dukler, A.E. 1969 Analysis and Prediction of Minimum Flow Rate for the Continuous Removal of Liquid from Gas Wells, J. Petroleum. Technology, 21, 1475.
Wallis, G.B. 1962 The Transition from Flooding to Upward Concurrent annular Flow in a Vertical Pipe, UKEA Report AEEW-R142.
Zabaras, G., Dukler, A.E. and Moalen-Maron, D. 1986 Vertical Upward Concurrent Gas-Liquid Annular Flow , AIChE Journal, 32 (5), 829-843.

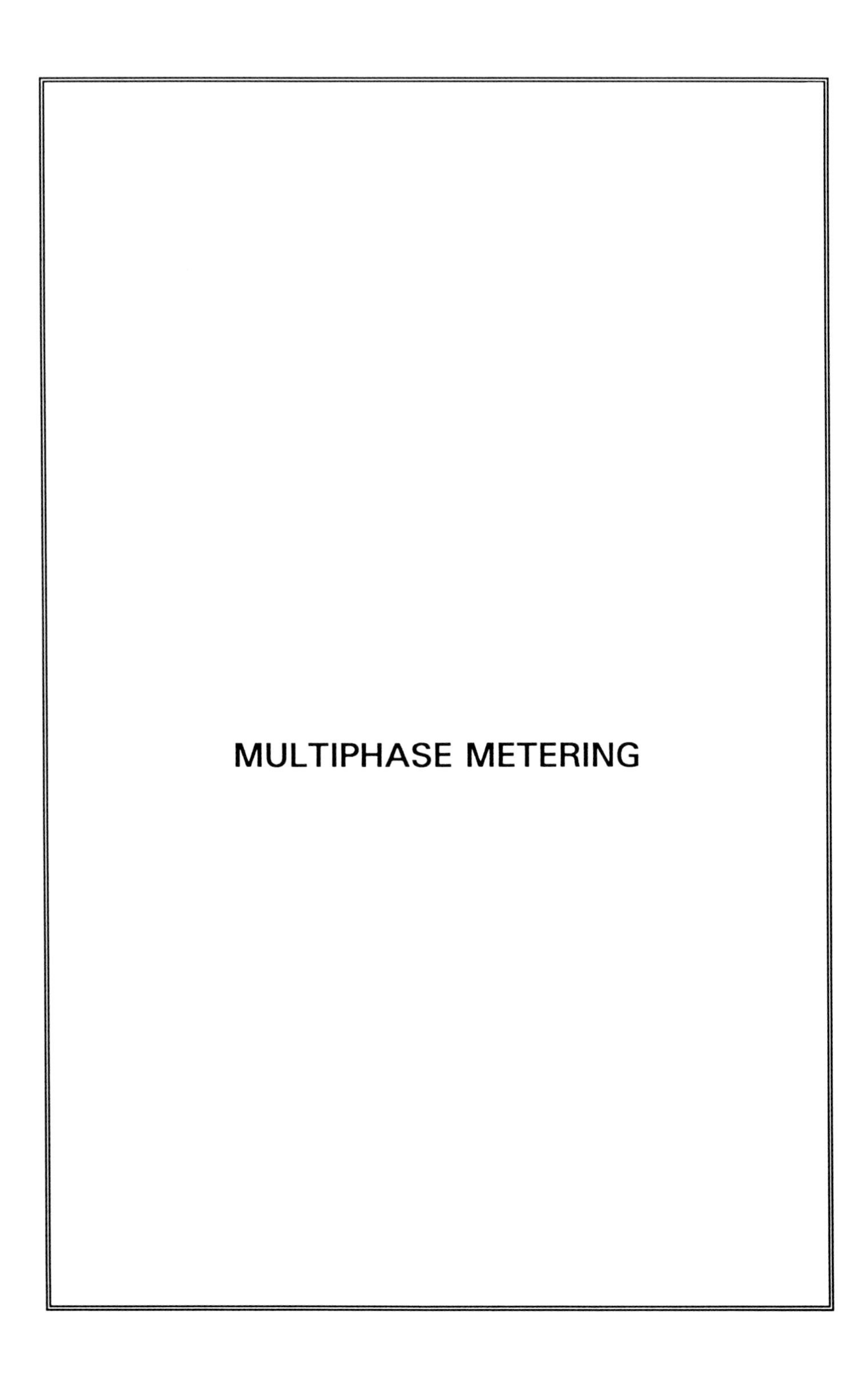

MULTIPHASE METERING

A REVIEW OF THE UNCERTAINTIES ASSOCIATED WITH THE ON-LINE MEASUREMENT OF OIL/WATER/GAS MULTIPHASE FLOWS

Dr B C Millington, NEL Executive Agency

SUMMARY

The paper describes the present state of multiphase flow metering technology from the point of view of measurement uncertainties in the individual gas, oil and water mass flow measurements. The functional relationship between uncertainty and flow composition is derived for a simplified calculation procedure, and it is shown that achieving 10 per cent uncertainty in single-shot mass flowrate measurements for all flow compositions will be very difficult using existing approaches to multiphase metering.

It is concluded that in the short term, the application of averaging procedures is the most practical method of reducing measurement uncertainties. This limits the application of multiphase meters to essentially throughput over a significant period of time. However, this is the major application. Pursuit of flowrate measurements taken over short periods - less than 15 seconds - is not considered worthwhile bearing in mind the transient nature of the flows. Opinions as to the direction of future research and development are briefly alluded to.

NOTATION

m_g	*mass flowrate of gas*
m_o	*mass flowrate of oil*
m_w	*mass flowrate of water*
Q_t	*total volumetric flowrate at metering conditions*
α_g	*gas phase fraction at metering conditions*
α_o	*oil phase fraction at metering conditions*
α_w	*water phase fraction at metering conditions*

β	*liquid water content*
μ	*liquid fraction of the multiphase mixture*
ρ_g	*gas density at the metering conditions*
ρ_o	*oil density at the metering conditions*
ρ_w	*water density at the metering conditions*
ρ_t	*total multiphase density at the metering conditions*
ρ_L	*liquid density at the metering conditions*

1 BACKGROUND

With the ever increasing demand to reduce crude oil production costs, the late 1980's and early 1990's has seen the rapid emergence of multiphase metering technology which offers the potential for major reductions in capital costs, as well as significantly improved reservoir management.

The technology is still developing, and the author is aware of nearly 20 independent research and development programmes currently in progress worldwide. The key short, medium and long term applications of multiphase flowmeters are respectively:

a To replace bulky separation equipment presently used to separate the flow into its individual components before single phase metering technology is applied;

b To use multiphase flowmeters subsea either on each individual well, or in a manifolded arrangement with perhaps four or five wells being serviced by one flowmeter. Improvements in reservoir management and allocation are the main goals in this application.

c To employ multiphase meters for fiscal purposes, although this is probably at least 5-10 years in the future.

Apart from the improved management of existing production systems, the economic gains to be made will be reflected in the improved viability of deep water and marginal fields.

In the haste to achieve operational multiphase meters, some of the fundamental measurement difficulties have often been overlooked. It is the purpose of this paper to rectify this situation, giving both an insight into the technical challenges and a balanced assessment of the likely measurement uncertainties involved in this type of metering.

2 PRESENT TECHNOLOGY

2.1 Multiphase flowmeter designs

In order to put the level of present technology into perspective, it is worthwhile considering the possible generic multiphase flowmeter designs:

Type 1 The ideal multiphase flowmeter would make 3 independent and simultaneous measurements: one of the oil mass flowrate, one of the water mass flowrate, and one of the gas mass flowrate. Each of the separate measurements would not be influenced by the fact that the other two phases are present. Such a meter is not yet technically possible, and is likely to remain so for at least the next 10-15 years.

Type 2 The next level down from the ideal multiphase flowmeter is one which provides measurements of the oil, water and gas velocities coupled with individual measurements of the area-based phase fractions. With PVT density information, these measurements would allow the mass flowrates to be derived. Although not yet achievable in its entirety, a number of the existing multiphase metering research programmes are tackling some of the technical problems involved.

Type 3 The next level down - and what could be considered as entry level - applies some form of flow conditioning to make all three phases flow at equal velocity. Measurements of the total flowrate (typically volumetric), the water content in the liquid and the gas fraction in the total flow, are sufficient to allow the individual mass flowrates to be derived. Most of the present research programmes fall into this category. A typical example of such a metering system configuration is shown in Fig.1.

2.2 Individual mass flowrate calculation procedure

Taking the Type 3 design, if the flow is to be gas-liquid with a liquid fraction of μ, then the total density can be written:

$$\rho_t = (1-\mu)\rho_g + \mu\rho_L \qquad (1)$$

Similarly, if the water content in the liquid is denoted by β then the expression for the liquid density is

$$\rho_L = (1-\beta)\rho_o + \beta\rho_w \qquad (2)$$

Substituting the expression for the liquid density into equation (1) gives an alternative expression for the total density

$$\rho_t = (1-\mu)\rho_g + \mu(1-\beta)\rho_o + \mu\beta\rho_w \qquad (3)$$

where the coefficients of the gas, oil and water densities are the volumetric phase fractions of each component, therefore

$$\alpha_w = \mu\beta \qquad (4)$$

$$\alpha_g = 1 - \mu \tag{5}$$

$$\alpha_o = \mu\,(1 - \beta) \tag{6}$$

So if μ and β can be measured or calculated then the volumetric flow composition can be derived. In the system shown in Fig.1, one method of determining the water content in the liquid is by measuring the liquid density, and applying equation (2)

$$\beta = \frac{\rho_L - \rho_o}{\rho_w - \rho_o} \tag{7}$$

where the oil and water densities are known from the well characteristics. Likewise from equation (1) the liquid fraction μ can be found by re-arranging to give

$$\mu = \frac{\rho_t - \rho_g}{\rho_L - \rho_g} \tag{8}$$

where the total density is typically measured using a gamma densitometer, and the gas density is known from the well characteristics.

From the values of μ and β, the individual mass flowrates of each phase can be found by obtaining a measurement of the total volumetric flowrate, by using a venturi for example

$$m_g = \alpha_g \rho_g Q_t \tag{9}$$

$$m_o = \alpha_o \rho_o Q_t \tag{10}$$

$$m_w = \alpha_w \rho_w Q_t \tag{11}$$

This simplified calculation procedure which omits parameters such as temperature and pressure measurement, is mapped out in Fig.2. The purpose of this figure is to highlight the links between various stages in the calculation route and to illustrate that the measurement uncertainties of some of the parameters can have significant cumulative effects. Incorrectly measuring the oil density, for example, not only causes the liquid water content and phase fractions to have higher uncertainties, but directly increases the uncertainty in the final oil mass flowrate calculation.

3 UNCERTAINTY CALCULATIONS

3.1 Procedure

The magnitude of the total random uncertainties in the measurements of oil, water and gas mass flowrates are determined using the Root Sum of Squares technique:

$$e^2(y) = \left[\frac{\partial y}{\partial x_1}\right]^2 e^2(x_1) + \left[\frac{\partial y}{\partial x_2}\right]^2 e^2(x_2) + \ldots + \left[\frac{\partial y}{\partial x_n}\right]^2 e^2(x_n) \tag{12}$$

where y is a function of n variables, x_1 to x_n. This calculation does not include any systematic effects, it represents only random effects, and can therefore be positive or negative in sense.

For a rigorous analysis of multiphase metering uncertainties the variable list would include primary measurement parameters such as temperature and pressure, however, to avoid over laborious mathematics the variable list will be restricted to those described in Section 2.2.

To calculate the uncertainty in the individual mass flowrates, start from the equation set (9-11). Take for example the oil mass flowrate, then the partial derivative (sensitivity) coefficients are:

$$\frac{\partial m_o}{\partial \alpha_o} = \rho_o Q_t \tag{13}$$

$$\frac{\partial m_o}{\partial \rho_o} = \alpha_o Q_t \tag{14}$$

$$\frac{\partial m_o}{\partial Q_t} = \alpha_o \rho_o \tag{15}$$

and the combined absolute uncertainty in the mass flowrate is

$$e^2(m_o) = m_o^2 \left[\frac{e^2(\alpha_o)}{\alpha_o^2} + \frac{e^2(\rho_o)}{\rho_o^2} + \frac{e^2(Q_t)}{Q_t^2}\right] \tag{16}$$

which can be written

$$E^2(m_o) = E^2(\alpha_o) + E^2(\rho_o) + E^2(Q_t) \tag{17}$$

where $E(x)$ is the relative uncertainty in x.

The last two terms on the RHS will be known from calibrations; the first term must be calculated from equation (6)

$$e^2(\alpha_o) = (1 - \beta)^2 e^2(\mu) + \mu^2 e^2(\beta) \tag{18}$$

and therefore

$$E^2(\alpha_o) = \left[\frac{1-\beta}{\alpha_o}\right]^2 e^2(\mu) + \left[\frac{\mu}{\alpha_o}\right]^2 e^2(\beta) \tag{19}$$

where the absolute uncertainties in μ and β are derived using the same RSS technique applied to equations (7) and (8)

$$e^2(\mu) = \left[\frac{1}{\rho_L - \rho_g}\right]^2 [e^2(\rho_t) + (1-\mu)^2 e^2(\rho_g) + \mu^2 e^2(\rho_L)] \tag{20}$$

$$e^2(\beta) = \left[\frac{1}{\rho_w - \rho_o}\right]^2 [e^2(\rho_L) + (1-\beta)^2 e^2(\rho_o) + \beta^2 e^2(\rho_w)] \tag{21}$$

It is a straightforward task to substitute these expressions into equation (19), and to use the result to compute the oil mass flowrate uncertainty using equation (17) with appropriate values for the relative uncertainties in ρ_o and Q_t.

The above procedure is a considerable simplification of a rigorous uncertainty analysis starting from first principles and including all potential sources of uncertainty, however, the method remains the same. Examples of more detailed procedures can be found in References 1-3.

3.2 Example calculations

Let the densities of the oil, water and gas be 830 kgm^{-3}, 1000 kgm^{-3} and 50 kgm^{-3} respectively, and consider three flow conditions which have the following volumetric phase fractions

---- Phase Fractions ----						
Oil	Water	Gas	μ	β	ρ_t	ρ_L
0.1	0.8	0.1	0.9	0.8889	888	981.1
0.1	0.3	0.6	0.4	0.7500	413	957.5
0.8	0.1	0.1	0.9	0.1111	769	848.9

where the last four columns of data are derived using equations (1-6).

Using the uncertainty data in Table 1, the equations derived in Section 3.1 are used to produce the following data

$e^2(\mu)$	$e^2(\beta)$	$E^2(\alpha_o)$	$E^2(m_o)$	$E(m_o)$
4.652×10^{-4}	1.662×10^{-4}	1.40×10^{-2}	1.44×10^{-2}	0.12
4.882×10^{-4}	1.600×10^{-4}	5.61×10^{-3}	6.01×10^{-3}	0.08
6.319×10^{-4}	1.662×10^{-4}	9.91×10^{-4}	1.39×10^{-3}	0.04

The figures in the righthand column can be multiplied by 100 to give the percentage oil mass flowrate uncertainties. Two factors are important: firstly even using very conservative levels of uncertainty in each parameter (Table 1), the level of uncertainty in a single measurement can be 10 per cent or more, and secondly the uncertainty is appreciably affected by the flow composition. Similar calculations can be performed for the water and gas phases, with identical conclusions.

The primary causes of the variation in the relative uncertainty are the terms on the RHS of equation (19). Dividing through by α_o causes the square of the sensitivity coefficient to approach infinity as the oil fraction tends to zero. This is the fundamental problem for all multiphase flowmeters which do not directly measure the individual mass flowrates.

4 PRACTICAL EXPERIENCE

A programme of work to develop a multiphase flowmeter from mainly proven technology has recently been completed at NEL - the system is shown schematically in Fig.1. The project provided a considerable amount of useful information and experience regarding the most suitable instruments for present multiphase metering - ie the Type 3 design in section 2.1.

The typical performance of this multiphase meter is shown in Fig.3. The predicted increase in uncertainty as the phase fraction (of oil) decreased is clearly evident in this data. Some systematic biasing was also present, but this was due primarily to gas entrainment in the oil/water sample stream from the separator. The decrease in the liquid density measurement was reflected in an apparent reduction in the liquid water content - see equation (7).

5 IMPLICATIONS

For the particular flow conditions described in Section 3, the course of action to reduce the uncertainty in the oil mass flowrate would be to reduce the uncertainty in μ so that the product of the sensitivity coefficient and the uncertainty in μ had a lesser effect.

Practical and technological problems prevent this being achieved at the present time. Firstly, bypass sampling systems suffer from two major difficulties: oil/water separation and consequently time-varying water content measurements and, as described in Section 4, gas entrainment in the oil/water sample stream, especially with emulsion-forming oils. Secondly, the other common method adopted for oil/water measurement is dual energy nucleonic densitometry, which uses the difference between two count rates to infer the liquid composition. This is quite acceptable, in terms of measurement uncertainties, if the volumetric phase fraction of the component of interest is high, but for low phase fractions taking the difference of two very similar quantities to infer an appreciably smaller quantity is less than ideal.

In the short term, the simplest method of reducing the mass flowrate uncertainties, is to increase the number of measurements and to average the data. With present, predominantly nucleonic based technology, the time required for each multiphase flowrate measurement is of the order of 30 seconds, or more. Consequently, a sufficient number of mass flowrate measurements to reduce uncertainties to acceptable levels, may require upwards of 10 minutes, depending on the flow composition.

To reduce this time frame down to a more practical 1-5 minutes for each updated measurement, the use of a rolling average multiphase density can be considered. The nucleonic system would continuously log the multiphase density, building up an average measurement with a continuously increasing time-base. The multiphase meter would read the density information as required. The rolling average technique does have drawbacks in that the density information required to interpret the signals from other instruments may be considerably different to the density of the mixture passing through the flowmeter at any one time. The measurements of individual flowrates will therefore be in error when taken as single-shot measurements, but time-averaging can be used to reduce this effect. Care must be taken in the use of such procedures.

Assuming a rolling average multiphase density is taken, multiphase metering will be restricted to applications requiring the measurement of the total mass throughput of each phase over a significant period of time. Fortunately this is the greatest area of application, as short term flowrate measurements - updated say every 15 seconds - are of very limited practical value because of the transient nature of many of the important flow patterns.

Having considered the mathematical constraints and practical difficulties posed by the flows, it is the author's view that future research should be focused on developing new techniques for multiphase density and water content measurement, along with improved velocity detection equipment. Additionally, moving away from systems that employ upstream mixing towards systems operating on the natural flow patterns, must also be a prime objective.

5 CONCLUDING COMMENTS

The main points arising from the study of multiphase flowmeter uncertainties are:

5.1 Present designs are far from ideal, because they do not measure the individual mass flowrates directly. Mass flowrates are inferred from a large number of measurements.

5.2 Some of the measured parameters are required in several calculations, consequently significant cumulative uncertainties can result from one measurement being slightly in error.

5.3 Ignoring the difficulties posed by transient flows, the mathematics presented explains the difficulties in obtaining individual mass flowrate measurements with an acceptably low uncertainty - 5 per cent of reading is an often quoted figure. For a single measurement, the random uncertainty in individual mass flowrate measurements may approach, and even exceed, 10 per cent even assuming very low levels of uncertainty in the measured parameters. The experimental data presented confirms this. It is found from more detailed analyses that when factors such as the uncertainties in mass absorption coefficients are considered, the random uncertainties in the individual mass flowrate measurements can exceed 20 per cent.

5.4 Because the phase fraction is an important determining factor in the level of uncertainty, it is essential that multiphase flowmeter uncertainty data is quoted along with the phase fraction at which it applies. For this reason, a standardised method of quoting the multiphase metering uncertainties needs to be developed in the near future which conveys the maximum amount of information to the user.

5.5 In the short term, it is concluded that improvements to the levels of uncertainty associated with the mass flowrate measurements will best be obtained by the application of averaging procedures.

ACKNOWLEDGEMENT

This paper is published by permission of the Chief Executive, NEL Executive Agency, and is crown copyright.

REFERENCES

1 MILLINGTON, B. C. An uncertainty analysis for a multiphase flowmeter. NEL Flow Measurement Memo 417. East Kilbride, Glasgow: National Engineering Laboratory, June 1991.

2 MILLINGTON, B. C. An uncertainty analysis for the proposed design of the NEL/SGS multiphase flowmeter. NEL Flow Measurement Memo 418. East Kilbride, Glasgow: National Engineering Laboratory, September 1991.

3 MILLINGTON, B. C. An uncertainty analysis for the alternative design of the NEL/SGS multiphase flowmeter. NEL Flow Measurement Memo 419. East Kilbride, Glasgow: National Engineering Laboratory, August 1991.

LIST OF TABLES

LIST OF FIGURES

TABLE 1

THE UNCERTAINTY IN THE PARAMETERS USED TO CALCULATE THE INDIVIDUAL MASS FLOWRATES

	Absolute	Relative (per cent)
Liquid density	2 kgm^{-3}	0.2 (approx)
Multiphase density	20 kgm^{-3}	Varies
Oil density	1 kgm^{-3}	0.13
Water density	1 kgm^{-3}	0.1
Gas density	2 kgm^{-3}	4.0
Total volumetric flowrate	Varies	2.0

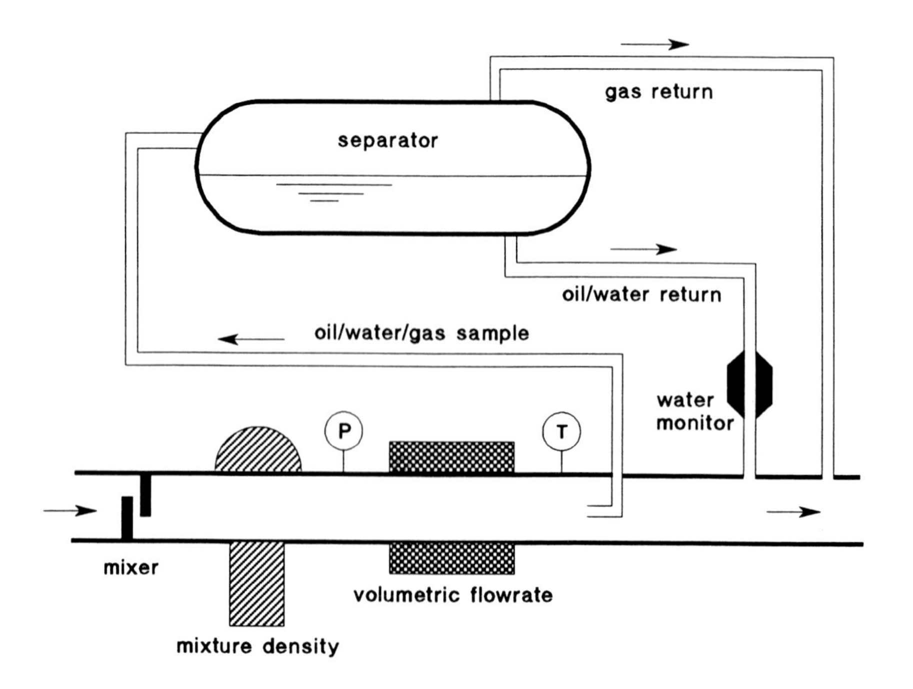

Fig.1 Schematic Diagram of the Multiphase Metering System

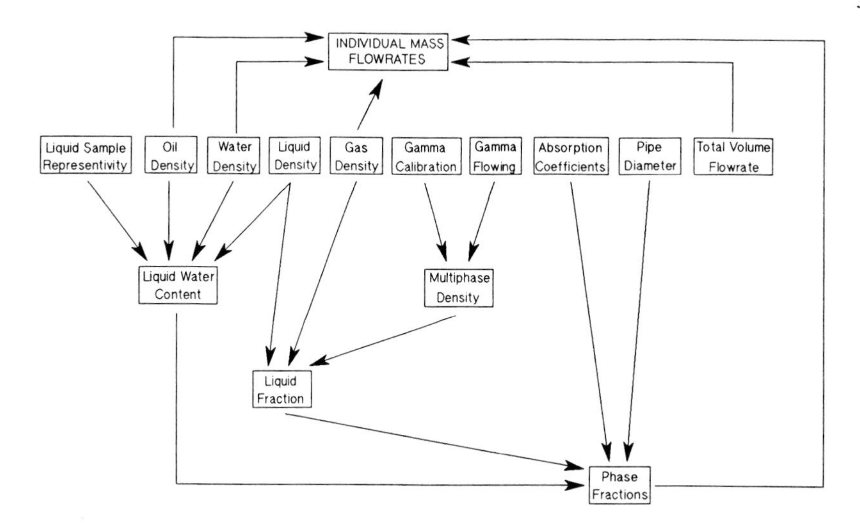

Fig.2 A Multiphase Flowrate Calculation Procedure

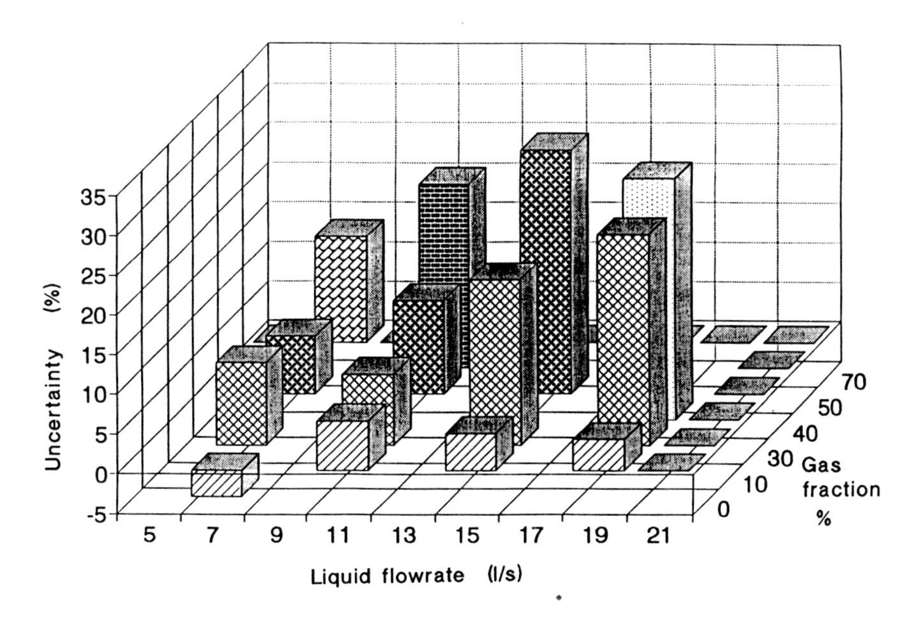

Fig.3 Oil Mass Flowrate Uncertainty With 50 per cent Water Content

DEVELOPMENT OF A MULTIPHASE FLOWMETER FOR SUBSEA WELL TESTING

P. Andreussi, University of Udine, Italy
P. Ciandri, C. Talarico, TEA, Italy
F. Giuliani, SNAMPROGETTI, Italy
W. L. Loh, BHR Group, UK
A. Mazzoni, AGIP, Italy

ABSTRACT

The multiphase flowmeter developed by TEA (Centro per le Tecnologie Energetiche ed Ambientali) for metering hydrocarbon production in presence of water and gas is described in its main features. The flowmeter has been adopted as a component of the Subsea Well Testing System (SWTS) developed by TECNOMARE, SNAMPROGETTI, BHR Group and AGIP and consists of an impedance meter coupled with a γ-densitometer and a Venturimeter. A prototype of the flowmeter has been produced by TEA and tested at the BHR Group Laboratory. The objectives of the experimental campaign have been:
- the calibration of the integrated flowmeter,
- the performance analysis of the metering system from the mechanical, hydraulic and electronic point of view;
- the development of model equations for dispersed three-phase flow in order to relate the output from the instrumentation to the volume fractions;
- the assessment of manufacturing critical issues.

The experimental results show that it is possible to determine the water content in the mixture through simple impedance measurements and the total mass flow by the Venturimeter. This is possible also in presence of a gaseous phase, at least in the percentages of gas envisaged by the SWTS project.

1. INTRODUCTION

The measurement of hydrocarbon production from a well is required for well testing, field management and fiscal purposes. The presence of natural gas and water which often accompanies crude oil, leads to a complex multiphase mixture whose composition and total flow rate must be monitored. Currently this can be achieved after separation of the phases, followed by metering of the individual constituents. In the case of well testing, the flow from each well is periodically diverted to a three phase test separator for this purpose. This method, especially in the exploitation of subsea reservoirs, may however be inconvenient, not feasible or very expensive. The main objective of the Subsea Well Testing System developed by TECNOMARE, SNAMPROGETTI, BHR Group and AGIP is to remove the need for a dedicated test line, by installing an underwater metering system for multiphase production /1/.

The SWTS operates as follows (Fig. 1): production fluids enter the two phase separator after which the gas flow rate is measured by conventional flowmeters. Liquid metering requires an innovative monitoring system consisting of flowrate, average density and water-cut meters integrated through a microcomputer for data collection and analysis. Two parallel metering lines are used to accommodate a wide range of flow rates. If less than 15% gas is present in the inlet flow, the test-separator is by-passed, and the total flow passes through the liquid metering equipment. Under these circumstances the liquid monitoring system operates on-line, i.e. it allows to measure each flowrate without separation of the phases. For high volumetric flows, the gas-liquid separator behaves as a partial separator and it is assumed that up to a maximum of 15% of gas (in volume) be entrained in the two liquid phases. Also in this case the metering system operates in presence of three phases. This is the innovative aspect of the method when compared with existing techniques.

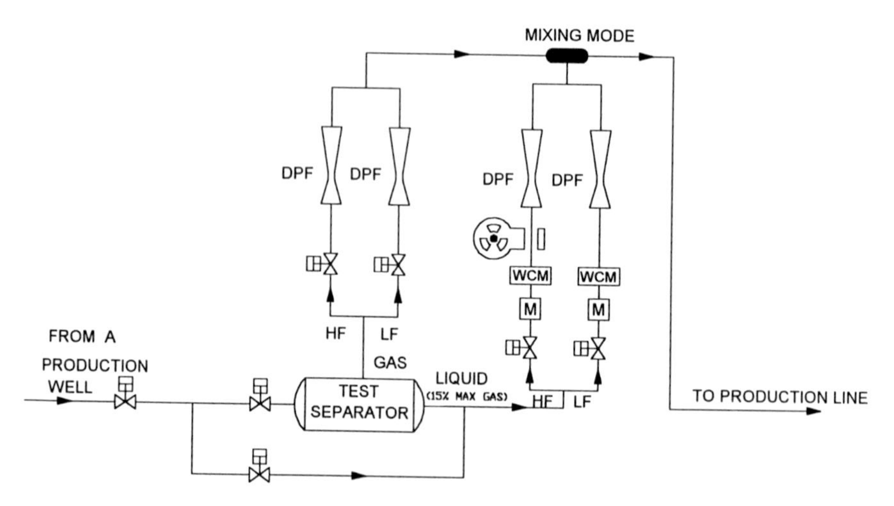

Fig 1 : Scheme of the Subsea Well Testing System

2. Description of the Instrumentation

The multiphase flowmeter is a compact measurement system composed by a set of instruments and a remote computing apparatus for data acquisition and analysis, designed for on line and real time measurement of gas, water and oil flow rates. The main instruments are an impedance meter, a Venturimeter and a single beam γ-densitometer. As an alternative to the γ-densitometer, the static head before the Venturi is also measured.

As shown by previous experimental analyses /2/, in order to ensure that the multiphase mixture forms a dispersion, the measurement system must be preceded by a static mixer or the speed of the multiphase mixture in the pipe must be higher than 3-4 m/s. The mixer - impedance probe - average density meter have been coupled with a Venturimeter to measure the flowrate of the three phases as well. The flowmeter is mounted vertically with the flow directed upwards.

As shown in Fig. 2, the three-phase oil/water/gas mixture first passes through the mixer (1) to ensure a homogeneous mixture. The LCR (inductance-capacitance-resistance) meter (3) converts the output of the water-cut meter (2) into total impedance -capacitance and conductance. The signal is sent to the data processing unit (4) integrated in the computer (5). After the water-cut meter the flow passes through a γ-densitometer (6) which sends a signal directly related to the average density of the three-phase mixture, to the processing unit. Knowing the density of the individual phases, it is possible to obtain a real time measurement of oil, water and gas flow rates from the impedance and average density values. The signal from the γ-densitometer can be replaced by the measurement of the head between two pressure taps located in the line between the mixer and the Venturimeter (static head density

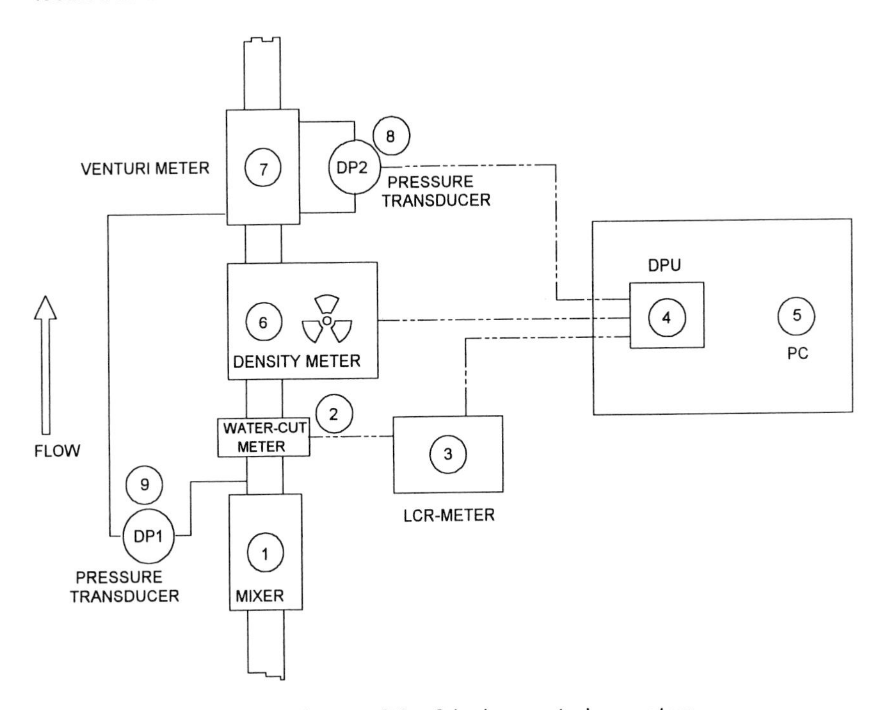

Fig 2 : scheme of the 3 inches metering system

meter (9)).

Two multiphase flowmeters are needed to cover the flow range expected within the well testing system. The main design characteristics of these flowmeters and the working conditions are reported in Table 1.

Table 1. Design characteristics

pipe diameter (")	2	3
expected fluids	oil+ water+15% vol. HC vap	oil+water+15% vol. HC vap.
length (m)	1.43	1.75
max operative pressure (bar)	100	100
max temperature (°C)	90	90
flowrate (m^3/h)	5 - 25	20 - 120
water-cut (%)	0 - 100	0 - 100
gas ratio (%)	0 - 15	0 - 15
system configuration	vertical upvards	vertical upwards

The computing apparatus is based on a PC with acquisition boards mounted; the data of each instrument are analyzed by means of specific software packages.

2.1 Venturimeters

Standard type Venturimeters have been adopted. The mechanical design is based on the ISO 5167 code. The pressure drops through the Venturi and in the line preceding the Venturi have been measured with Honeywell ST 3000 smart transmitters (Differential Pressure Model STD 120).

2.2 Water-Cut Meter

The water-cut probe, shown in Fig. 3. is a total impedance probe which consists of two parallel rings mounted flush to the pipe wall. In the present test the impedance between the two rings is measured using a Hewlett Packard 4285A LCR Meter. This instrument will be replaced with dedicated electronics in the near future The electrodes physically touch the process fluid and are electrically insulated from each other and from the pipe wall which is earthed. Similar probes have been developed by Andreussi *et al.* /3/ /4/ for the measurements of the time varying liquid hold-up in gas-liquid flow systems.

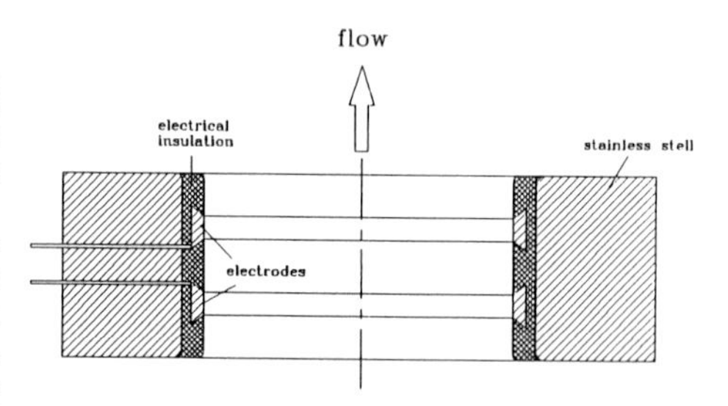

Fig. 3 - Probe section

The probe can be considered electrically as a three-terminal device: ring (1), ring (2) and screen (3). We can assume that between these terminals there are three impedances: Z_1, Z_2 and Z_{12}. The impedance meter measures only one of them, Z_{12}, whose value is directly related to the fluid present in the probe. The effects of Z_1 and Z_2 can be disregarded. The parameters affecting the modulus and phase of Z_{12} are:

- nature of the fluid,
- distance between the electrodes,
- frequency at which the impedance is measured.

Although the probe gives resistive, inductive and capacitive components, for frequencies of the order of 10 MHz it can be compared to a capacitance and resistance operating in parallel. By simultaneously reading capacitance and conductance it is possible to distinguish between water dispersed in electrically non-conductive liquid hydrocarbons and hydrocarbons dispersed in water.

The relationship between the dielectric constant of the mixture and the percentage of water is theoretically given in the literature /5/ and has been experimentally verified. This relationship assumes that there is a dispersion of oil in water or water in oil. The transition between the two conditions is determined by a sharp change of both the conductance and the capacitance. The experimental phase of probe calibration provides the data to replace Maxwell or Bruggeman equations (see /3/) with best-fit curves.

2.3 Static Mixer

As regards the low viscosity multiphase mixture, turbulent flow condition can be assumed. Nevertheless, the turbulence level may be too low to give rise to homogeneous flow conditions. In order to form a dispersion, a five-elements Sulzer SMV static mixer has been adopted.

2.4 γ-Densitometer

In principle, the measurement of the mean density of a fluid mixture by a γ-densitometer is a straightforward industrial operation, in particular when the flowrate is varying slowly with time and the mixture presents uniform characteristics over the cross section. The γ-densitometer used in the instrumentation test is a Krohne densitometer and is mounted onto a purpose-built pipe. Signals from the pick-up unit is transmitted to the signal converter which indicates the density of the mixture flowing through the pipe.

2.5 Head Meter

The static head meter instrumentation simply consists of a differential pressure transmitter (Honeywell STD 120). The measured linear pressure drops can then be appropriately related to the mixture density.

3. The Water-cut and Total Flowrate Calculation

Output signals from the instruments are elaborated by means of an original software package which requires PVT data and physico-chemical properties of the fluids as well. The algorithm to calculate the volume fractions may be summarized as follows:

- From the average density value, knowing the density of the three phases, it is possible to calculate the gas content for a water-cut trial value:

$$H_w = 1 - \frac{C_w - C_{dd}}{C_w - C_h} \cdot \left[\frac{C_h}{C_{dd}} \right] \tag{1}$$

$$H_w = \frac{C_h - C_{dd}}{C_h - C_w} \cdot \left[\frac{C_w}{C_{dd}} \right] \tag{2}$$

$$\alpha = \frac{H_w \cdot (\rho_w - \rho_h) + \rho_h - \rho_m}{H_w \cdot (\rho_w - \rho_h) + \rho_h - \rho_g} \tag{3}$$

where:

H_w = water-cut,
α = void fraction,
C_w, C_h, C_g = measured capacitance values when only water, only liquid-state hydrocarbons only gas are present,
C_{dd} = measured capacitance value of the multiphase mixture,
ρ_w, ρ_h, ρ_g = water, liquid hydrocarbon and gas density respectively,
ρ_m = measured average density of the three phase mixture.

Eq. (1) applies to water-in-oil dispersions, while Eq. (2) applies to oil-in-water dispersions. By reading the capacitance and the conductance it is possible to distinguish between the two patterns. On the basis of the gas content, the water-cut reading provided by the impedance probe is corrected. Such correction consists first in the calculation of the dielectric constant of the liquid phase only, C_m, given by:

$$\frac{C_g - C_{dd}}{C_g - C_m}\left[\frac{C_m}{C_{dd}}\right]^{1/3} = 1 - \alpha \tag{4}$$

then in the actual computation of the water-cut value, using different equations in case of oil continuous phase or water continuous phase, again according to the capacitance or conductance value:

$$H_w = 1 - \frac{C_w - C_m}{C_w - C_h} \cdot \left[\frac{C_h}{C_m}\right] \tag{5}$$

$$H_w = \frac{C_h - C_m}{C_h - C_w} \cdot \left[\frac{C_w}{C_m}\right] \tag{6}$$

The experimental phase of flowmeter calibration makes possible to replace equations (1) to (6) with best-fit functions.

As regards density measurements, the γ-densitometer output from the pick-up unit can be directly related to the average density of the three phase mixture; no data processing is therefore needed. The head measurement, on the other hand, has to be suitably analyzed to get the average density of the fluid. The differential pressure value Δp_l provided by the pressure transmitter is a function of the linear pressure drops and the static head ρgh:

$$\Delta p_l = h \cdot g \cdot (\rho_f - \rho_m) - \Delta p_a \tag{7}$$

where

h = test section length,
ρ_f = density of the dp cell filling fluid,
Δp_a = frictional pressure drops.

In Eq. (7) the static head term is, in general, much larger than frictional losses. It is then possible to use for Δp_a a correlation given in literature /6/ and to iteratively calculate the average density.

The analysis of Venturi data requires the density of the multiphase mixture as external input. Venturimeter measurements can be corrected for the presence of gas by means of original relationships that link the flowrate to pressure, quality and throttling ratio /7/. Nevertheless, as a first approach the homogeneous multiphase density may be used, yielding:

$$\Gamma = C_v \cdot \varepsilon \cdot A_t \cdot \sqrt{\frac{2\Delta p_v \cdot \rho_m}{1-\beta^4}} \quad (8)$$

where

Γ = mass flowrate,
C_v = discharge coefficient,
ε = compressibility coefficient,
A_t = throat cross-section area,
Δp_v = pressure drop through the Venturi,
β = throat-pipe diameter ratio.

The compressibility coefficient ε is equal to 1 for incompressible flow; for compressible flow it is less than 1 and depends upon the isoentropic coefficient of the gas, the ratio p_2/p_1 and β. The discharge coefficient C_v depends upon the device geometry, is very close to 1 and is generally obtained by calibration. For a classical Venturi tube, most of the national codes (ASME, DIN, UNI) give $C_v = 0.995$ in the following parameter ranges:

$$50 < D < 250 \text{ mm}$$
$$0.4 < \beta < 0.75$$
$$10^5 < Re < 10^6$$

Considering the errors involved both in the measurement of the water-cut and of the mean density, the expected precision in the measurement of the gas fraction is low. However, it should be noted that an appreciable uncertainty in the measurement of the gas fraction has only limited consequences on the value of the total mass flowrate, at least when the volumetric gas fraction is as small as expected.

4. Test Facility

Present measurements have been obtained by TEA, BHR Group and SNAMPROGETTI in a low pressure rig available at the BHR Laboratory in Cranfield (U.K.). The operating fluids in this rig are air, water and a dielectric oil containing a red dye to aid visualization. Oil and water are held in different storage tanks. The two fluids are pumped by two positive displacement pumps through their respective metering systems prior to mixing. Water is metered by a magnetic flowmeter while oil is metered by a turbine flowmeter. After metering, the fluids are mixed in a y-piece. Air from a compressed air source is first metered by a gas turbine flowmeter, before being injected into the oil/water mixture through four radially spaced inlets, which is further downstream of the y-piece. After passing through the test spool, the 3-phase mixture enters a 3-phase separator where separation takes place. Air after scrubbing is exhausted to the atmosphere outside the laboratory while 'oily-water' and 'watery-oil' enters the water and oil coalescers, respectively, for final clean up before returning to their storage tanks. The facilities can operate in any combination of single and multiphase flow.

5. Test Campaign and Experimental Results.

Objective of the tests is the calibration and the performance analysis of the liquid metering system in presence of a three phase mixture. High flowrate and low flowrate lines have been separately tested at pressures up to 0.35 MPa. The experimental campaign consisted of a series of tests in which each phase volume fraction has been varied by means of a variation of the volume flowrate. Accordingly, the water-cut and the void fraction ranged from 0% up to 100% and from 0% up to 20% respectively. The volume flowrate of the 2" and 3" lines ranged up to 40 m^3/h. Repeatability tests have also been performed.

The experimental results on each line have been analyzed taking into account first the behaviour of each instrument and then the whole integrated system. The system output has been compared to the actual flow conditions metered by the laboratory instrumentation. Obviously, all the measuring equipment was calibrated before starting the experimental campaign.

In case of two-phase flow, the impedance probe is able to provide the composition of the mixture. The analysis of the probe behaviour with a liquid-liquid (water-oil) flow has therefore been performed plotting the dimensioneless capacitance versus the water-cut. As shown in Fig. 4, the impedance probe output only depends on the water content; it is not a function of the pipe diameter and the total flowrate. The dimensionless capacitance can then be related to the water-cut by means of a single calibration curve [Eqs. (5) and

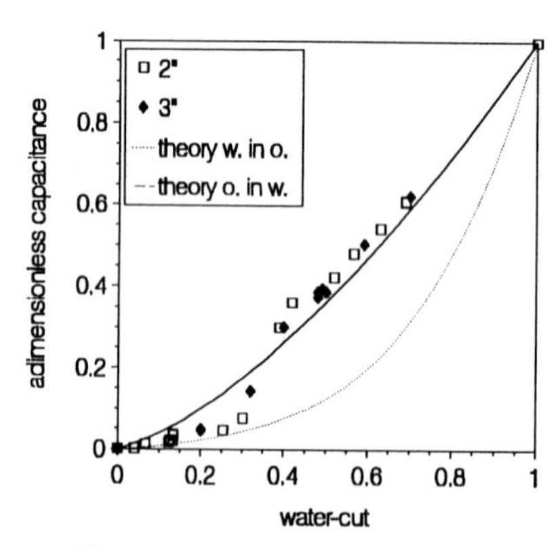

Fig. 4 - 2" and 3" adimensional capacitance

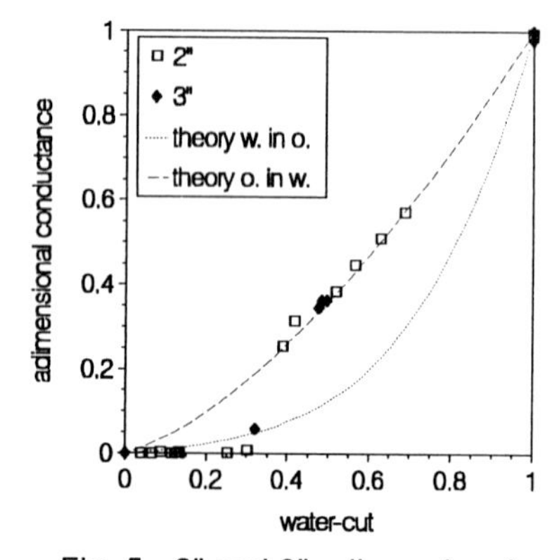

Fig. 5 - 2" and 3" adimensional conductance

(6)]. Taking into account the dimensionless conductance, a similar behaviour can be found. The sharp transition between oil continuous phase and water continuous phase has been experimentally verified. It takes place in the water-cut range of 0.30 ÷ 0.35.

As previously underlined, the average density measurements have been carried out by means of two different tecniques: γ-rays attenuation and static head meter. In both cases the experimental results have been compared with the homogeneous (no slip) model (Fig. 6). The static head tecnique is less accurate. Nevertheless it is based on a very simple and economic system.

Fig. 7 indicates that the homogeneizing effects of the mixer persists throughout the test section or, at least, that the flow can be considered as homogeneous for flowrate calculation by means of Eq. (7).

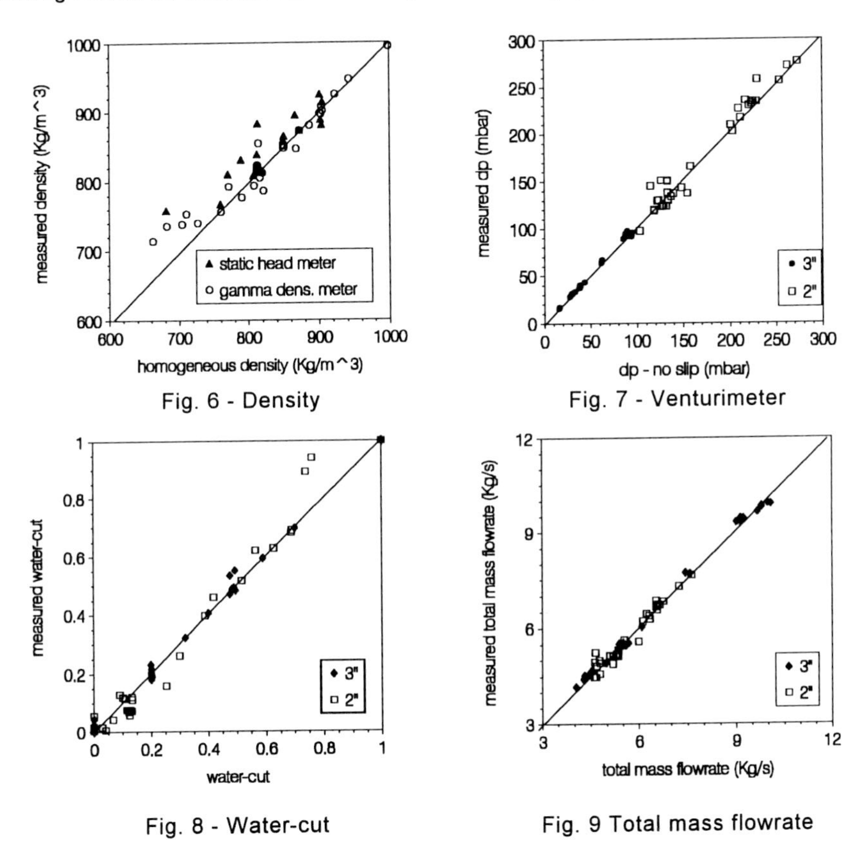

Fig. 6 - Density

Fig. 7 - Venturimeter

Fig. 8 - Water-cut

Fig. 9 Total mass flowrate

Final results are plotted in Figs. 8 and 9 as points representing the flowmeter readouts, mass flowrate and water-cut, against effective quantities. As the diagonal fits quite well the experimental data, the multiphase flowmeter behaviour appears to be satisfactory. The resulting errors in the water-cut and flowrate predictions are reported in Figs. 10 and 11 as a function of water-cut and flowrate, respectively. Most of the experimental data points are comprised within a ± 5% band. Only 3 points out of 71 are clearly out of this band. It seems likely that these large errors be due to mistakes in the set-up of the instruments, rather then to more substantial reasons

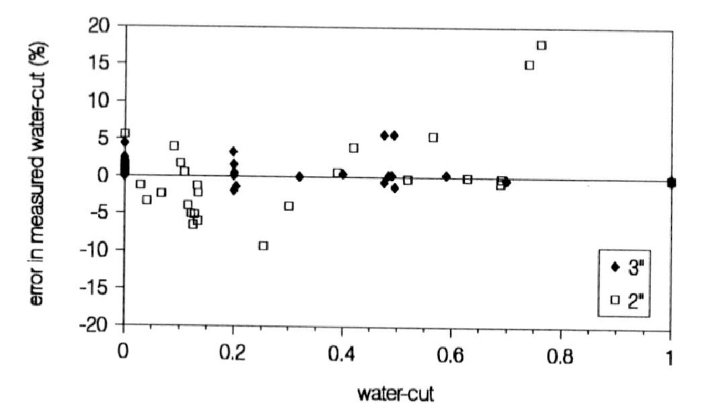

Fig. 10 - Water-cut error

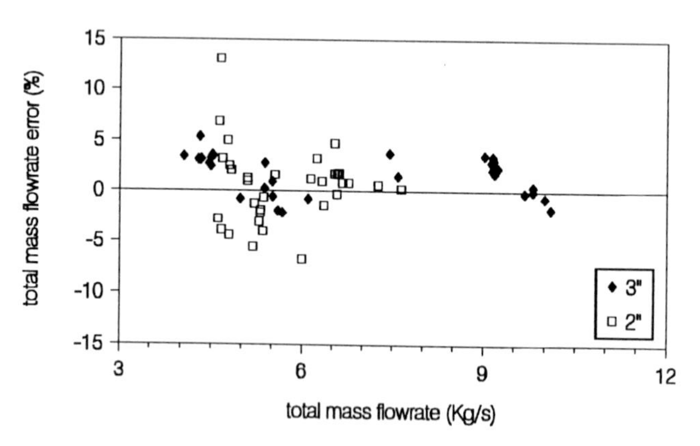

Fig 11 - Total mass flowrate error

6. CONCLUSIONS.

The multiphase flowmeter developed by TEA as a component of a subsea well testing system based on partial separation of the gas phase from the two liquid phases (oil-water) has been thoroughly tested at the BHR Group laboratory.

The tests performed have shown that the flowmeter matches the required specifications with an error less than 5% in the measurement of the total flowrate and the water cut. This result can be attribuited to the intrinsic robustness of the metering system, for which appreciable uncertainties in the estimate of the gas fraction lead to small errors in the measurement of the important parameters, the oil and water flowrates.

Among the positive results of the experimentation, it should be pointed out that the measurement of the pressure drop in the vertical line preceding the Venturi meter is able to give a fairly good estimate of the mixture density, allowing the replacement of an awkward and expensive instrument with a very simple component.

As the specification on the maximum volumetric gas fraction flowing through the metering system is fairly low (15%), it has been possible to match this value with a limited modelling effort. It is believed, and preliminary test confirm this point, that the same set-up, based on the use of a static head meter to measure the mixture density, can be adopted to measure the total flowrate and the water-cut with the required precision, in presence of a volumetric gas fraction up to 30%. To this end improved models or correlations are required both for the Venturi meter and the water-cut meter. In a parallel project, also supported by AGIP and based on the use of a two-beam γ-densitometer to measure the mixture density and the water-cut, in series with the impedance meter, the specification given to the gas fraction arrives to 50%.

REFERENCES.

/1/ MAZZONI A., K. BASSITI, M. BENETTI and F. GIULIANI. (1991) The development of a subsea well testing system. In *Multiphase Production*, A.P. Burns editor, Elsevier Applied Science, London.

/2/ ANDREUSSI P., R. FACCO, C. TALARICO, S. PINTUS, F. GIULIANI and M. BENETTI, (1992) A flowmeter for subsea well testing, *2nd International Congress on Energy, Environment and Technological Innovation*, October 1992, Roma.

/3/ ANDREUSSI P., A. DI DONFRANCESCO and M. MESSIA, (1988) An impedance method for the measurement of liquid hold-up in two-phase flow. *Int. J. Multiphase Flow*,14, 777-785.

/4/ ASALI J. C., T. J. HANRATTY and P. ANDREUSSI, (1985) Interfacial drag and film height for vertical annular flow. *AIChE Jl* 31, 895-902.

/5/ BRUGGEMAN, D. A. G., (1935) Calculation of different physical constants of heterogeneous substances. *Annln Phys.* 24, 636 - 679.

/6/ MALNES, D., Slug flow in vertical, horizontal and inclined pipes, IFE/KR/E-83/002, Institute for Energy Technology, Kjeller, Norway, (1983).

/7/ SANCHEZ SILVA F., P. ANDREUSSI and P. DI MARCO, (1991) Total mass flowrate measurement in multiphase flow by means of a Venturimeter. In *Multiphase Production*, A.P. Burns editor, Elsevier Applied Science, London.

NON-INTRUSIVE MULTIPHASE METERING USING ARTIFICIAL NEURAL NETWORKS [1]

N Beg, J G McNulty - CALtec Limited, Cranfield, Bedford, England

C Sheppard, A Frith - EDS-Scicon Limited, Pembroke House, Pembroke Broadway, Camberley, Surrey, England

ABSTRACT

Transducers monitoring multiphase flow in pipelines produce complex signals containing a wealth of information on the nature of the flow. In order to extract information on flowrates for individual phases from these signals, a considerable degree of signal processing is required. The non-linear nature of the multiphase flow makes its analysis by conventional signal processing techniques difficult. This is due to the high level of understanding of the signals required to implement the processing algorithms.

CALtec and EDS-Scicon, on behalf of a consortium of oil/gas companies and UK Health and Safety Executive, are undertaking a joint venture to predict multiphase flowrates by employing artificial neural networks. A neural network architecture has been developed which when trained on responses from a gamma ray densitometer is able to predict individual phase flow rates of a multiphase fluid mixture.

This paper describes the neural network techniques employed and presents some preliminary results on tests carried out on two-phase (gas/liquid) flow data. Finally the paper reviews the current work programme and indicates where further work is required to successfully exploit the technique.

---

[1] *Presented at the 6th International Conference on Multiphase Production, Cannes, France, June 1993.*

INTRODUCTION

The exploitation of many marginal and deep water reserves requires a move to subsea developments tied to floating, central or land based production facilities if they are to be economically viable. The application of multiphase flow technology to full well-stream transportation is often cited as a financially attractive solution. However for production systems based on minimum processing and on full well-stream transportation to be a practical proposition a high level of confidence in the application of multiphase flow technology will be required. The design, operation and economics of these developments that need to be understood includes transportation, pumping and flow metering. Multiphase flow metering is one of the main technical challenges facing its use is the subject of this paper.

MULTIPHASE FLOW METERING

Measurement of hydrocarbon production is important for reservoir management, production allocation and optimisation of total oil during the producing life of a field. The conventional approach offshore often involves using test separation equipment mounted on fixed or floating platforms. This requires separation of the fluids into single-phase oil, gas and water from where the flow rates can be measured using proven meters such as turbine, positive displacement meters and orifice plates. This approach is not acceptable for many future developments such as unmanned satellite platforms and seabed well completions. Instead novel metering techniques will be required.

One such technique involves metering the fluid as a multiphase flow with, in the short term, accuracies sufficient for reservoir management.

While there have been several developments that address multiphase metering with mixed or segregated flow, there have been few attempts to measure a multiphase flow in its natural state and over a variety of flow patterns. The difficulty is that velocity and phase composition of a cross-section for all three phases needs to be determined. In the course of its work on developing multiphase design methods CALtec has performed a large number of experiments using sensors such as capacitance probes, gamma-ray densitometers, acoustic and pressure transducers. These transducers have the advantage of being non-intrusive, making installation easier and avoiding problems of flow disruption, sensor erosion and pipeline maintenance. However, transducers of these types produce complex signals which inherently contain a wealth of information. While this information is difficult to understand, it does intrinsically contain the characteristics of the natural flow. An important challenge of multiphase metering is therefore establishing an effective means of extracting the flow characteristics from the signal and relating these to flowrates of individual phases.

Conventional filtering and signal processing techniques find such complex signals difficult to analyse without loss of accuracy because of the high level of understanding of the signal required to implement the processing algorithms. Expert system approaches suffer in a similar way in that comprehensive rules which govern the flow characteristics are difficult to specify and ultimately depend on engineering knowledge of the processes. It is precisely for problems of this type that artificial neural networks promise a significant breakthrough.

ARTIFICIAL NEURAL NETWORKS

Unlike conventional data processing techniques which require complex programming, artificial neural networks develop their own solutions to problems through exposure to examples. In effect, neural networks are trained rather than programmed. Although it is only recently that neural networks have been successfully applied to practical data processing problems the technology increasingly shows significant advantages over traditional techniques, particularly when applied to complex non-linear problems.

A neural network comprises a large number of simple interconnected processing elements (neurons) operating in parallel. Each neuron typically combines the weighted outputs from other neurons, passes this through a non-linear transfer function and feeds the result on into the network. Although many different network architectures and training algorithms are in common use, these can be categorised into two basic types; supervised learning systems and unsupervised learning systems. Figure 1 illustrates a popular multi-layer feed forward network.

In supervised learning schemes (shown schematically in Figure 2), the network is presented with sets of training pairs comprising known inputs and corresponding desired outputs. The inputs are entered into the network's input layer of neurons and the neuron activations are fed forward through the network. The activation on the output layer of neurons forms the network output. For multiphase metering the input (problem observation) is the sensor response history while the outputs (known solutions) include the known gas and liquid flowrates. Training of the network is achieved by modifying weights on the internal connections within the network according to the difference between the current output and the desired output for all available examples. In this way, the network learns to produce the required response to each input.

The characteristics of neural networks of particular relevance to multi-phase flow metering include:

- The ability to learn by example; whilst the theoretical understanding of multiphase flow is limited, there exists a wealth of information in data which can be exploited by neural technology.

- The ability to cope with complex non-linear problems; the characteristics of multiphase flow, particularly in the transition regions between flow patterns, display a high degree of non-linearity which is amenable to a neural solution.

- The ability to abstract essential information from noisy data; non-intrusive sensors are characteristically noisy and neural networks have the ability not only to extract signals from the noise but also build up an understanding of the sensor characteristics.

- The ability to generalise from examples; neural networks can interpolate and, to some extent extrapolate from the experience generated from a limited set of examples.

- The ability to fuse data from a number of sources; neural networks excel at the combination and assimilation of data of disparate types and therefore offers the potential to resolve the limitations of single sensor systems.

- The ability to construct effective solutions quickly; being trained rather than programmed, neural networks are less reliant on expert domain experience.

- The ability to work in real time; a trained neural network is highly computationally efficient and ideally suited to modern processing technology.

TWO-PHASE FLOW EXPERIMENTS

CALtec and EDS Scicon are currently engaged in a work programme to employ neural network technology to solve some of the outstanding issues for multiphase flowmetering. An early phase of this work involved assessment of the feasibility of the technique on two-phase (air/water) flows. This involved training and evaluating several EDS-Scicon neural network designs customised for this application, on a set of experimental data generated over a range of different gas and liquid flow rates. The data was generated on a two-phase (air/water) pipeline 2" in diameter and 367m long. The flow loop was horizontal and instrumented with pressure transducers, thermocouples, slug detection probes and gamma densitometers. The data had been collected as part of another project and was not ideal for training and testing. However it did provide a good basis for judging feasibility.

Gamma ray densitometer and pressure time series data recorded after flow was stabilized was provided for 42 air/water test points covering a wide range of gas and liquid flowrates and flow patterns. Measurements of the single-phase air and water flow rates were taken at both the inlet and the outlet of a 16" diameter and 400m long pipeline (shown on the Taitel and Dukler flow pattern map in Figure 3).

The performance of the neural network in predicting gas and liquid flowrates from a combination of pressure and density measurements is shown in Figures 4, 5 and 6. The measured flowrates have been plotted against the neural network predicted flowrates using a jackknifing test technique.

The technique involves training the network on all but one of the available data points and subsequently testing on the single left-out unseen data point. Each data point is omitted from the training set and used for testing in turn. Thus, for the 42 data sets used here, 42 runs were carried out for a complete set of results. The results given are therefore a true indication of the network's ability to calculate flowrates on unseen data.

Figure 4 illustrates the network's predicted mean flow against the measured flow rates, in the form of a set of 42 vectors (one for each of the test points) for a network trained on pressure and gamma densitometer responses. The vectors provide a good graphic illustration of the prediction accuracy. Where the vectors are short the performance is good, where the vectors are long and horizontal the air prediction is poor, and where the vectors are long and vertical the water prediction is poor.

Figures 5 and 6 present the same data on graphs of predicted against measured flowrate for water and air respectively.

In general, the technique has been shown to produce very good gas and liquid flow rate predictions except where the training data is sparse - at the extreme gas and liquid flow rates. For the other points, mean errors of less than 10% of the known flow rates are achieved for both the gas and liquid phases. The mean percentage error is defined as:

$$mean\ \%\ error = \frac{100}{N} \sum_{i=1}^{N} \frac{(Pi-Ai)}{Ai}$$

Where Pi is the predicted and Ai is the actual phase flowrates for the ith test data point ; and N is the total number of data points tested.

SUMMARY

This work has shown the potential value of employing neural network techniques to enhance or, eventually replace some signal processing techniques used in multiphase flowrate prediction. However, the accuracy of the neural network prediction was limited by the amount of the data available.

The work has shown the importance of good data, correct neural network design and knowledge of the application (in this case multiphase flow metering). Neural networks are not a 'black-box' solution to metering, but, when properly applied, they appear to be able to resolve some of the problems associated with modelling the flow.

The main problem with any sensor used to infer flow information is that the response of the sensor is flow pattern dependent and, in some cases, the presence and extent of a second, liquid phase, can affect its performance. Although, flow pattern is often a problem for sensors when used in a conventional manner, for neural network enhanced sensors, it is potentially a useful feature. In the test data used here the main errors in measurement accuracy where not so much due to flow pattern as limited data on a flow pattern on which to train the neural network.

Recognising the successes and limitations of the results so far, EDS-Scicon and CALtec are involved in a continuing programme of work supported by a consortium of oil and gas companies to develop technology which will ensure its successful application. The programme is currently addressing a number of areas which will maximise the way in which neural networks can be employed. This work includes:

- Demonstrating the technical feasibility to predict flowrate in three-phase flows
- Validating the neural network design against field data
- Identifying training data and sensor requirements

The next phase of work will include:

- Preparing a prototype system(s), comprising sensors, neural network designs, etc
- Developing scaling laws for prototype system(s)
- Field validation trials

CONCLUSIONS

1. Neural networks can be successfully employed with conventional sensors to extract unique information about multiphase flow rates.

2. The technique can be used to interpret the information from single sensors.

3. Measurement accuracies better than $\pm10\%$ are possible. However performance is critically dependent on the availability of a good quantity of well distributed training data.

4. The success of the technique is dependent on the selection and optimisation of the neural network architecture and on the pre-conditioning of the sensor measurement information to provide the network with a small number of relevant features.

ACKNOWLEDGEMENT

The authors would like to thank the sponsors of the 'Neural Network Enhanced Multiphase Flowmeter' consortium for allowing this paper to be published. However, the views in the paper are those of the authors and do not necessarily represent those of the sponsors.

c:\JGM\cmc\gerry\neurnet

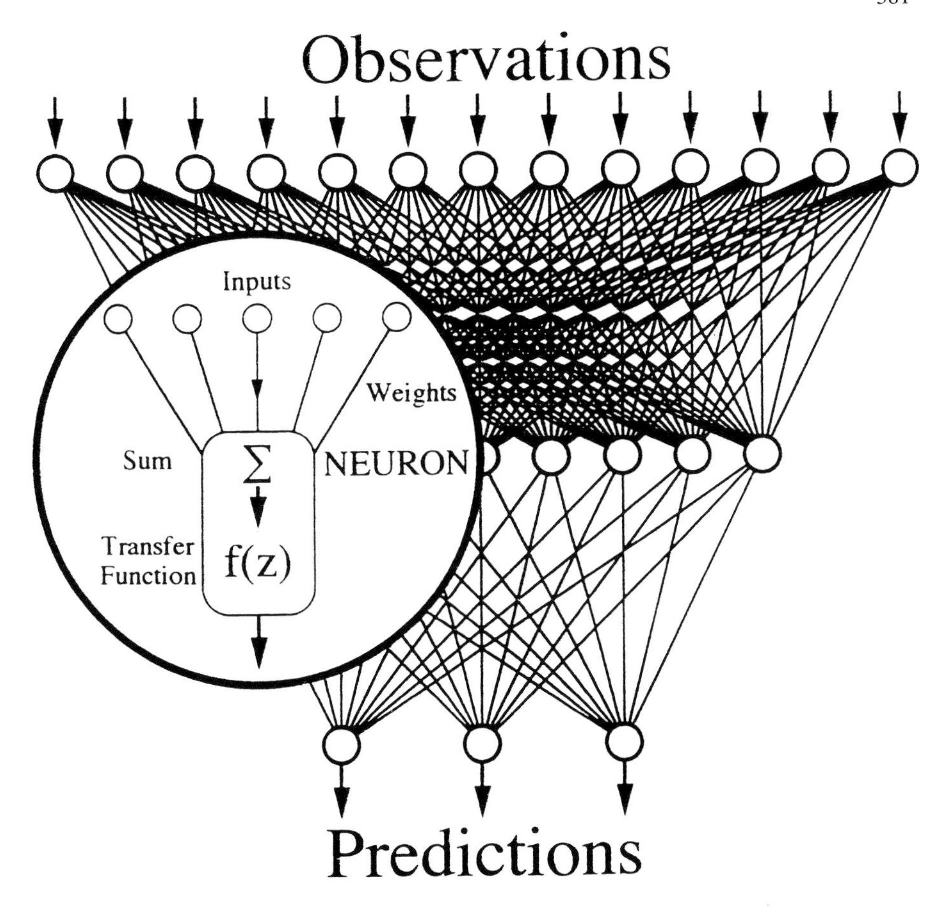

FIGURE 1 MULTI-LAYER FEED FORWARD NEURAL NETWORK

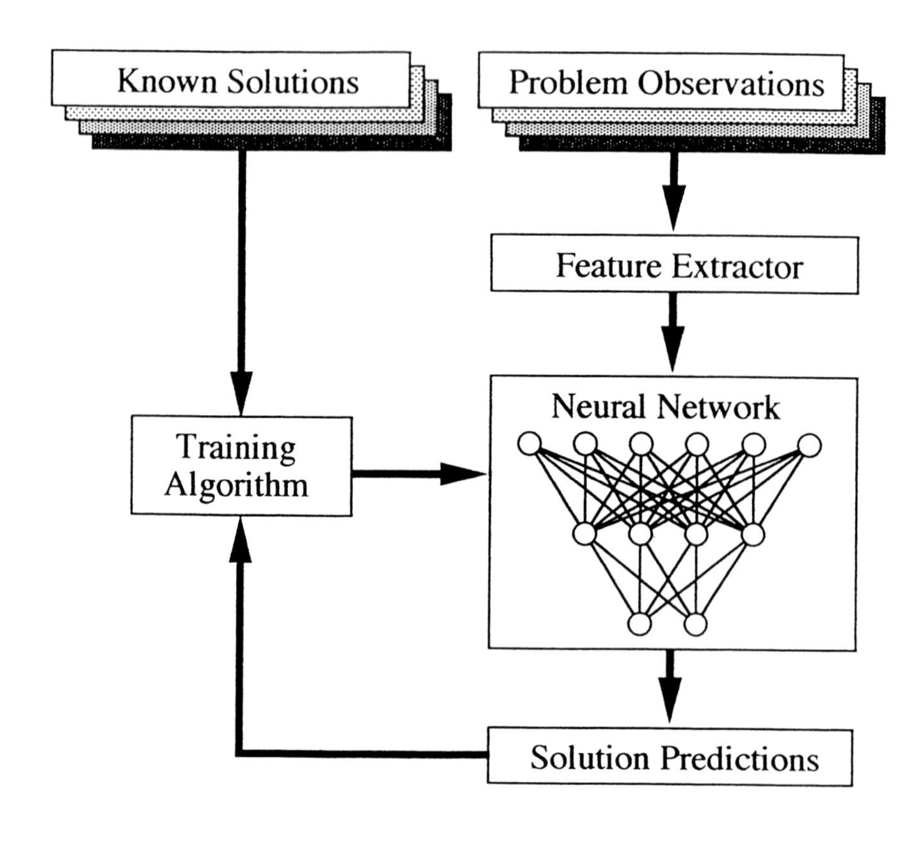

FIGURE 2 SUPERVISED LEARNING NEURAL NETWORK

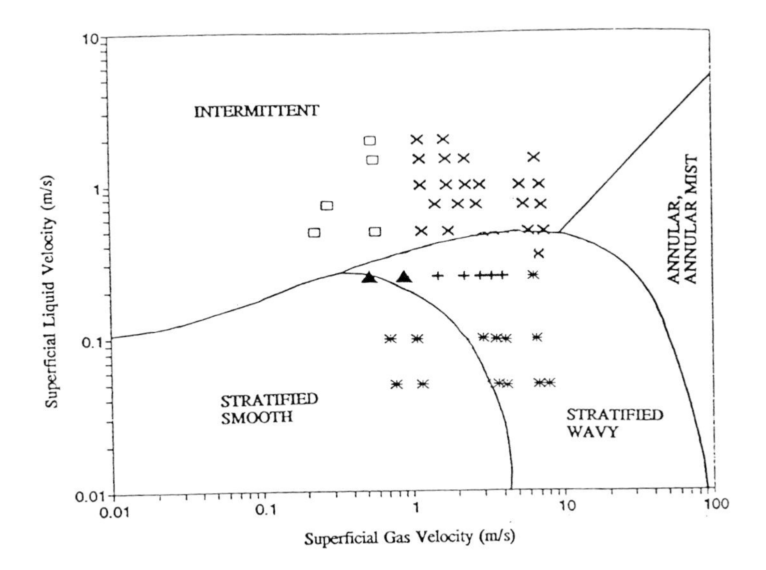

FIGURE 3 DATA SET SHOWN ON FLOW PATTERN MAP

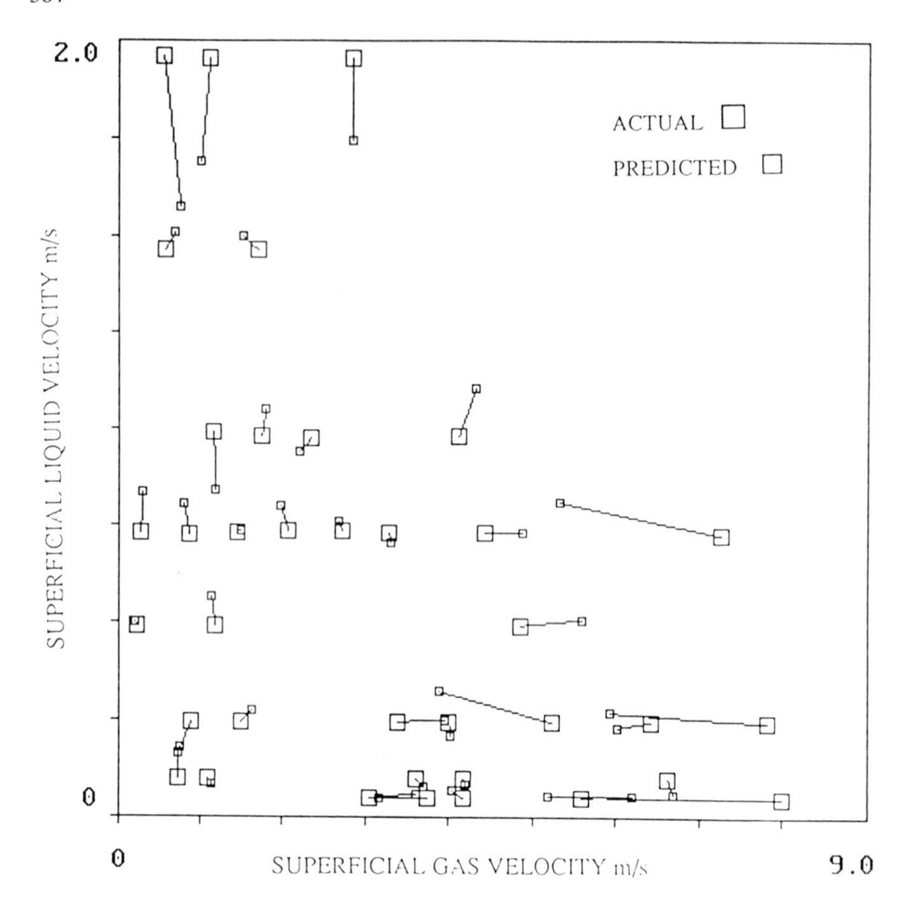

FIGURE 4 VECTOR PLOT OF GAMMA AND PRESSURE PREDICTIONS

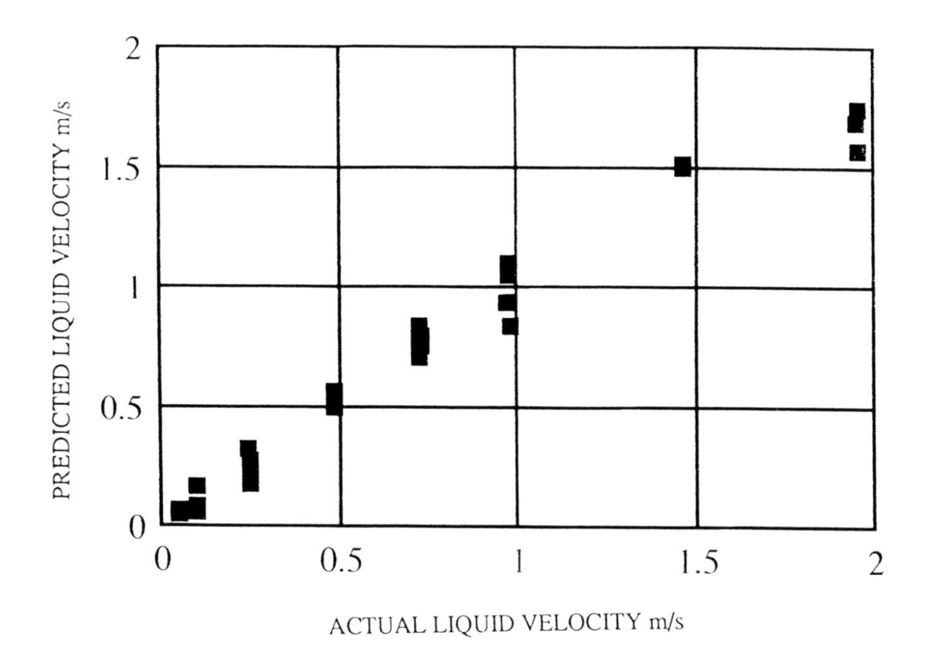

FIGURE 5 WATER VELOCITY PREDICTION ACCURACY

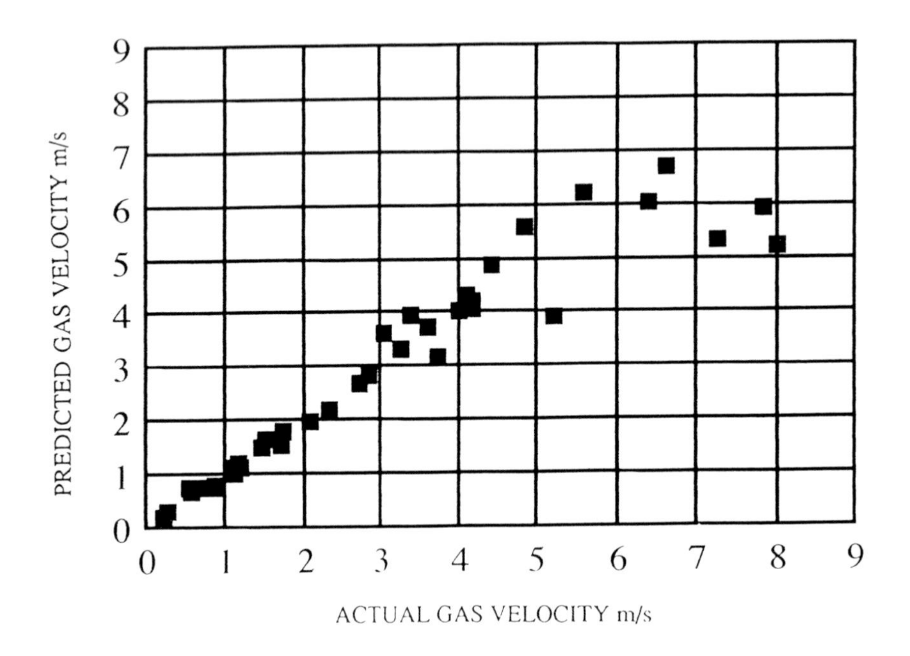

FIGURE 6 AIR VELOCITY PREDICTION ACCURACY

MULTIPHASE BOOSTING

STUDIES IN MULTIPHASE PUMPING WITH SCREW PUMPS

Wietstock, P. Witte, G.

GKSS-Forschungszentrum Geesthacht GmbH; Geesthacht, Germany

Multiphase technology has been developed mainly for the oil industry. The multiphase pump is an essential part of a multiphase transport system . Especially designed for application in multiphase transport systems are modified centrifugal pumps and rotating positive displacement pumps.

At the GKSS-Forschungszentrum Geesthacht GmbH a research programme began 1988 ,in co-operation with industrial partners, to investigate the static and dynamic behaviour of multiphase pumps .

This report presents some selected hydraulic energetic aspects on different pump systems and fundamentals for a characterisation of multiphase pumps.

For pressure generation in multiphase flow systems positive displacement pumps are best suited. To meet today's requirements, multiphase pumps are tested with high gas fractions at the pump inlet .

The GKSS - Multiphase - Pump - Test - Facility is designed for

- pumps of original size (actually up to 1000 kW)
- real stationary working conditions (capacity up to 500 m^3/h and Δp up to 70 bars)
- simulated unstationary conditions (slug-flow)

The GKSS - Multiphase-Model-Test-Facility allows operation of

- fundamental research on multiphase flow
- pumps of scaled size (actually up to 100 kW)
- simulated stationary and unstationary working conditions.

Both facilities allow tests to be carried out with a high accuracy and at high reproducibility.

Improvement of screw pumps for multiphase transport requires detailed information on the conditions of flow inside the screw chambers, especially in relation to the pressure increase. Pressure distributions at fixed measuring positions will be presented.

Moreover results related to the behaviour of flow capacity, power consumption and slug simulation tests will be discussed.

1. Introduction

Multiphase technology has been developed mainly for the oil industry /1 - 4/, to provide technical-economical solutions for exploitation of hydrocarbon fields which have an output which is not economically efficient or technically possible today. The goal of this new technology is an ecologically beneficial and robust new system. An essential part of a multiphase transport system is the multiphase pump.

Besides the application of hydrocarbon transport the use of multiphase pumps in chemical process engineering is also possible / 5 /.

In the past few years several contributions were published / 6-9 /, which describe the progress on multiphase pump developments. However, there is very little informations on the transformation of knowledge about the behaviour of single phase transport with pumps and compressors to multiphase transport applications. Furthermore, there is no comparison of hydraulic energetic aspects of different pump systems, which can be used in multiphase transporting.

At the GKSS-Forschungszentrum Geesthacht GmbH "multiphase pumps" have been investigated since the 1970's ,beginning with pumps used in the field of nuclear power generation. With this background a research and development programme started in 1988 in co-operation with industrial partners to investigate the static and dynamic behaviour of multiphase pumps for hydrocarbons. With these test results the analytical description of the behaviour of multiphase screw pumps was started. The activities were completed in close co-operation with the University of Hannover. The result was a numerical simulation model for screw pumps.

This paper presents results related to the behaviour of flow capacity, power consumption and also pressure distributions on fixed measuring positions as well as the integral pressure increase along the screw length. Dependencies between pressure values, gas content, inlet and differential pressure are illustrated and conclusions concerning the condition of flow and separation effects inside the pump will be discussed.

2. Pumps for multiphase transport technology

2.1 Goals, tasks and different multiphase pump systems

The unprocessed multiphase well stream fluid (oil, gas, water, solids) will be pumped by a multiphase pump from a location of sensitive environmental conditions over a long distance through a multiphase pipeline to a processing plant with a minimum of environmental side-effects. This new technology provides an economical and safe alternative for many problems in off-shore and onshore-field developments. The essential component of a multiphase-flow-transport-system is the multiphase pumping unit.

Sufficient viscous fluids convert flow energy into thermal energy during the transport in a pipeline; energy input has to compensate for the decreasing flow energy for transport over long distances. The present state of technology provides methods for

determination of the necessary power input for one phase fluids by pumps and compressors (partially with solids).
For multiphase flow in the hydrocarbon industry different boosting systems are used. In this report only two features of such systems are examined, namely

- rotodynamic pumps
- positive displacement pumps

Centrifugal pumps are fluid dynamic systems. To this kind of machinery belong all devices, which are passed by a fluid or with solid contaminated fluid, which transform mechanical energy into fluid energy. This changing of fluid direction is typically done by a vane.

Gas - liquid flow in centrifugal pumps may result in severe head degradation due to stratification in the impeller. In this case the energy is mainly transferred to the liquid phase. To avoid the head degradation, the machinery should keep the two-phase flow as homogeneous as possible throughout the impeller and diffuser. Nevertheless, head degradation will occur with increasing gas fraction.

Pump head is proportional to the density of pumped fluid and the specific pump work: (Symbols used are explained in the nomenclature at the end of the paper).

$$p_D - p_S = \Delta p = \rho * Y_p \tag{2.2.2 - 1}$$

For a horizontal pump neglecting the kinetic energy difference between suction and discharge side can be stated

$$Y_p = Y \tag{2.2.2 - 2}$$

$$Y_{Sch} = Y * \eta_h \tag{2.2.2 - 3}$$

and

$$Y_{Sch} = u_2 * c_{2u} - u_1 * c_{0u} \tag{2.2.2 - 4}$$

$$\Delta p = \frac{\rho}{\eta_h} * (u_2 * c_{2u} - u_1 * c_{0u}) \tag{2.2.2 - 5}$$

Consequently the pressure difference is a function of pump construction, pump speed and density of the pumped fluid.

This means that a rotodynamic pump can reach appreciable pressure differences in the case of multiphase pumping if
- the pumped fluid is homogeneous
- pump speed is very high
- density of pumped fluid is not arbitrarily low.

To get a constant differential pressure by the multiphase pump in the case of varying gas fractions the pump speed must be variable.

Positive displacement pumps work on the principle of enclosing a defined volume on the low pressure side and moving it to the high pressure side by adding energy. The pumped discharge flow of a positive displacement pump is independent of the density of the pumped fluid. The pumped flow is proportional to the pump speed:

$$Q_{th} = V_{Geom} * n \qquad (2.2.2 - 6)$$

The differential pressure of the positive displacement pump is independent of the pumped flow. Therefore the density of the pump flow can be arbitrarily low and the pump speed may be constant for different gas fractions on the pump inlet side.

Fig. 1 shows the correlation of working points for a multiphase rotodynamic pump and a multiphase positive displacement pump. It clearly shows, that constant working conditions can be reached more easily by a positive displacement pump than by a rotodynamic pump.

The knowledge of the behaviour of rotodynamic and positive displacement pumps results from single-phase fluid pumping. The determination of energy input on multiphase fluids requires more complex considerations.

2.2. Characterisation of multiphase pumps

For a description of the fluid dynamic state at an arbitrary point of the transport line (for example the suction side of a pump) more parameters for multiphase flow are needed than for single phase flow.

2.2.1. Flow, pressure and temperature

The capacity of centrifugal and positive displacement pumps is defined as the effective time averaged capacity at the discharge side of the pump. Considerable changes in density require a conversation to inlet conditions.

For a description of the real volume flow of a fluid gas mixture it is necessary to provide input data of gas fractions as well as statements on the pressure and temperature conditions at pump inlet and outlet.

2.2.2. Gas fraction , Gas Oil Ratio

There are different possibilities for describing the gas fraction (gas content). In this paper the description of two phase flow uses the definition:

gas fraction = volume flow of gas / overall volume flow (2.3.2 - 1)

$$\alpha = \frac{Q_G}{Q_G + Q_{Fl}} = \frac{Q_G}{Q_{Mix}} \qquad (2.3.2 - 2)$$

That means that the volume flow is given by:

$$Q_{Mix} = \alpha * Q_{Mix} + (1 - \alpha) * Q_{Mix} \qquad (2.3.2 - 3)$$

More exactly equation (2.3.2-2) should be written as:

$$\alpha = \frac{Q_{G,p}}{Q_{G,p} + Q_{Fl}} \qquad (2.3.2 - 4)$$

The volume flow of gas at a given pressure is related to the volume flow of the mixture at the same pressure.

In the oil industry, gas volumes are expressed normally in standard cubic feet (scf), that is the volume of the gas measured at 0 ^{o}C and one atmosphere (1 scf = 0.02832 m^3). Oil is measured normally in barrels (1 barrel = 0.15899 m^3). A commonly used term in the oil industry is the Gas Oil Ratio (GOR), which is the number of scf of gas per barrel oil. The GOR varies enormously from hydrocarbon field to hydrocarbon field.

$$GOR = \frac{Q_{Gas\,(1bar)}}{Q_{Oil}} \qquad (2.3.2 - 5)$$

$$Q_{Gas\,(pbar)} = Q_{Gas\,(1bar)} * \frac{p_0}{p} \qquad (2.3.2 - 6)$$

$$\alpha = \frac{1}{(1 + \frac{p}{p_0\,GOR})} \qquad (2.3.2 - 7)$$

Fig 2 and fig. 3 show the correlations between a, GOR and gas fraction, respectively, under standard conditions.

Example:

$GOR = 28\ \frac{scf}{bbl} = 5\ \frac{m^3}{m^3}$ Pump inlet pressure = 20 bars

In this case the gas fraction for the real inlet pressure is $\alpha = 0.20$. Converted into 1 bar condition yields $\alpha_{1\,bar} = 0.83$, but this is a theoretical figure, which does not describe the reality. Speaking of volumetric gas flow rate or gas fraction in the following means α and not $\alpha_{1\,bar}$.

2.2.3. Solid fuel content

Besides the fluid phase, a solid phase can also occur. Information on realistic values for sand content, for example, are rare. Known values vary from 43 mg solid / l oil up to 210 mg solid / l oil /2, 11/.

2.2.4. Efficiency

For a comparison of similar technical products, efficiencies are needed. The efficiency is normally defined as:

$$\text{Efficiency} = \frac{\text{actual power}}{\text{input power}} \tag{2.3.4 - 1}$$

or

$$\text{Efficiency} = \frac{\text{power of ideal machine (process)}}{\text{power of real machine (input power)}} \tag{2.3.4 - 2}$$

It is possible to define two different powers for ideal machines:

$$P_{Id} = Q_{Mix} * \rho_{Mix} * (Y_p)_{rev} \tag{2.3.4 - 3}$$

$$P_{Id} = Q_{th} * \rho_{Mix.} * (Y_p)_{rev} \tag{2.3.4 - 4}$$

This paper uses definition (eq. 2.3.4 - 3).

In this case $(Y_p)_{rev}$ is the specific work of a adiabatic isentropic compression. This is the minimum work for a loss-free machine which is impervious to heat.

With that the isentropic power of a pump with multiphase flow is:

$$P_{Id} = Q_{Mix} * \rho_{Mix} * (Y_p)_{rev} = \alpha * Q_G * \rho_{G.} * (Y_p)_{rev\,G} + (1-\alpha) * (\rho)_W * Q_W * (Y_p)_{rev\,W} \tag{2.3.4 - 5}$$

$$\pi = \frac{p_2}{p_1} \tag{2.3.4 - 6}$$

$$P_{Id} = Q_{Mix} * p_1 * \left[\alpha_{In} * \frac{\kappa}{\kappa-1} \left[\pi^{\frac{\kappa-1}{\kappa}} - 1 \right] + (1-\alpha) * (\pi - 1) \right] \tag{2.3.4 - 7}$$

It is also possible to take the power of the ideal machine as the isotherm power. Then the efficiencies become smaller.

The real machine should have the power:

$$P_{re} = Q_{Mix} * p_1 * (\pi - 1) \tag{2.3.4 - 8}$$

Then an isentropic efficiency can be defined:

$$\eta = \frac{\alpha_{In}}{\pi - 1} * \frac{\kappa}{\kappa - 1} * \left[\pi^{\frac{\kappa - 1}{\kappa}} - 1 \right] + \left(1 - \alpha_{In} \right) \qquad (2.3.4 - 9)$$

This efficiency is for $\alpha_{In} = 0$

$\eta = 1$

and for $\alpha_{In} = 1$ follows

$$\eta = \frac{1}{\pi - 1} * \frac{\kappa}{\kappa - 1} * \left[\pi^{\frac{\kappa - 1}{\kappa}} - 1 \right] \qquad (2.3.4 - 10)$$

Fig. 4 shows $P_{id} / Q_{mix}*p_1$ and $P_{re} / Q_{mix}*p_1$ for different gas fractions.

2.4. Need of experimental research

Besides a comparison of hydraulic energetic aspects of different multiphase pump systems which is possible without experiments, the application of multiphase pumps needs exact knowledge about the behaviour under all working conditions and the influences of construction details.

For screw pumps exact knowledge is required for
- whether the clearances are fluid filled or not,
- in what manner the clearances change under working conditions and the influence on clearance loss,
- in what manner the hydraulic losses can be predicted satisfactorily and minimised,
- optimisation of the number of chambers.

The mentioned above questions are only part of other questions which can only be answered by experimentation.

3. Experimental possibilities with multiphase fluids at GKSS

3.1. Multiphase pump facility

The multiphase pump test facility (MPV = Mehrphasen-Pumpen-Versuchsstand) was conceived for testing original size plants and works with a water nitrogen mixture.

The possibilities of experimental investigations are multiple, the test facility can be used for
- fundamental investigations of the behaviour of multiphase flow in pumps and valves
- testing the integral stationary behaviour of pumps (experimental determination of pump characteristics) and valves

- testing the integral unstationary behaviour of pumps and valves (experimental determination of plant response in the case of slug flow)
- determining the oscillating properties of pump plants and valves, for minimising sound emission and for monitoring purposes
- testing measurement devices for application under real working conditions.

The MPV(Fig.5) was built as a closed loop system. The loop is divided into two different systems for variation of gas fractions at the pump inlet between 0 and 100 % and for measurement of the two single phases. Only single phase measurements of mass flow secure accurate measurements for experimental purposes.

The facility can be operated alternatively with mixtures of air or nitrogen and water. Maximum working pressure is 96 bars and 180 ^{O}C between the discharge side of the pump and the separator. The pressure on the suction side of the pump is limited to 20 bars and also 180 ^{O}C. All components coming in contact with the test fluid are made of stainless steel resp. plated with stainless steel. It is possible to use different materials for the test pump.

A mixing unit mixes both fluid components (liquid - gas) just before the fluid enters the test pump. The two phase flow mixture now passes the test pump, a check valve directly at the pump discharge side and finally a throttle section. The throttle section consists of a throttle valve and throttle flap. The check valve prevents back-flow of compressed fluid after shut down. Shut down may be possible after a failure of power supply. Back-flow of compressed fluid can raise the pump speed in a wrong direction and destroy the pump.

The throttle valve provides the possibility to throttle the two phase mixture between 0 - 64 bar. After throttling a separation of the mixture flow takes place in the separator. In the separation unit the liquid phase is extracted through the bottom part while the gas is removed through the upper part.

The pump is driven by a speed controlled induction motor with a variable power up to 1000 kW. Different heat exchangers installed in both water and gas loops extract the heat energy generated by the system. These heat exchangers have been designed in such a way that regardless of operational conditions the temperature of the mixture at the inlet can be set to below 50 ^{o} C.

GKSS - Multiphase - Test - Facility also allows simulation of slug conditions at the inlet. In this case, water from a water reservoir is injected just before the pump inlet and when the loop is operated with a high gas content by means of an extra gas reservoir.

Dynamic tests, such as frequency analysis, complete the test capabilities. All important data of the pump and the test loop are recorded online and can be processed in a various manner.

Past experiences with several multiphase pumps have shown that the concept of the facility to test pumps without automatically controlled flows (valves) and pressures leads to stable operational conditions within the test loop. Variation of flow and pressure is performed by remote controlled valves.

GKSS - Multiphase - Test - Facility can also be used for investigations on other components, as long as the original multiphase mixture and its flow conditions are of secondary importance and only the components performance is of concern. For installation of test components pipe diameters from 300 to 500 mm can be operated.

3.2. Multiphase model facility

The multiphase model facility should be used for examinations of detailed problems in the field of multiphase pumps and for the development of methods for multiphase measuring techniques. The main parts of the facility are made of glass.

The multiphase model test facility is a combination of a closed water circuit and an open air circuit. In the water circuit a booster pump is installed for tests on pumps over a wide range of inlet conditions (two quadrant operation).

Water and air are mixed before entering the test pump for measuring mass flow in single phase conditions and for varying the gas fraction between 0 an 100 %. After passing the test pump the fluid flows through a pressure-proofed stainless steel pipe and a throttle valve. The performance data in this part of facility are DN 100 and PN 40.

Behind the throttle valve the glass part of the test loop begins. The first part is built out of a horizontal DN 100 pipe of about 10 meters and a short vertical pipe. The latter part is used for observing different kinds of flow patterns and for the development of multiphase measuring devices.

The flow sheet of the facility is shown in fig. 6.

4. Conducted tests on multiphase pumps at GKSS

4.1. Test with JHB-pumps

In 1988 Bornemann (JHB) and GKSS started a co-operation for developing multiphase screw pumps.

In 1989 a first prototype of a multiphase screw pump from Bornemann ("Tritonis" pump) was tested with a water-nitrogen mixture at the GKSS multiphase pump test loop. Due to time limitations the first test series only included some function and running tests. After successful test runs at GKSS the pump "Tritonis" started onshore in situ operation in Tunisia (Sidi El Itayem).

The second Bornemann pump "Pegasos", a pump of the same design and construction as the "Tritonis" pump was exhibited in Norway and Germany in 1990. After that the pump was installed on the GKSS-Multiphase-Test-Facility (fig. 7) and prepared for extensive scientific examination of pump behaviour, and included experimentation on the most difficult working conditions /9, 12-16/.
Performance data of JHB- pumps are:

max speed	=	1800	rpm
theoretical inlet flow	=	500	m^3 / h
max. gas inlet ratio	=	96	%
max. power	=	1000	kW
differential pressure	=	64	bar
min. inlet pressure	=	2	bar
max. inlet pressure	=	40	bar

4.2. Tests with Leistritz-pump

Prototype L4H - main data:

max inlet pressure	=	25	bar
differential pressure	=	60	bar
capacity at $\nu = 1 mm^2/s$	=	225	m^3/h
max speed	=	3000	U/min
power	=	550	kW
Total pump efficiency	=	68	%
gas fraction	=	0 - 95	%
max solid size	=	100	µm

5. Test results

5.1 Flow behaviour at different gas fractions

For the "Pegasos"-pump Fig. 8 shows the normalised (related to the theoretical flow of the pump) gross inlet flow as a function of the normalised (related to the maximum design differential pressure) differential pressure at a pump speed of 1200 rpm for four different gas fractions. The pump inlet pressure was 3 bar; the fluid was a water-nitrogen mixture.

The flow-differential pressure behaviour of a screw pump is determined essentially by the construction of the inlet chamber region. The difference between theoretical and real flow volume is determined by the back-flow between the first chamber and the suction room. The back-flow is a function of the pressure distribution along the screw between inlet and outlet pressure. As shown in figs. 17 and 18 the distribution is a function of gas content and inlet pressure. As shown the flow differential pressure characteristic approaches the theoretical characteristic when the gas fraction increases, and this means that the pressure difference between the suction region and the first screw chamber becomes smaller.

This correlation shows that the experimental determination of the different chamber pressures for different screw pumps together with the experimental determination of the flow differential pressure characteristic gives valuable informations on construction parameters of the pump.

Fig. 9 shows the flow behaviour of the "Pegasos"-pump for different pump speeds. The gas fraction in this case was 20 %. The relative back-flow decreases with increasing pump speed. The absolute flow loss is only a function of the pressure difference between first chamber and suction room of the pump. This characteristic does not differ from the known characteristic single-phase flow behaviour (Fig.10).

5..2 Power supply and efficiency

Fig. 11 shows the normalised power supply of a multiphase screw pump ("Pegasos" pump) as a function of the normalised differential pressure for different gas fractions. The influence of the inlet gas content on power supply is only very small.

Fig. 12 shows the known normalised power supply as a function of normalised differential pressure for different pump speeds. Gas fraction was 0 % in this case.

5.3 Feeding screw pressure measurements

5.3.1. Measuring technique

Seven pressure gauges were mounted along one of the double thread feeding screws with 4 1/2 threads each (fig. 13). These gauges give a pattern of the pressure increase in every feeding chamber, because a half length of the feeding chamber passes the measuring position twice per revolution. As a consequence the measuring position in the chamber changes periodically in axial direction. The gauges were mounted in boreholes through the double-walled casing in such a way that gauges´ membranes were flush with the inside of the inner casing. First measurements with longer borings were tried, but no reproducible results were achieved. Cavity resonances appeared due to the unavoidable accumulation of gas in the measuring channels. Their amplitudes completely covered the pressure pulsations to be measured. The used pressure gauges were small and robust piezoresistive systems. All these signals were stored digitally on a magnetic tape together with a revolution trigger and the "static" operating data of the pump. In addition, a two channel on-line analysis in the time and frequency domain was made for control purposes.

Averaging, triggered by screw revolutions, gave an excellent smoothing of the very noisy original data. This was due, on one hand, to the unfavourable signal to noise ratio (the static pressure is much higher than the dynamic values to be measured) and, on the other hand, to the very noisy background as a consequence of oscillations of gas bubbles and collapsing cavities. There signals were transformed into the frequency domain (FFT) for analysing the shape. The results showed the formation of harmonics which was strongly dependent on the operating parameters, such as gas content, differential pressure and on the measuring position. This can be expected from the pressure patterns shown in the following. Measurements made during very low inlet pressures and low gas contents gave problems. In these cases extreme oscillations of the dynamic pressure signals were observed. Periodic pressure oscillations in the feeding chamber appeared together with slower variations of the mean value (fig. 14). This seems to be a phenomenon of the test loop itself due to an inhomogeneous gas-water-mixture at lower inlet pressures. The decreasing influence of this inhomogenity with increasing gas contents can be explained by the increasing elasticity of the total system.

5.3.2. Measured pressure patterns

Fig. 15 shows averaged pressure patterns p_{dyn} at different measuring positions (see also fig. 13) during a half revolution of the double thread feeding screw for different gas contents (80, 90, 95%), a constant speed (3000 rpm) and a constant inlet and differential pressure (5 resp. 30 bars). The static pressure (DC-part of the signal) is hereby omitted. Notice the expected pressure increase when the feeding chamber passes the measuring position, up to a maximum value when the tooth of the screw reaches this position. A pressure decrease down to the initial value follows along the width of the tooth. The sum of all measured pressure increases per feeding chamber corresponds to the total head (differential pressure) of the pump. In doing so it should be considered, that only half lengths of the feeding chambers could be observed. The different relationship between the various maximum values is remarkable in this figure 15, depending on the measuring position and on the gas content. Also the shape of the pressure increase or decrease depends on both these parameters.

Fig. 16 shows pressure patterns similar to those in fig. 15, but for an another inlet pressure (10 bars) and with an additional gas content of 60%.

Fig. 17 shows the pressure pattern related to the differential pressure for different measuring positions and gas contents at a constant speed and constant inlet pressure. This normalisation is allowed with respect to the inaccuracy of the measurement. It leads to a reduction in the number of measuring parameters and to a simplification of the following evaluation of pressure increase along the screw.

5.3.3. Pressure increase along the screw

Fig. 18 shows the measured pressure increase Δp_{dyn} (amplitude of the pressure pattern) versus the normalised measuring position. This is the distance from the screw inlet in relation to the total screw length. The speed (3000 rpm) and the inlet pressure (10 bars) are constant, the total head (about 30 to 60 bars) and the gas contents (60, 80, 90, 95%) are changed.
As a measurable pressure increase could not be monitored before the middle of the screw with the above mentioned gas contents, the normalised length starts with 0.45 for both figs. 18 and 19.
Standardising the measured pressure increase Δp_{dyn} with the differential pressure Δp and additionally with the accompanying related screw length $\Delta L/2L$ (ΔL = observed screw length) leads to

$$\Delta p_{norm} = \frac{\Delta p_{dyn}}{\Delta p} * \frac{\Delta L}{2L} \qquad (5.1.3\text{ -1})$$

Fig. 19 shows the normalised pressure Δp_{norm}. The integral of Δp_{norm} for a constant gas content corresponds to the total head and is in this normalised form equal to "1".

$$\int_0^1 \Delta p_{norm}\, d\left(\frac{1}{L}\right) = 1 \qquad (5.1.3\text{ - }2)$$

5.3.4. Discussion of the measuring results

In a liquid with a gas content of 0% the pressure increase is constant versus the screw length (meaning "1" in normalised form). The measured results show that the pressure increase along the screw length moves to the outlet side with increased gas content. This effect, which is well known from theoretical considerations, causes a remarkable compression at the very end of the screw if the gas content is for example 80%. Furthermore, with higher gas contents (i.e. 95%) the maximum of the pressure increase moves to a position in front of the screw outlet. This is due to gas leakage in the unavoidable gap between the screws and between screws and housing. There is not enough liquid available to "seal" these gaps.

The measured results show that an optimisation of the number of feeding chambers as a function of the gas content is possible.

Another aspect is the possibility of indirectly determining of the gas content by measuring the pressure increase. Conventional methods for the determination of the gas content are difficult at minor liquid portions. It is just here, that the pressure increase is very sensitive. This measuring method depends on the geometry of the pump, and for its application a calibration at different inlet pressures and speeds is necessary at the beginning. Using a computer for measuring conditioning and for comparison of the calibration curves with actual values, an exact determination of the gas content should be possible, which could fulfil the high demands of a fiscal measurement.

5.4 Remarks on slug behaviour

A typical attribute of a slug flow is the periodical sudden change of liquid flow into gas flow and vice versa. This liquid frontier leads to strong loads on structures and results in pressure changes if this frontiers must change the direction.

The principle of the screw pump does not lead to any special requirements in power supply aside from the possibility of structure load in the case of changes in the direction of flow in the suction part of the pump. Therefore, no changes of flow or power can occur in a pipe without throttle valve in the case of slug flow. This has been experimentally confirmed.

6. Outlook and acknowledgements

The presented results represent only a small part of the total work. The aim of this presentation is to give a small contribution for application of multiphase pumps in hydrocarbon field technology.

The experimental work was performed with pumps from Bornemann and Leistritz. The participating companies are gratefully acknowledged.

7. Nomenclature

Δl	part of the screw
l	screw co-ordinate
L	screw length
n	pump speed
p	static pressure
Q	pump capacity
P	power
U	revolution
Y	specific pump work
α	gas fraction - gas content
κ	adiabatic coefficient
π	pressure ratio
ρ	density
η	efficiency

Indices:

D	discharge side
Fl	fluid
G	gas
Geom	geometrical
h	hydraulic
Id	Ideal
max	maximal
Mix	mixture
S	suction side
p	pressure
P	pump
re	real
rev	reversibel
Sch	vane
th	theoretical
W	water

8. Reference

/1/ Daniels, L. C.
Multiphase Flow and Offshore Oil Production
Conference Documentation
Multiphase Operations Offshore; 11/12 December 1991, The London Marritt Hotel

/2/ Goodridge, R.A.
Operational Experience of a Multiphase Pump on Live Well Production
Conference Documentation
Multiphase Operations Offshore; 11/12 December 1991, The London Marritt Hotel

/3/ Bratu, Ch.; Gie, P.; Durando, P.; Buvat, P.
Operational Multiphase Pumping Integration
Conference Documentation
Multiphase Operations Offshore; 11/12 December 1991, The London Marriott Hotel

/4/ Bratu, Ch.; Gie, P.; Durando, P.; Buvat, P.
Operational Multiphase Pumping Integration
Conference Documentation
4th EC Symposium; Berlin 3-5 November 1992

/5/ Prang, A. J.
Rotary Screw Pumps for Multiphase Products
Conference Paper
12th International Pump Technical Conference, 17 - 19. April 1991, London

/6/ De Donno, S.; Senni, S.; Benvenuti, E.
SBS Project - System Evaluation and Multiphase Pumps Testing for a Subsea Booster Station Development
Conference Paper
Offshore Multiphase Production, 14 - 15 September 1988, London

/7/ Sigmundstad, M.
Multiphase Boosting - Status of a Current Project
Subsea Separation and Transport II, 1 - 2 March 1989, Oslo

/8/ Sigmundstad, M.
Multiphase Boosting - Problem Areas
Multiphase Flow - Technology and consequences for field development
8 - 9 May 1989, Stavanger, Norway

/9/ Neumann, W.
Efficient Multiphase Pump Station for Onshore Application and Prospects for Offshore Application
Proceedings of the Eighth International Pump Users Symposium, S. 43 - 48

/10/ Goodridge, R.A.
Multiphase pump advancements prompt commercial applications
Live-well tests prove reliability factor
Offshore, November 1990, Seite 47 - 52

/ 11 / Karge, V.:
Two-Spindle Screw Pumps for Handling Multiphase Mixtures
Reprint of
Pumps - Vacuum Pumps - Compressors '88, VDMA, Frankfurt am Main, Germany (1988)

/ 12 / Wietstock, P.; Seeliger, D.; Biemann, P.; Kolbe, G.; Bein, H.; Möller, H.; Vogt, F.; Schafstall, H.G.:
Mehrphasentransport in der Unterwassertechnik
Sonderdruck aus " GKSS Jahresbericht 1989, S. 62 - 74, GKSS-Forschungszentrum Geesthacht Gmbh

/ 13 / Bein, H.; Wietstock, P.:
Multiphase Pumps according to Twin Screw Design for High Gas Fractions and Differential Pressure
SMM ´90 Conference, September 25 - 27, 1990, Hamburg, Germany

/ 14 / Beckmann, J.; Felten, H.; Neumann, W.:
Überwachung einer Schraubenspindelpumpe zur Rohölförderung
VDI Berichte Nr. 846, Düsseldorf, Germany (1990)

/ 15 / Neumann, W.:
Efficient Multiphase Pump Station For Onshore Application and Prospects For Off-shore Application
Proceedings of the Eighth International Pump Users Symposium, Houston 1991

/16/ Brandt, J.U.; Wietstock, P.
Screw Pumps for Multiphase Pumping
4th EC Symposium Berlin 3-5 November 1992

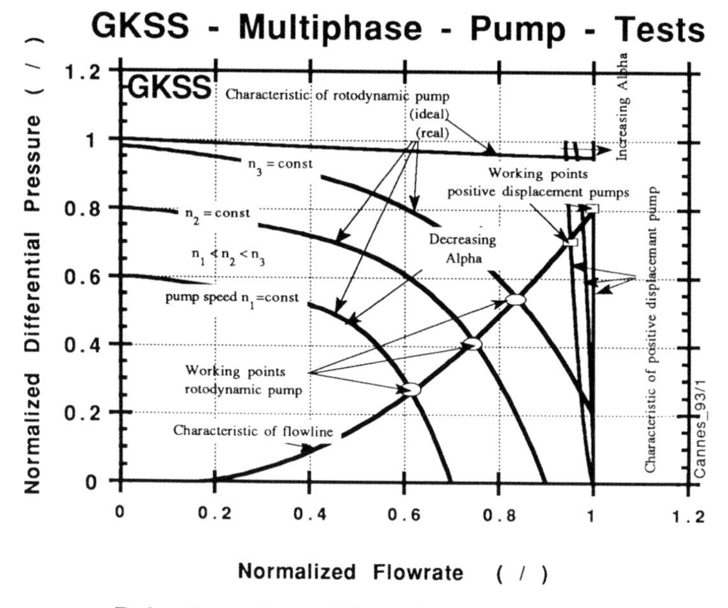

Fig. 1: Behavior of positive displacement pumps and rotodynamic pumps in a circuit

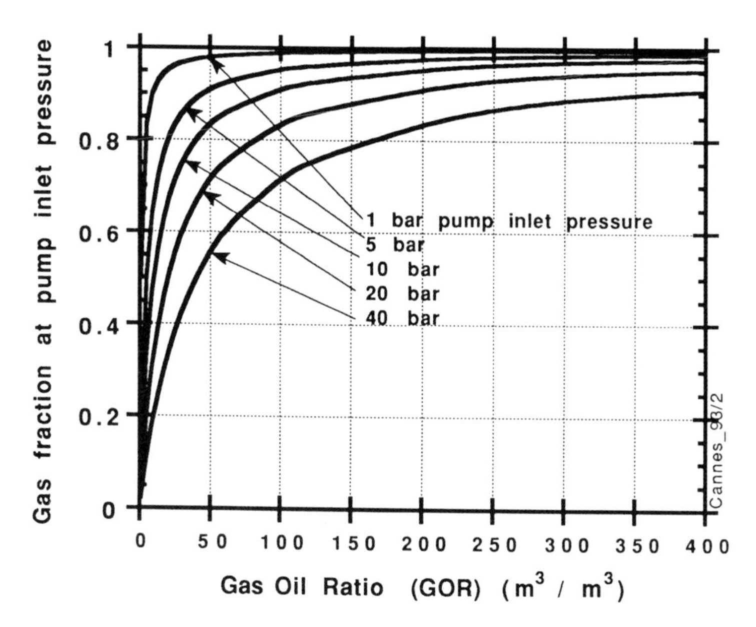

Fig. 2 : Gas fraction at different GOR

GKSS - Multiphase - Pump - Tests

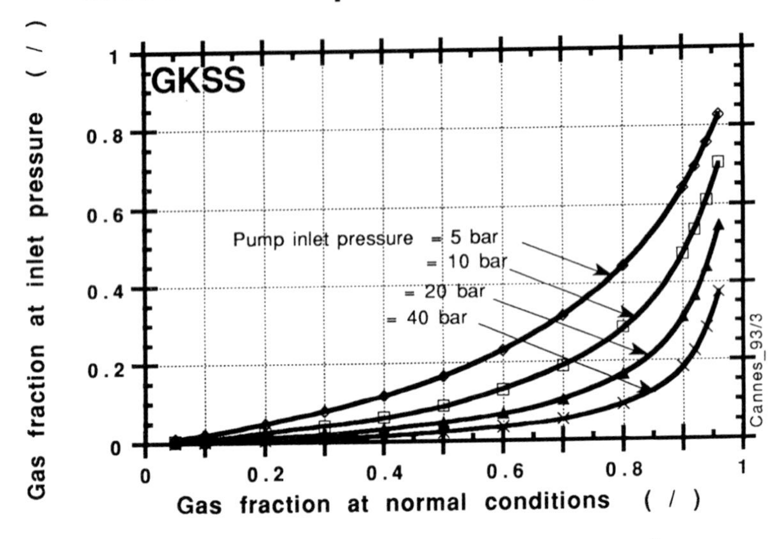

Fig. 3: Gas fraction at inlet conditions of multiphase pumps

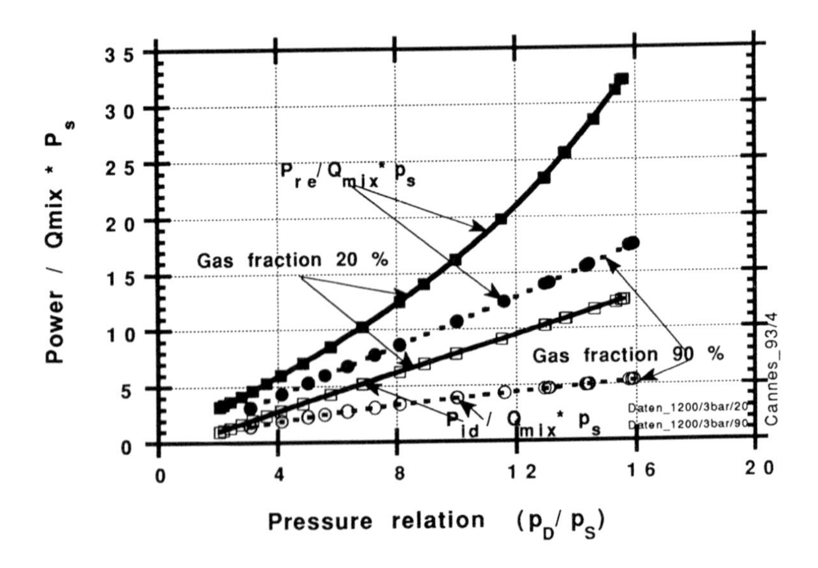

Fig. 4 : Normalised Power as a function of pressure relation

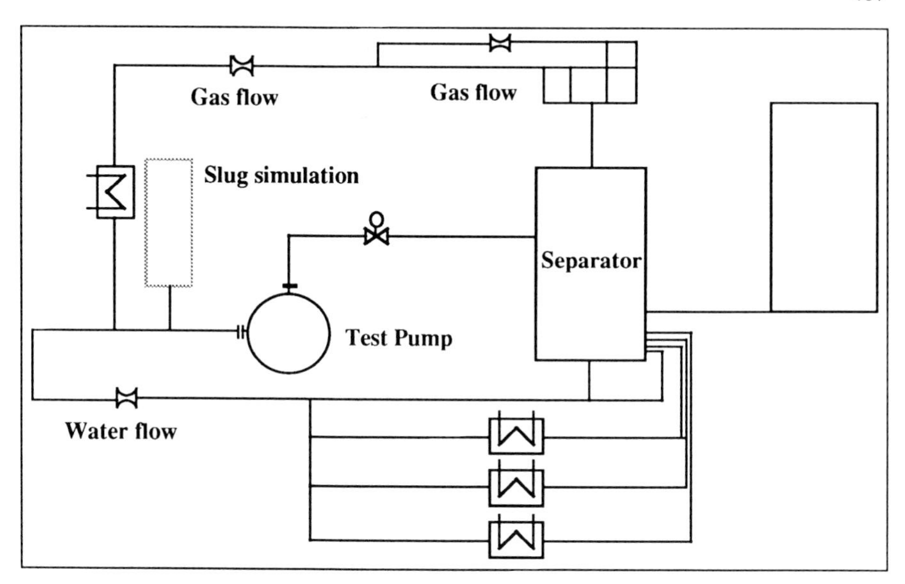

Fig. 5 : Multiphase pump test facility without auxiliary systems

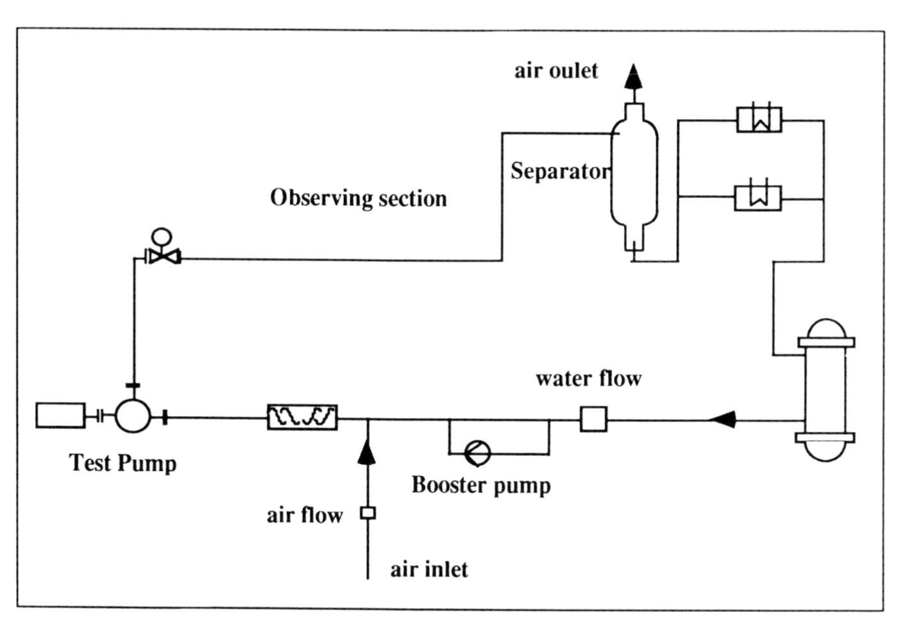

Fig. 6 : Multiphase model test facility

Fig. 7 : View on GKSS test facility

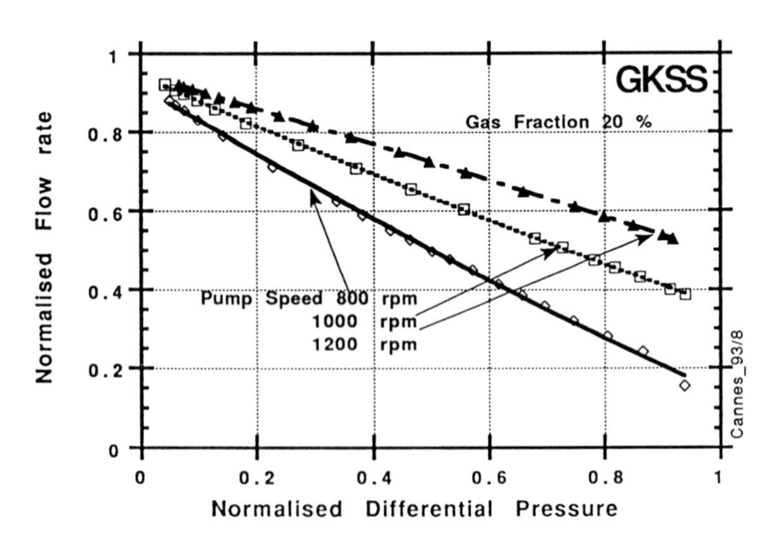

Fig. 8 : Normalised flow rate as a function of normalised differential pressure

GKSS - Multiphase - Pump - Tests

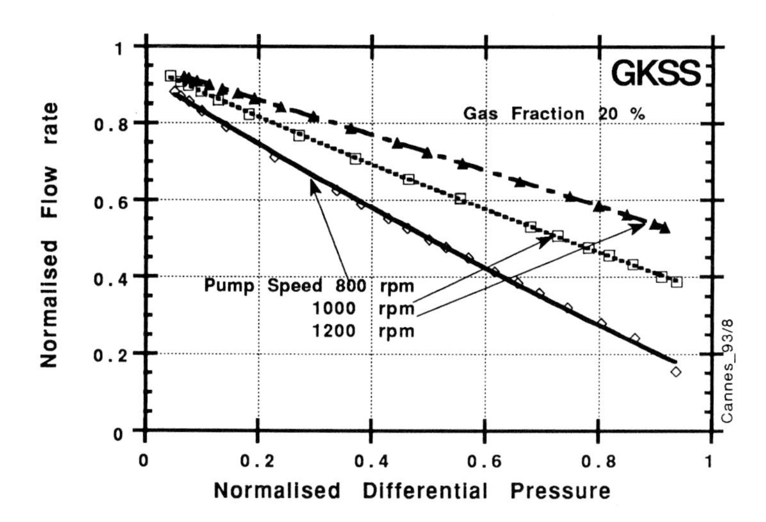

Fig. 9 : Normalised flow rate as a function of normalised differential pressure - Influence of pump speed without gas fraction 20 %

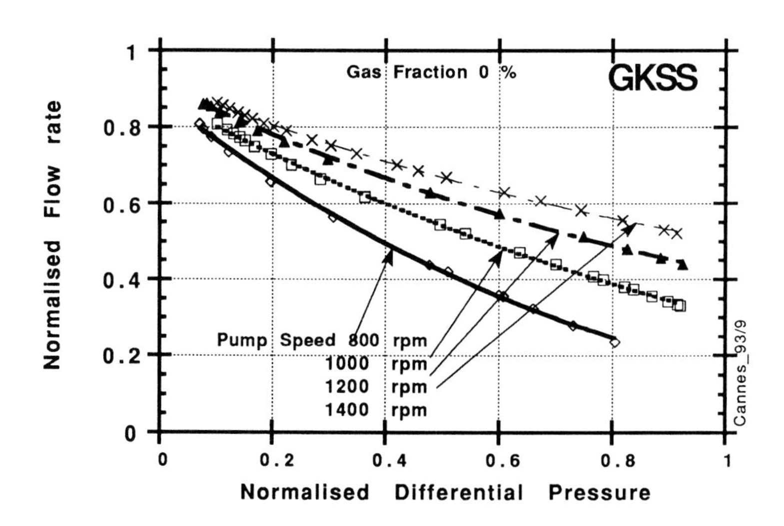

Fig. 10 : Normalised flow rate as a function of normalised differential pressure - Influence of pump speed without gas fraction

GKSS - Multiphase - Pump - Tests

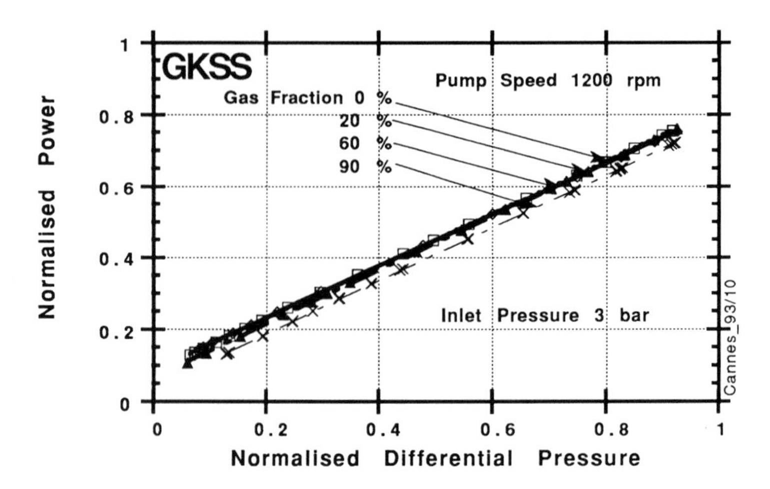

Fig. 11 : Normalised power as a function of normalised differential pressure - Influence of gas fraction

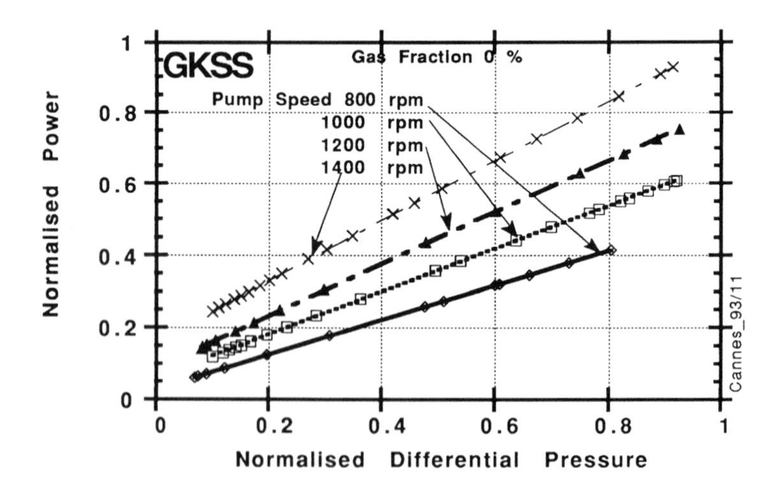

Fig. 12 : Normalised power as a function of normalised differential pressure-Influence of pump speed without gas fraction

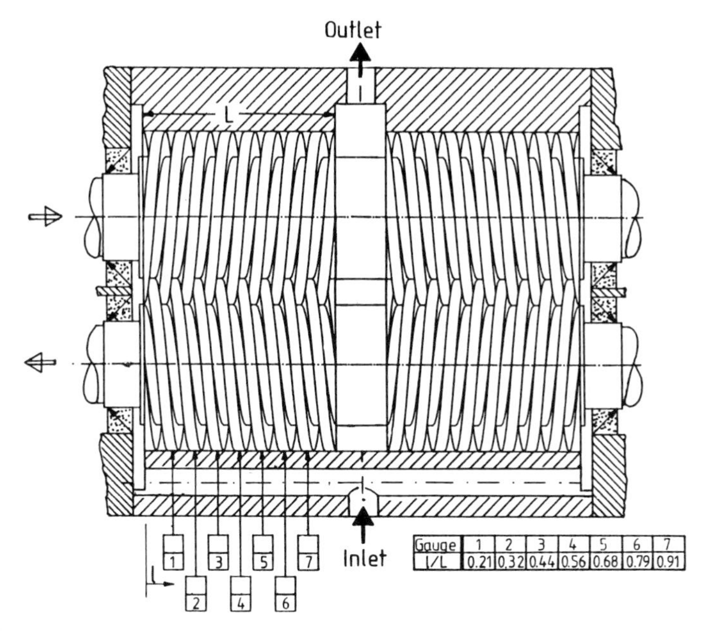

Fig. 13 : Arrangement of pressure gauges

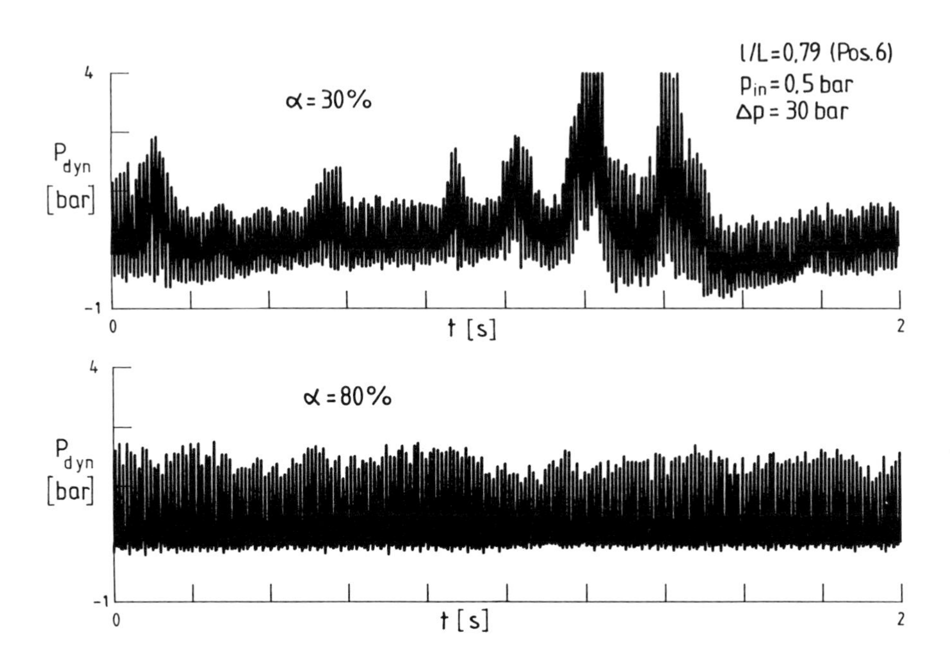

Fig. 14 : Pressure oscillations at low inlet pressure

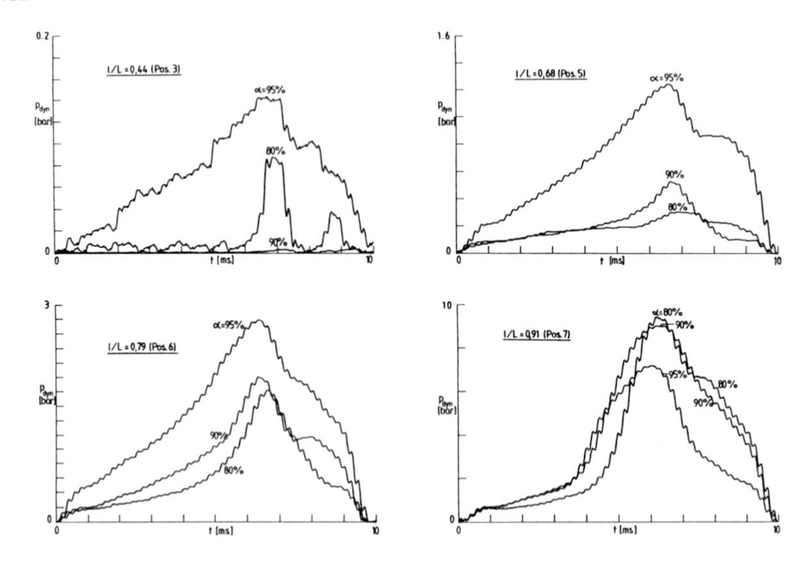

Fig. 15 : Pressure pattern as a function of gas content (p_{in} = 5 bar)

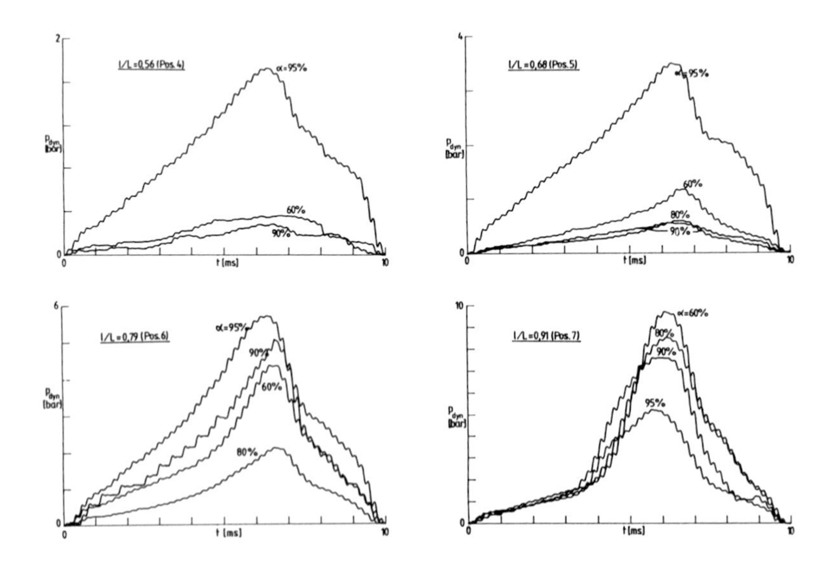

Fig. 16 : Pressure pattern as a function of gas content (p_{in} = 10 bar)

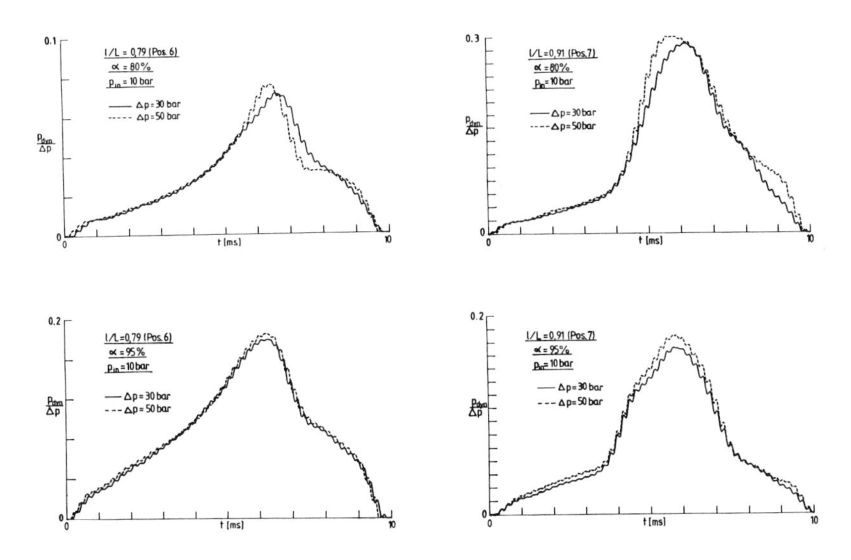

Fig. 17 : Pressure pattern as a function of differential pressure

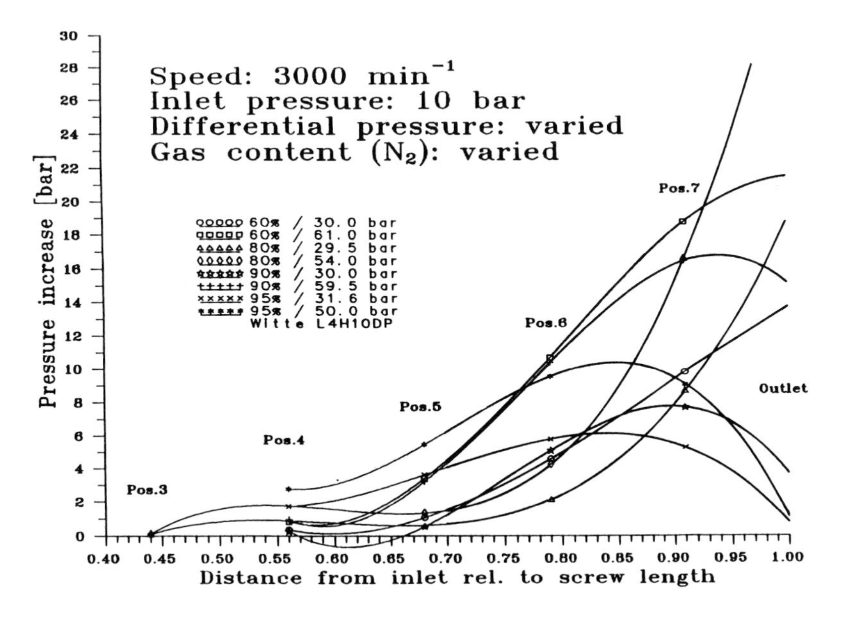

Fig. 18 : Pressure increase as a function of differential pressure and gas content

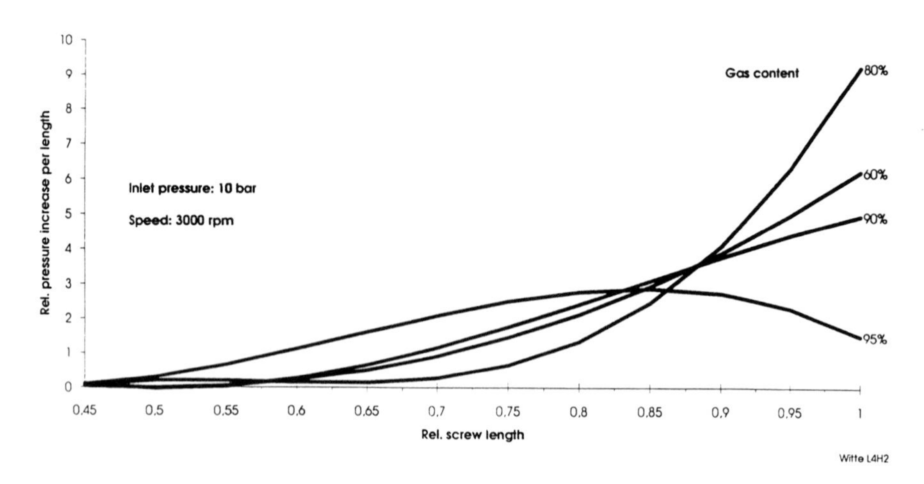

Fig. 19 : Relative pressure increase per length as a function of gas content

Test Results from the Integration Testing of the Kvaerner Booster Station at the SINTEF Multiphase Flow Laboratory

K.O.Stinessen, S.Gustafson (Kvaerner Energy, Oslo,Norway) and
O.Lunde (SINTEF,Trondheim, Norway)

SUMMARY

The paper presents performance data of a subsea pump/compressor installation aimed for pressure boosting in a multiphase pipeline in order to achieve increased hydrocarbon production.

The unit denoted KBS (Kvaerner Booster Station) has been developed by Kvaerner Energy, Norway, with Esso Norway a.s and Saga Petroleum a.s as sponsoring companies.

The prototype that has been tested consists of an integrated separation, pumping and compression unit. A multiphase mixture of oil and gas enters the KBS separator. Gas enters the suction side of the compressor, and liquid in equilibrium with the gas enters the suction side of the pump. The gas compressor and the liquid pump provide the necessary pressure increase, determined by transport distances and back pressure.

During the period September - December 1992, the KBS unit has been tested at controlled conditions, accomplished by an integrated test at the SINTEF Multliphase Flow Laboratory. The principal objective of the integration testing is to prequalify the KBS for operation on subsea wellstreams. To meet this objective a series of tests has been carried out.

Important issues addressed in the paper are operational experience of the integrated KBS at various inlet flow conditions, and verification of the control system. Finally an evaluation of the tests results so far are given.

During the integration testing the SINTEF Multiphase Flow Laboratory has proven to work excellent for the purpose of creating realistic wellhead flow conditions for the integrated KBS unit and also for component testing in general.

The function and mechanical status of the main KBS modules have been established to be satisfactory according to design requirements and standards before they were assembled for integration testing.

The tests have finally confirmed the satisfactory function and operation of the integrated KBS unit.

1. Introduction

Development of deepwater marginal fields by direct well stream transfer from subsea installations directly or via small wellhead platforms to central platforms or onshore terminals have a great potential both in terms of cost savings and increased operating flexibility . The full well stream transfer concept is very attractive because it eliminates the need for investing in huge platforms in deep waters.

A key concern connected to the well stream transfer is the loss of pressure along the line caused by pipe friction and elevation differences. Knowledge of the multiphase flow behaviour is required in order to predict these pressure losses.

The pressure gradient along the flowline and the distances between a subsea well and a wellhead platform or a central processing unit offshore or onshore may be detrimental for the production life of a well. In the worst case the transportation of well fluid will not be possible without some sort of subsea pressure boosting. Pressure boosting may also increase production rate and may prolong the production life of a field.

Production boosting can be described as the combination of the suction created by the booster on the upstream well heads and the transportation pressure added to the pipeline flow downstream. The upstream suction effect makes it possible to significantly increase both the production rate and the total production . The downstream pressure increase allows transport of the wellstream to distant existing process platforms or to onshore processing facilities.

The multiphase boosting concepts being developed so far are either based on multiphase pumping alone or on separation of gas and liquids at or near wellhead ; with the liquids being pumped in single phase back to shore, or to a remote production facility, and separated gas flowing back to the same location under wellhead pressure.

The Kværner Booster Station (KBS) compose a third and versatile alternative to multiphase pressure boosting where also subsea gas compression is included. Gas compression may typically be required when a reservoir reaches production decline and the available wellhead pressure drops below the required transport pressure. From the beginning of the development programme in 1985, Kværner , with Esso Norway a.s. and Saga Petroleum a.s. as sponsoring companies, selected separation followed by single-phase pressure boosting as basic concept for the KBS development. The KBS is further based on established technology and as far as possible built up of proven technology. The concept of separation followed by pumping and compression in parallel gives the flexibility to boost the whole range of gas oil ratios from pure liquid to pure gas. The three main modules , separator, pump and compressor, can be individually sized to adapt to the capacity needs of the actual field. After the pressure boosting, the liquid and gas can either be transported in separate pipelines to their respective terminals, or they can be commingled into multiphase flow in a single pipeline.

During the period September 1992 - January 1993, the KBS has been tested at controlled conditions, accomplished by an integrated test at the SINTEF Multiphase Flow Laboratory. The principal objective of the testing was to prequalify the KBS for operation on subsea wellstream. In order to meet this objective several types of tests have been carried out.

The current paper gives a brief technical description of the KBS prototype. In addition the SINTEF Multiphase Flow Laboratory test facilities as an integrated part of the test unit, used to generate realistic wellhead flow conditions, is presented.The various types of tests, test conditions and test results are further given. An evaluation of the test results are finally given before conclusions are drawn.

2. The KBS concept

2.1 General background

The unit, denoted KBS (Kværner Booster Station) has been developed by Kværner Energy with Esso Norway a.s. and Saga Petroleum a.s. as sponsoring companies. The main features of the KBS is presented in a paper by Stinessen and Queseth (1992). A brief description of the KBS concept is given in the following.

The complete Kværner Booster Station System consists of the following main components and modules , ref. figure 2.1 :

- Primary separator and gas scrubber
- Pump module with water filled electric motor
- Compressor module with gas filled electric motor and gear transmission
- High voltage power supply system for topside installation and frequency control
- Computerised electro-hydraulic control system
- Lube oil system
- Intermodule piping systems

The KBS is designed for a multiphase mixture of oil, water, gas and sand entering the KBS separator. Gas enters the suction side of the compressor, and liquid in equilibrium with the gas enters the suction side of the pump. The gas compressor and the liquid pump provide the necessary pressure increase, determined by transport distances and back pressure.

The main control objectives of the KBS are to maintain a specified liquid flowrate at a separator liquid level within certain limits and also to keep compressor out of surge conditions.

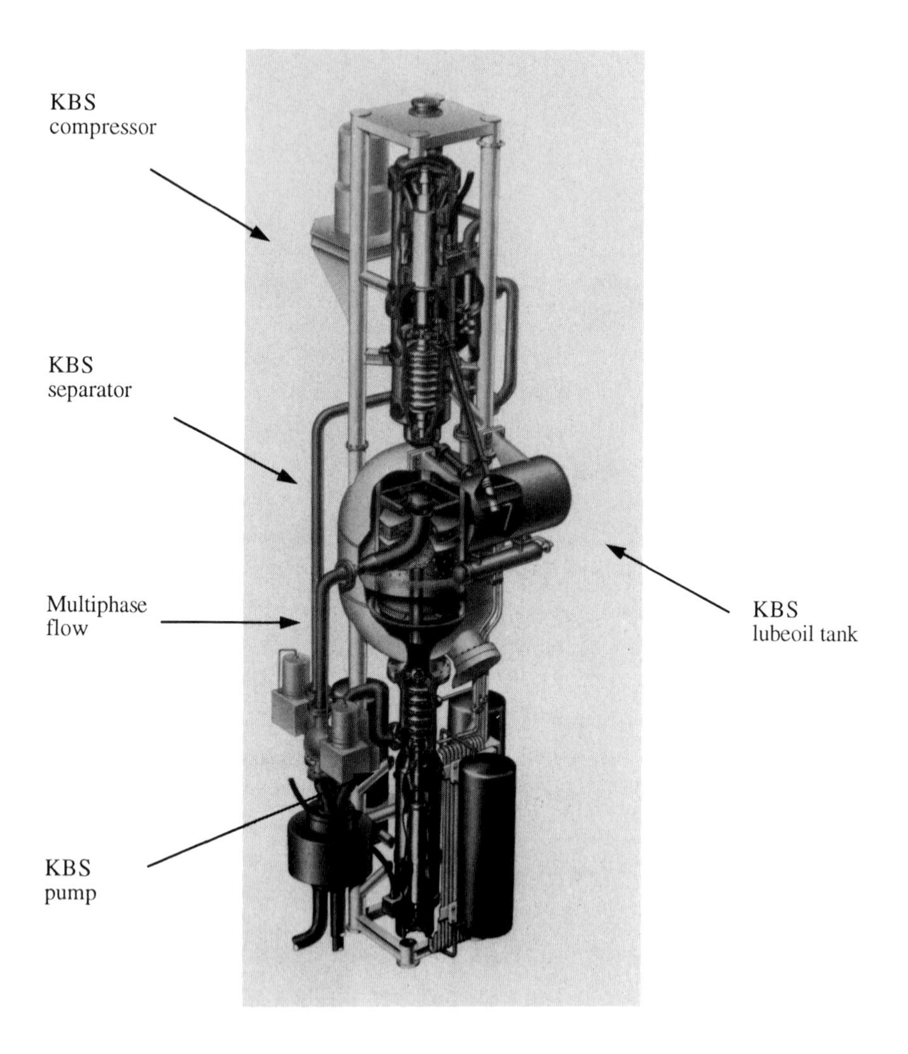

Figure 2.1 : KBS prototype - Main components and modules.

2.2 Design basis - KBS prototype

The prototype design is based upon typical North Sea Oil and Gas properties. The dimensions and performance data of the prototype KBS are defined as follows :

Liquid flowrate	:	10 000 BPD (77 m^3/h)
GOR (gas/oil ratio)	:	100 - 120 Sm^3/Sm^3

<u>KBS separator</u> :

Operating Pressure	:	10 - 55 bar
Operating Temp.	:	80°C
Design Pressure	:	60 barg

<u>Compressor</u> :

Max. Pressure Ratio	:	4 :1
Power	:	830 kW
Max.Discharge Pressure	:	73 barg
Discharge Temperature	:	250 °C

<u>Pump</u> :

Max.Pressure Rise	:	45 bar
Power	:	300 kW

Most of the major prototype components are designed for deep water installation at up to 300 meters.

The above flowrates are for a live oil stream.

3. The SINTEF Multiphase Flow Laboratory as modified for the KBS Integration Testing

3.1 The process flow diagram

The KBS has been tested at controlled conditions, accomplished by an integrated test at the SINTEF Multiphase Flow Laboratory. A simplified flow diagram of the test facilites is shown in figure 3.1. This will be used as reference for the further description of the process systems involved.

The KBS unit was submerged in a huge water tank in order to obtain required cooling for the electromotors. Gas and liquid were circulated in a closed loop where gas and liquid were mixed before entering the KBS unit. Gas and liquid were then separated in the KBS separator. The gas was pressurized by the gas compressor, and the liquid pressure increased by the KBS pump. Gas and liquid were then routed separately back to the existing process facilities as shown in figure 3.1. Gas and liquid flowrates were metered before remixing.

The geometry of the two-phase pipeline, diameter 8 inch, is shown as part of the instrument diagram in figure 3.2. Gas and liquid enters the two phase line through a mixing "Y". 345 m downstream the mixing element the horizontal line is connected to a 180° curved horizontal line. Downstream of the curved line the two phase mixture enters a 138 m long 8 inch diameter 1° upward inclined pipe, and finally the two phase mixture arrives at the KBS unit.

The total length of the two phase pipe applied for the current tests was approximately 720 m.The gas and liquid were separated in the KBS, and single phase pressures increased to specified discharge pressures by variable speed control pump and compressor drive. The specified discharge pressures were set by pressure control valves downstream the KBSunit

Downstream the booster the single phase gas flow was routed to the 8 inch diameter 1/2° upward inclined pipe, connected to the existing 50 m vertical riser, entering the top separator VS-01. From the top separator the gas was routed back to the gas scrubber, VO-02. The gas was further routed through the gas cooler, E-01, for temperature control. Downstream the gas cooler, a vortex meter station was applied for flow control. The gas was routed back into the two phase pipe through the Y-mixing section.

The liquid outlet from the booster was connected to the 8 inch diameter 1° upward inclined pipe entering the horizontal separator VS-02. The liquid from VS-02 was routed through the heat exchanger E-03 for temperature control. Then the liquid entered a turbine meter station for flow control, and was routed back to the two phase line via the Y-mixing section.

3.2 Instrument description

An instrument diagram of the multiphase flow loop is shown in figure 3.2.
It should however be noticed that instruments part of the main process system are not indicated on the aforementioned figure.

The following fluid flow parameters were decided as a minimum to be identified during each test run :

- Absolute press.(Pa) at specified locations upstream and downstream the KBS unit
- Pressure gradient (Pa/m) along the multiphase flowline
- Temp.(°C) at inlet and outlet of the multiphase flowline (allthough heat traced)
- Single phase flowrate (m^3/h) at the entrance of the multiphase pipe
- Superficial gas / liquid velocities (m/s) at specified locations along the multiphase flowline
- Single phase flowrates (kg/s and m^3/h) of gas and liquid downstream the KBS
- Flow regime identification at the 1° inclined entrance to the KBS
- Liquid holdup (-) at specified locations along the multiphase flowline

In addition to the flow parameters here being identified, a computer based slug flow analysis was performed for each test , ref. section 5. The slug flow analysis was based on signal input from the gamma densitometers as shown in figure 3.2.

(Two broad beam gamma densitometers denoted Fast Volume Weight Meter (FVWM), covering the total pipe cross sectional area, and two single beam gamma densitometers, SBGD, with a source strength of 0.3 Curie, Cs 137, ionisation chamber and with a response time of 0.1 seconds were used for the liquid holdup measurements and for generating input to the slug flow analysis).

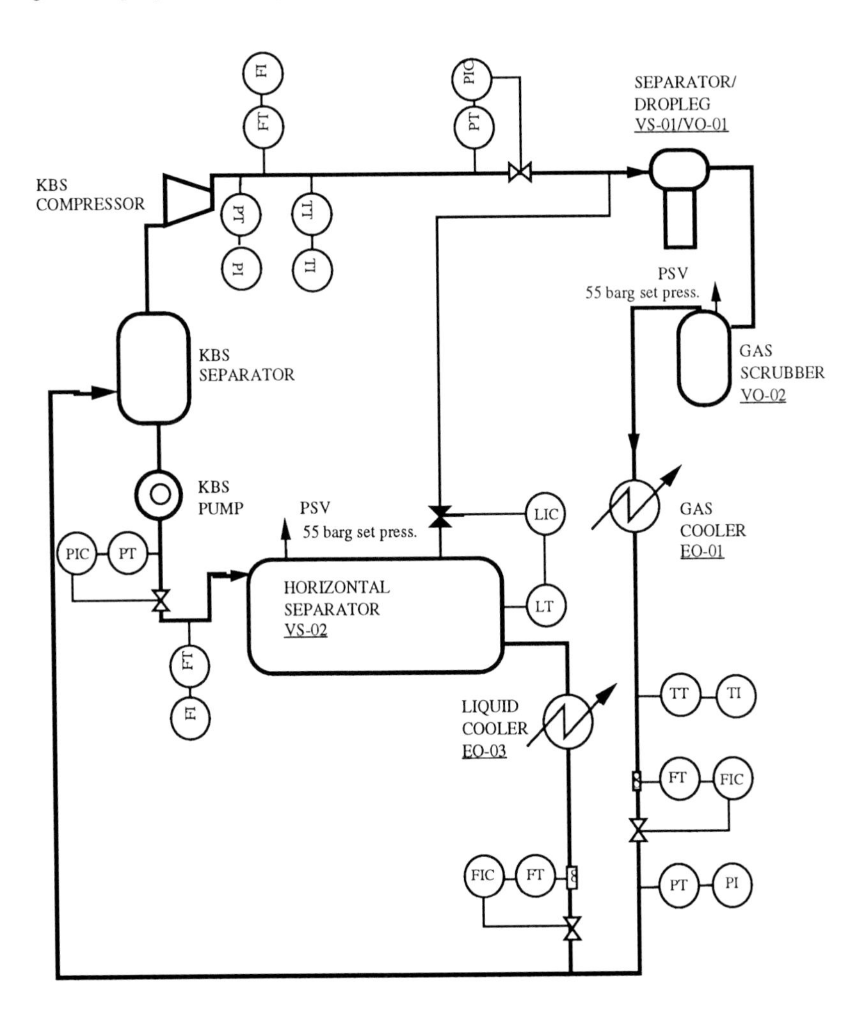

Figure 3.1 : A simplified flow diagram of the KBS test facilities.

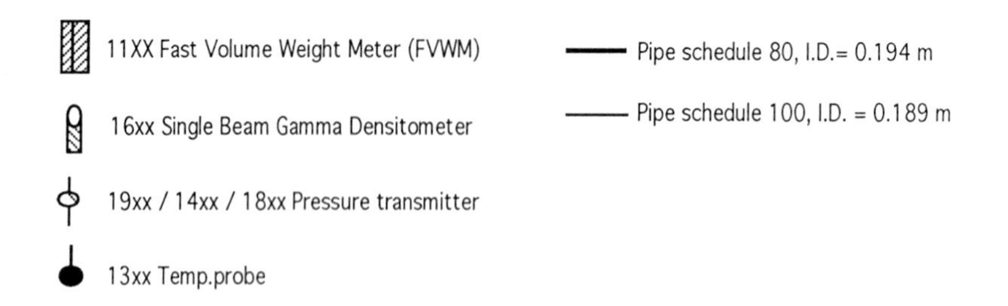

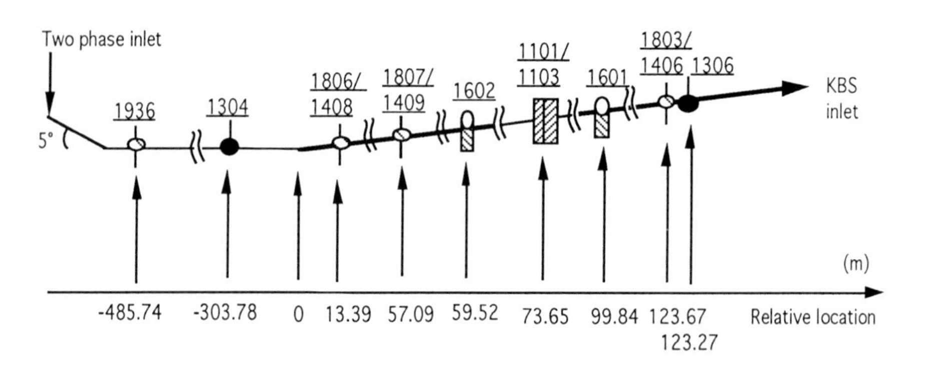

Figure 3.2 : An instrument diagram of the multiphase flow loop

4. Types of tests and test conditions

4.1 Types of tests

As a result of the philosophy used during component and module testing, ref. Stinessen and Queseth (1992), the final Integration Testing at the SINTEF Multiphase Flow Laboratory has three main objectives :

- To confirm the integration and functioning of the whole system over the total range of operation.
- To test the ability of the control systems to monitor and to act through control commands on the process and to verify the correct ESD (Emergency Shut Down) actions by the control system when trip situations are simulated.
- To start the reliability verification work through a duration test.

In order to meet these objectives, several groups of tests have been carried out. The tests were organized as follows :

1. Commissioning tests
2. Steady State tests
3. Ramping tests
4. Endurance tests
5. Slug tests

Commissioning tests
The Commissioning tests were carried out in order to verify that all equipment comply with the mechanical requirements which are given in the design specification and that all systems and components function according to specifications.

Steady State tests
The objective of the Steady State tests , i.e. the combined separation, pumping and compression process,was to verify that the booster process is properly coordinated by the control system within the specified area of operation of the KBS.

The KBS is intended to be tested at different pressure levels ranging between 5 and 50 bar, i.e. separator pressure. At each pressure level the compressor and the pump will be tested at different combinations of pressure ratios and flow rates ("high-medium-low").The pressure ratios and flow rates are constant for each test, ref. table C.1, Appendix C.

Ramping tests
In the Ramping tests the flow rates into the two phase pipe (and thereby in average into the KBS separator) and outlet pressures of the compressor and the pump should be constant. The setpoint of the KBS separator pressure controller will be varied as a "saw-tooth" function, ref. figure 4.1. This is a dynamic test where the KBS controllers do not have an unlimited period of time to reach steady state as in the Steady state test. They have to follow the period of the set point variation in the "saw-tooth" function. The Ramping tests are therefore a special case of and a more "advanced" test for the KBS control system than the Steady State tests. The Ramping tests will be performed at a nominal separator pressure of 15 and 40 bar.

Slug tests
The purpose of the Slug tests is to study how the control system of the KBS will handle slugs with increasing length. The practical accomplishment of these tests is to fill some oil in the two phase pipe upstream, and then use the gas pressure to push the slug into the KBS separator. The operation is done quite manually by opening/closing different valves.

Endurance tests
The Endurance tests should verify that for a given set of operational parameters the booster process is kept stable over a long period by the process modules (i.e. separator, compressor and pump) coordinated by the control system. The performance of the pump and compressor modules over a long period shall also be verified.

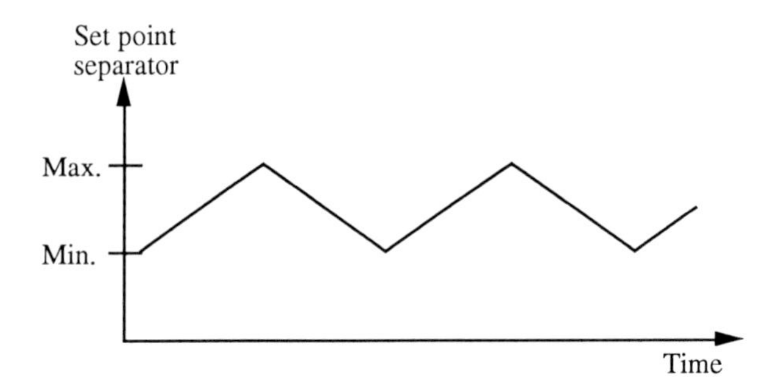

Figure 4.1 : Ramping test schematic

4.2 General test conditions

The flow behavior of the KBS feed line determines the entrance conditions of the KBS and was therefore of key concern for the tests. Thus, by integrating the KBS into the SINTEF Multiphase Flow Laboratory, realistic wellhead conditions could be simulated.

The liquid level in the KBS separator is controlled by a pump variable speed drive. When the liquid level increases, the pump speed will increase and bring the level down again. When the pump is operating at full capacity and the liquid level still continues to rise, the separator pressure is automatically increased to reduce the (well) flowrate. If the liquid level even still continues to rise a high-high alarm would trigger a compressor shut-down and an anti-surge control would keep the compressor out of surge conditions.

A pressure regulator was installed downstream the KBS compressor. The valve was operated to give a constant back pressure on the compressor. A pressure regulator was likewise installed on the liquid line between the KBS pump and the horizontal separator VS-02. The purpose was to keep a constant backpressure on the pump outlet.
Diesel and nitrogen were used as model fluids for the tests. Fluid physical properties are

given in Appendix A. - Previous two phase flow experiments with diesel and nitrogen have been published by Linga (1991), and was used as reference in the current test planning.

Each test was defined by a set of variables given by the above mentioned control system. Standard forms to be filled in during test execution, ref. Appendix B, were used to make records of the variables involved in a KBS tests run. Anticipated results, determined by using a dynamic simulator of the KBS integrated test facilities, were included as part of the test records. The simulator was also used to generate a realistic set of operating conditions.The dynamic simulator in concern was developed by SINTEF Automatic Control by use of the CADAS[1] program system.

A test matrix covering the hole range of test conditions is shown in table C.1, Appendix C.

(1) CADAS (Computer Aided Design, Analysis and Synthesis of industrial processes) is a general dynamic simulation and design system. The CADA system is used for dynamic simulation tasks, test of process design, design and test of control strategy, test of safety system, analysing of start-up and shut-down, optimization and operator training. CADAS is distributed by Simrad Albatross a.s.

5. Test results - some examples

In addition to the test records as mentioned in the previous section, data listings and plots visualizing the various flow conditions at the KBS entrance were developed. Time series representing typical 600 seconds sampling periods were plotted for most of the scientific instruments.

An advanced condition monitoring system, supported by Marintek, Norway, was also incorporated in the KBS integration test facility. Vibrations and temperatures in bearings were continuously measured and logged as well as temperature in motor windings and pressures at various positions.

Unfortunately not all the tests described above were finished at the time when this paper was issued. However the test results so far , including all the commissioning tests and an endurance test, were completed by December -92 and will provide a basis for the conclusions later being drawn.

As part of the commissioning phase , the KBS has been tested by a series of steady state tests. The gas / liquid velocity combinations at which the KBS has been tested are shown in figure 5.1. For the tests here illustrated, nominal system pressures were varying between 10 and 20 bara, however constant for each test. For all the velocity combinations the flow regime was interpreted as slug flow. This is from a KBS operational point of view considered to be a worst case concept. The KBS was observed to operate satisfactory for all the gas/liquid velocity combinations and pressures here given.

In addition to the aforementioned tests a series of single phase gas / liquid tests were carried out by gas/liquid circulation performed by the KBS. Finally the endurance test , composing all the tests /work hours on the KBS compressor and pump machinery , documents a roughly 1000 hours contiuous operation of the KBS. - An evaluation of the various test results is given in section 6.

In table 5.1 below a series of flow parameters given by the computerised flow analysis are listed, thus illustrating some typical entrance conditions at which the KBS has been tested.

Table 5.1 : Typical flow conditions at which the KBS has been tested.

Flow parameters	Test no.9522	Test no. 9516	Test no. 9500
USL :	0.376 m/s	0.452 m/s	0.320 m/s
USG :	0.827 m/s	0.447 m/s	1.256 m/s
Abs.press. :	10.2 bar	20.3 bar	13.9 bar
Mixture velocity :	1.203 m/s	0.900 m/s	1.576 m/s
Slug front velocity :	2.231 m/s	1.946 m/s	2.581
Average holdup :	0.502	0.660	0.406
Average slug length / L / D :	3.9 m / 20.6	3.7m / 19.6	4.7 m / 24.9
Max. slug length/ L/D:	8.3 m / 43.9	8.2 m / 43.4	10.1 m / 53.4
Min. slug length/ L/D :	0.7 m / 3.7	0.6 m / 3.2	0.8 m / 4.2
St.dev. slug length :	1.4 m	1.9 m	2.2 m
Average time interval between slugs :	7.5 sec	4.6 sec	10.6 sec
Standard dev. of time interval between slugs:	2.5 sec	1.8 sec	3.6 sec
Average bubble length/ L / D :	11.9 m / 63.0	4.8 m / 25.4	21.0 m / 111.1
St.dev. bubble length :	5.2 m	2.2 m	8.4 m
Max. bubble length L / D :	23.2 m / 122.8	10.1 m / 53.4	38.5 m / 203.7
Min. bubble length / L / D :	1.6 m / 8.5	0.6 m / 3.2	1.0 m / 5.3
Average time interval between bubbles :	7.2 sec	4.7 sec	9.9 sec
St.dev. of time interval between bubbles :	2.5 sec	1.8 sec	3.4 sec

Corresponding time series for liquid holdup measured by the broad beam gamma densitometer denoted "Fast Volume Weight Meter" is shown in figure 5.2. As seen from the holdup curves, the liquid holdup in the slugs are approximately equal to one. Thus, when estimating effective slug length (100% liquid), correction for void fraction in the slug is not needed.

As far as flowrate and pressure are concerned, the values given in table 5.1 do reflect typical North Sea applications. The slug lengths and slug frequency are further within the same order of magnitude as reported in the literature, ref. Lunde (1989). Thus , by the operating conditions here outlined, it is confirmed that the KBS can operate satisfactory at a series of realistic field conditions.

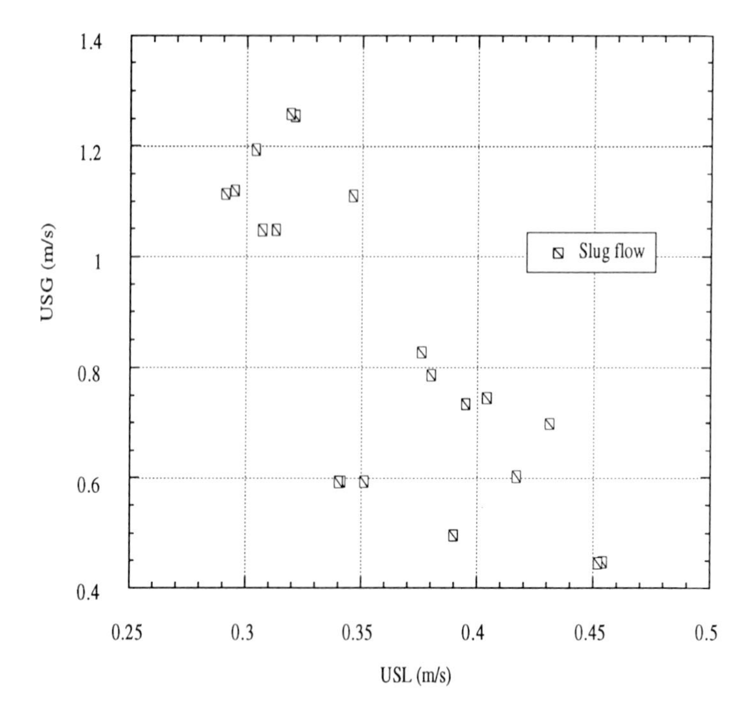

Figure 5.1 : KBS test conditions - Flow regime identification
Superficial gas velocity (USG) versus superficial liquid velocity (USL).

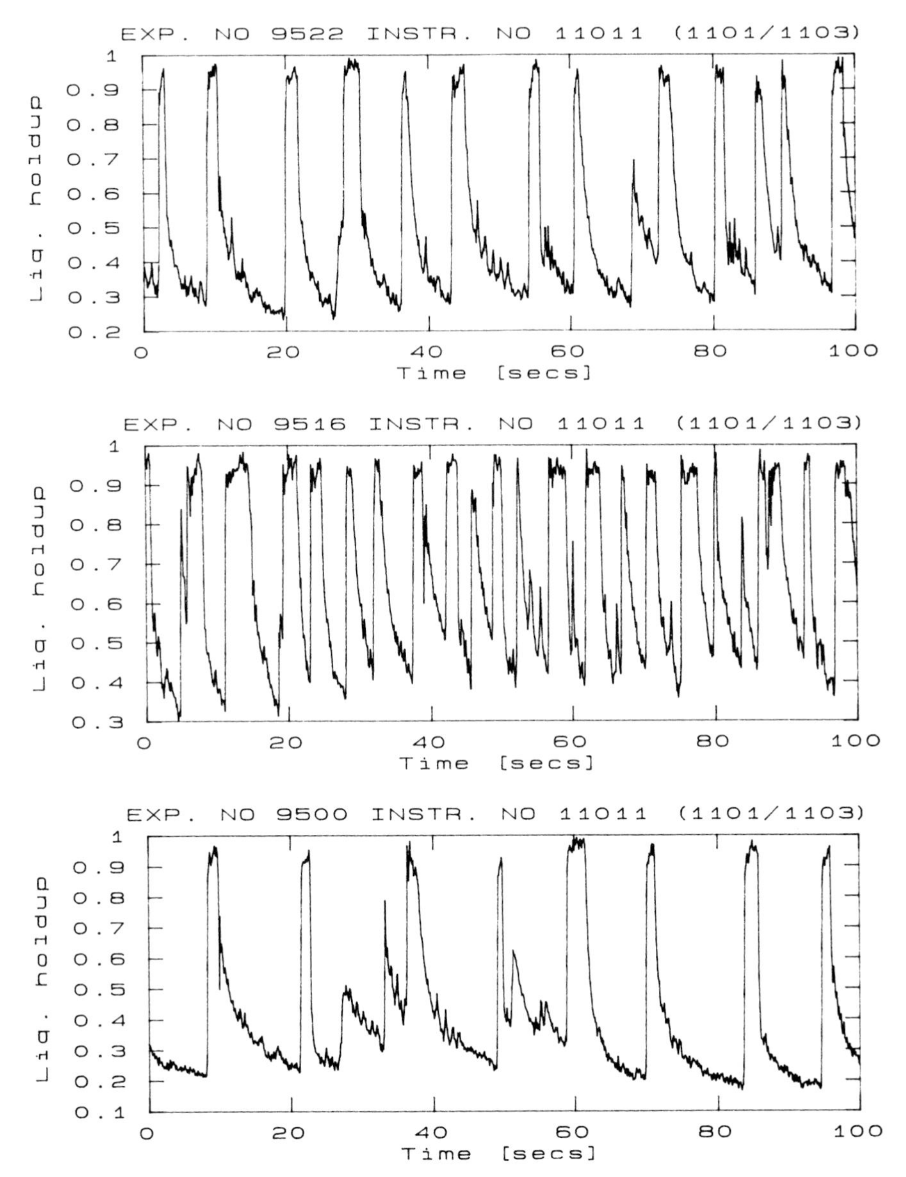

Figure 5.2: Time series of liquid holdup for the three different test conditions as given in table 5.1.

6. An evaluation of the KBS test results

In the following, test results with respect to the relationship between observed flow condition in the multiphase flow line upstream the KBS and design requirements including liquid level control of the KBS separator have been evaluated.

The main task of the KBS separator is liquid and gas phase separation. In addition the separator is designed to dampen the disturbances on liquid level caused by slug flow at the entrance of the KBS. As far as the liquid level in the KBS separator is concerned, the inclusion of an extra separator volume of 1.25 m^3 together with momentum absorbing process internals keeps the level under control. The diameter of the separator is 2.2 m, thus a volume of 1.25 m^3 corresponds to a variation of the liquid level of 300 mm.
An important part of the control philosophy for the KBS is that liquid level is allowed to oscillate within a deadband of 300 mm around the set nominal operating liquid level in the separator. Thus, when the KBS is operating under steady state conditions, the pump and the compressor are running at constant speed as long as oscillations of liquid level is within the defined deadband.

A volume of 1.25 m^3 corresponds to a liquid slug length (liquid core) of roughly 200 pipe diameters in the 8 inch entrance pipe upstream the KBS. This is a much longer slug length than normally found in horizontal or near horizontal multiphase flow lines. For horizontal flow hydrodynamic slugging will typically involve slug lengths in the order of 30 pipe diameters, ref. Lunde (1989).

In table 5.1 it is seen that the maximum slug length from these three tests is 53.4 pipe dimaeters (10.1 m) and the minimum length is 3.2 pipe diameters (0.6 m).

It is assumed that the slugs enter the separator instantaneously. A slug of 53.4 pipe diameters would result in a sudden increase of the liquid level of approximately 80 mm.

In practice a slug does not enter instantaneously, but over a period of some seconds. From table 5.1 it is found that the slug front velocity is 2.58 ms^{-1}. The time the 10.1 m (53.4 pipe diameters) slug uses to travel into the separator is therefore approximately 4 seconds. The assumption of instantaneous entrance of slugs is therefore correct for design purposes because little of this additional liquid is pumped out of the separator during these 4 seconds. (The volume of the 10.1 m slug is approximately 0.3 m^3, and the volume of liquid pumped out of the separator while this slug is entering the separator is in the range of 0.04 m^3 or 13% of the slug volume. The calculated flow is based on a superficial liquid velocity (USL) equal to 0.32 ms^{-1}, pipe diameter 8 inches and a steady state flow condition, i.e. flow out of the separator equals flow into it.)

A gas bubble will follow a liquid slug. From table 5.1 it is seen that corresponding to the maximum slug there is a maximum gas bubble of 38.5 m (203.7 pipe diameters). Based on a mixture velocity of 1.576 ms^{-1}, the time this bubble uses to travel into the separator is approximately 24 seconds. Therefore, at separator inlet, the total duration for entrance of the maximum slug plus the maximum bubble is 28 seconds. During this hole period 0.28 m^3 of liquid volume is pumped out of the separator. This corresponds with satisfactory accuracy for this type of calculation to the stipulated volume of the liquid slug of 0.3 m^3.

It can accordingly be concluded that when the KBS is operating under steady state conditions, the liquid level variation due to slug flow shall be allowed to oscillate without any correcting action from the control system for pump and compressor speed. This is because the separator liquid level only oscillate around the set average level.

Because the hole period of level oscillation is short, maximum observed approximately 28 seconds, effort to suppress the natural oscillation around the desired average level, would probably result in the opposite of dampening the oscillations ; i.e. the response time of the rotating equipment is too slow to follow oscillation of some few seconds and large oscillations could result instead of dampening.

Based on the test results and the above considerations it can further be concluded that an extra volume of 200 entrance pipe diameters (1.25 m^3), to dampen the effect of slug flow and ease the liquid level control function, was a sound design assumption.

7. Conclusions

During the integration testing the SINTEF Multiphase Flow Laboratory has proven to work excellent for the purpose of creating realistic wellhead flow conditions for the integrated KBS unit and also for component testing in general.

Testing of the integrated KBS unit at the SINTEF Multiphase Flow Labortory represents the final stage of a thorough test program. Testing has been carefully carried out from component level to the main modules and systems, through to the final KBS integration test incorporating the control system.

Before commencement of the testing of the integrated unit, the main modules, i.e. separator, pump and compressor module, with their individual part of the control system were tested out during a commissioning period.

The function and mechanical status of the main modules were therefore established to be satisfactory according to design requirements and standards before they were assembled for integration testing.

The main objective of the integration testing has been to verify that the individual modules, by action of the control system, are coordinated to operate as an integrated booster unit, keeping the booster process under control over a wide range of operational conditions including the demanding test program.

The tests have confirmed the satisfactory function and operation of the integrated KBS unit.

An advanced condition monitoring system was incorporated in the KBS integration test facility. Vibrations and temperatures in bearings were continuously measured and logged as well as temperatures in motor windings and pressures at various positions. The condition monitoring has established that the mechanical status of the equipment is within acceptance criteria, and no condition deviations have been observed after approximately 1000 hours of operation.

8. References

(1992) Stinessen K.O., Queseth P.O., "Kvaerner Booster Station for Subsea Production Boosting - Testing and Test Results", Multiphase Transportation III - Present Applications and Future Trends, Norwegian Petroleum Society, Røros, Norway, 20-22 September, 1992.

(1991) Linga H., "Flow pattern evolution ; some experimental results obtained at the SINTEF Multiphase Flow Laboratory", BHR 5th Int.Conf.on Multi-Phase Flow, Cannes, France, 19-21 June, 1991.

(1989) Lunde O., "Experimental study and modelling of slug stability in horizontal flow", Dr.ing. thesis, The Norwegian Institute of Technology, Trondheim, Norway, 1989.

APPENDIX A

FLUID PHYSICAL PROPERTIES

A PVT analysis was performed on a sample of diesel taken 27 September 1990. The analysis was conducted by combining diesel and nitrogen in the volume ratio 1:1, at 90 bar and 30°C. The pressure was then decreased to 20 bar, keeping the temperature of the sample constant. The results are summarized in table A.1, while accuracy in fluid property data, as reported by the test laboratory is listed in table A.2.

Table A.1: Fluid properties of diesel and nitrogen at 30°C

Pressure (bara)	Liquid density (kg/m^3)	Liquid viscosity (Ns/m2$\cdot 10^3$)	Surface tension (N/m$\cdot 10^3$)	Gas density (kg/m^3)	Gas viscosity (Ns/m$^2 \cdot 10^5$)
90	824.6	2.45	18.2	99.7	1.3
20	825.3	2.34	21.8	22.3	1.0

Table A.2 : Accuracy in fluid property data

Liquid density (kg/m^3)	Liquid viscosity (Ns/m$^2 \cdot 10^3$)	Surface tension (N/m$\cdot 10^3$)	Gas density (kg/m^3)
+/- 0.5%	+/- 5.0%	+/ -5.0%	+/- 1.4%

The fluid properties here presented were used as input for the integrated KBS tests.

APPENDIX B

STANDARD FORM TO BE FILLED IN DURING TEST EXECUTION

KBS at Tiller	Steady state			Documentation			
Description of test:							
Estimated time consume: (min)							
Real time consume	Start:		Steady:		End:		Used:
Tiller setup	**Gpres**	**Gflow**	**P0nit**	**Lpres**	**Lflow**	**Oil mass**	
	(bar)	(kg/s)	(bar)	(bar)	(kg/s)	(kg)	
Nominal values	4) The numbers here will reflect the requirements from 1. Go to 5.						
Actual values	5) Fill inn the actual values and use them to calculate values in 6,7 and 8.						
KBS setup/test	**Ccrat**	**Gflowi**	**Psep**	**Pdelp**	**Lflowi**	**Tsep**	
		(kg/s)	(bar)	(bar)	(kg/s)	(C)	
Requested values	1) The values in this row will give the requested test for the KBS. When this is done, go to 2 and 3.						
Expected values	6) The values will be calculated from the actual Tiller setup.						
Actual values	9) Actual values when steady state is achieved						
Other related KBS values	**Cflow** (kg/s)	**Crpm** (rpm)	**Cpower** (kW)	**C-Tdis** (C)	**ASV** (%)	**Pflow** (kg/s)	**Prpm** (rpm)	**Ppower** (kW)
Corresponding to requested	2) The values in this row are calculated corresponding to the requested values from 1.							
Corresponding to expected	7) The values will be calculated from the actual Tiller setup.							
Actual values	10) Actual values when steady state is achieved							
Multiphase lab characteristics	**Pnom** (bar)	**ULS/ UGS** (m/s)	**Holdup**	**Pressure gradient** (bar)	**Flow regime**	**GOR**	**Slug length** (m)	**Slug front velocity** (m/s)
Corresponding to requested	3) The values in this row are calculated corresponding to the requested values from 1).							
Corresponding to expected	8) The values will be calculated from the actual Tiller setup.							
Actual values	11) Actual values when steady state is achieved							
Simulated by:		Sign:			Date:		
Run by:		Sign:			Date:		
Approved by:		**Sign:**			**Date:**		

List of symbols :

P0nit : Initial system pressure (bara)
Oil mass : Initial amount of oil (diesel) (m^3)
Gpres : KBS compressor outlet pressure (controlled by discharge throttling) (bara)
Gflow : Gas flow into the multiphase flowline (controlled by the vortex meter) (m^3/s)
Lpres : KBS pump discharge pressure (controlled by discharge throttling) (bara)
Lflow : Liquid flow into the multiphase pipeline (controlled by the turbinemeter)(m^3/s)
Pvs01 : Pressure in "buffer volume" (VS-01,VO-02) (bara)
Pnom : Nominal pressure multiphase flowline (bara)
ULS : Superficial liquid velocity (m/s)
UGS : Superficial gas velocity (m/s)
HoldUp : Liquid fraction two phase pipe (-)
Pdel : Pressure gradient (Pa/m)
FlowRegime : Type of flow in two phase pipe (slug,stratified etc.)
GOR : Gas/oil ratio (m^3/m^3)
LS : Length of liquid slug (m)
UF : Slug Front Velocity (m)
Psep : KBS separator pressure (bara)
Tsep : KBS separator temperature (°C)
Lsep : KBS separator level (mm)
Gflowi : Gas flow into the KBS separator (kg/s)
Lflowi : Liquid flow into the KBS separator (kg/s)
Cflow : Gas flow through the KBS compressor (kg/s)
Ccrat : KBS compressor pressure ratio (bar/bar)
Crpm : KBS compressor speed(rpm)
Cpower : KBS compressor power consumption (kW)
C-Tdis : KBS compressor discharge tempertaure (°C)
ASV : Position anti surge vlave (%)
Pdelp : Differential presure KBS pump (bar)
Pflow : Liquid flow through KBS pump
Prpm : KBS pump speed (rpm)
Ppower : KBS pump power consumption (kW)

APPENDIX C

KBS TEST CONDITIONS

Table C.1 : KBS test conditions

KBS test no.	Nom.press. two phase pipe, i.e. KBS sep. set press.	KBS compr. disch.press.	KBS pump discharge press.	USL (superficial liq.vel., 8" pipe)	USG (superficial gas vel., 8" pipe)
-	(bara)	(bara)	(bara)	(ms^{-1})	(ms^{-1})
1	5.0	16.0	25.6	0.713	0.892
2	5.0	13.0	25.6	0.713	3.207
3	5.0	15.0	42.1	0.427	2.566
4	5.0	16.0	46.3	0.142	0.892
5	5.0	13.0	46.3	0.142	3.207
6	10.0	26.0	30.6	0.712	3.207
7	10.0	30.0	47.1	0.427	2.566
8	10.0	32.0	51.3	0.142	0.892
9	15.0	48.0	35.6	0.712	0.892
10	15.0	39.0	35.6	0.712	3.207
Ramp.test 1	15.0	39.0	35.6	0.712	3.207
11	15.0	45.0	52.1	0.427	2.566
Ramp.test 2	15.0	45.0	52.1	0.427	2.566
12	15.0	48.0	56.3	0.142	0.892
Ramp.test 3	15.0	48.0	56.3	0.142	0.892
13	15.0	39.0	56.3	0.142	3.207
14	20.0	52.0	40.6	0.712	3.207
15	20.0	60.0	57.1	0.427	2.566
16	20.0	64.0	61.3	0.142	0.908
17	40.0	70.0	60.6	0.712	0.601
18	40.0	70.0	60.6	0.712	2.405
Ramp.test 4	40.0	70.0	60.6	0.712	2.405
19	40.0	70.0	77.1	0.427	1.844
Ramp.test 5	40.0	70.0	77.1	0.427	1.844
20	40.0	70.0	81.3	0.142	0.601
Ramp.test 6	40.0	70.0	81.3	0.142	0.601
21	40.0	70.0	81.3	0.142	2.405
22	50.0	70.0	70.6	0.712	1.604
23	50.0	70.0	87.1	0.427	1.283
24	50.0	70.0	91.3	0.142	0.445
Endurance test	15.0	45.0	52.1	0.427	2.566
Slug test	15.0	-	-	-	-